BIOMATERIALS
SCIENCE

An Integrated Clinical *and* Engineering Approach

BIOMATERIALS
SCIENCE

An Integrated Clinical
and Engineering Approach

Edited by

YITZHAK ROSEN
NOEL ELMAN

CRC Press
Taylor & Francis Group
Boca Raton London New York

CRC Press is an imprint of the
Taylor & Francis Group, an **informa** business

CRC Press
Taylor & Francis Group
6000 Broken Sound Parkway NW, Suite 300
Boca Raton, FL 33487-2742

First issued in paperback 2019

© 2012 by Taylor & Francis Group, LLC
CRC Press is an imprint of Taylor & Francis Group, an Informa business

No claim to original U.S. Government works

ISBN-13: 978-1-4398-0404-9 (hbk)
ISBN-13: 978-0-367-38125-7 (pbk)

Visit the Taylor & Francis Web site at
http://www.taylorandfrancis.com

and the CRC Press Web site at
http://www.crcpress.com

Contents

Foreword

Biomaterials science is a multi-disciplinary field. The book offers a good overview of biomaterials as medical devices, drug delivery and tissue engineering systems. The emphasis is on integrating clinical and engineering approaches. In particular, the book covers various applications of biomaterials in unmet clinical needs in a variety of fields which include tissue engineering of musculoskeletal and cardiovascular tissues, neurosurgery, hemocompatibility, Micro-Electro Mechanical Systems (MEMS), nanoparticle based drug delivery, dental implants, and obstetrics/gynecology. It also covers areas such as regulatory challenges and commercialization issues.

Robert Langer
Institute Professor
David H. Koch Institute for Integrative Cancer Research
Harvard-MIT Division of Health Science and Technology
Chemical Engineering Department
Massachusetts Institute of Technology

Summary

This book provides a comprehensive list of applications summarized as follows:

- **Hemocompatibility.** Overview of clinical and engineering integration and its role and importance; examples of stents and their challenges. This chapter discusses examples of special clinical states such as hypercoagulability in pregnancy and patient individual differences.

- **Nanoparticles.** This chapter provides a review of drug delivery methods, challenges, and complications. These include various nanoparticle-based systems and their functionalization with targeting molecules for various applications.

- **Neurosurgery/Neurology.** This chapter provides a review of examples of devices and their integration barriers and complications. The challenges from a clinician point of view are discussed.

- **Dental.** Odontological – Engineering Integration. This chapter provides an insightful review on the need to combine clinical and materials engineering to design new materials for dental applcations. Various materials are described with their impact.

- **BioMEMS.** Biological Micro-Electro-Mechanical-Systems. This chapter provides a technological review of devices. A number of examples are described, as well as microdevices, materials, and integration challenges.
- **Tissue Engineering.** Musculoskeletal description. This chapter provides methodologies for scaffolding in this area. Hydrogels are described for this purpose as well as the use of designated stem cells. Also, the use of electrospun nanofibers and supercritical CO_2 are described.
- **Tissue Engineering.** Cardiovascular application. This chapter provides methodologies for scaffolding. Hydrogels, polymeric porous scaffolding, biomaterial free tissue engineering and various stem cells are described.
- **Obstetrics and Gynecology.** Clinical integration. This chapter provides a comprehensive insight related to a number of issues in this field, including: fetal toxicity; understanding the histological, physiological aspects; design of new materials and devices. A number of cases are described, including an example of clinical and engineering integration with a copper intrauterine device releases copper ions into the endometrium.
- **FDA.** Regulation/Ethic. This chapter provides an overview of clinical trials and regulation. The differences between various regulation administrations in the world are described. Radiological applications are also discussed. Excellent case studies are used.
- **Commercialization.** Transition. This chapter provides an understanding of market needs and transitioning into the market. Diagrams are used to describe a useful process to achieve market endpoints.
- **Appendix.** FDA references. This appendix provides relevant references related to the regulatory processes.

Janet Zoldan
Research Scientist
David H. Koch Institute for Integrative Cancer Research
Chemical Engineering Department
Massachusetts Institute of Technology

Authors

Yitzhak Rosen, MD, is a graduate of the Tel Aviv University of Medicine. He is currently a visiting research scientist at the Institute for Soldier Nanotechnologies, Massachusetts Institute of Technology. He is also the president and CEO of Superior NanoBioSystems LLC, a biomedical company. He has served in the Israel Defense Forces (IDF) as a medical officer and physician in militarily active areas. He completed a medical internship at the Rabin Medical Center and has worked at the Oncology Institutes of both the Rabin and the Sheba Medical Centers in Israel. He has invented a microfluidic chip platform, funded by the Defense Advanced Research Projects Agency (DARPA), for effecting extremely rapid blood typing and cross-matching for mass casualties in collaboration with the MEMS and Nanotechnology Exchange. In addition, he is the inventor of several medical ultrasound technologies. At Johns Hopkins University, he has been an invited lecturer in the area of nanotechnology in medicine for several years at the Biology Department for full-time undergraduate students. He has also taught a full course at the Department of Materials Science and Engineering at Johns Hopkins in biomaterials science for full-time undergraduate students. He took part in several key humanitarian medical relief missions as a medical doctor in Haiti in January 2010, immediately after the earthquake, and then in May 2010. He also worked as a medical doctor with the Global Medical Brigades Chapter, School of Public Health, Johns Hopkins University, in several remote areas in Honduras in June 2010. He is the author of publications in the fields of clinical medicine and micro- and nanotechnologies.

Noel Elman, PhD, is currently a research scientist at the Institute for Soldier Nanotechnologies at the Massachusetts Institute of Technology (MIT). He leads a research group focused on biomedical technologies based on nano- and microtechnologies for both diagnostics and therapeutics. He received his BS and Master's degrees in Electrical Engineering from Cornell University, where he focused on the development of micro-opto-electro-mechanical systems (MOEMS). He received his PhD degree in Physical Electronics from the Department of Electrical Engineering at Tel Aviv University in 2006. His PhD thesis focused on the development of a new family of biosensors based on the unique integration of living whole cells with semiconductor, MOEMS, and nanotechnologies. His postdoctoral studies at the Department of Materials and Engineering at MIT focused on the development of biomedical microdevices based on MEMS and nanotechnologies for both therapeutics and diagnostics. He is also the founder of high-tech startups in the field of micro- and nanotechnologies. His current interests relate to translational research for biomedical and biotechnological applications. He is the author of publications in the fields of biomedical technologies, MEMS, materials science and engineering, and micro- and nanotechnologies.

Contributors

Yitzhak Rosen, MD
Visiting Research Scientist, Institute for Soldier Nanotechnologies
Massachusetts Institute of Technology
President and CEO, Superior NanoBioSystems LLC

Noel Elman, PhD
Research Scientist, Institute for Soldier Nanotechnologies
Massachusetts Institute of Technology

Prof. Emanuel Horowitz, PhD
Professor, Department of Materials Engineering, Johns Hopkins University
Co-Chair, Cell Signaling Committee, American Society for Testing and Materials (ASTM)/ International, Committee F04 on Medical and Surgical Materials and Devices (Surgical Implants)

Michael A. Huff, PhD
Director, MEMS and Nanotechnology Exchange

Thomas Moore
Department of Bioengineering
Clemson University

Elizabeth Graham
Department of Bioengineering
Clemson University

Brandon Mattix
Department of Bioengineering
Clemson University

Prof. Frank Alexis, PhD
Department of Bioengineering
Clemson University

Prof. Jack E. Lemons, PhD
Professor of Dentistry, Department of Prosthodontics,
University of Alabama at Birmingham School of Dentistry

Urvashi M. Upadhyay, MD
Department of Neurosurgery
Brigham Women's Hospital
Harvard University School of Medicine

David Shveiky, MD
Department of Obstetrics and Gynecology
Hadassah Ein Karem Hospital
Hebrew University of Jerusalem

Dr. Yael Hants, MD
Department of Obstetrics and Gynecology
Hadassah Ein Karem Hospital
Hebrew University of Jerusalem

Ayelet Lesman, PhD
Department of Biomedical Engineering,
Technion Institute of Technology

Prof. Shulamit Levenberg, PhD
Department of Biomedical Engineering,
Technion Institute of Technology

Michael Keeney, PhD
Department of Orthopedic Surgery
Stanford University

Li-Hsin Han, PhD
Department of Orthopedic Surgery
Stanford University

Sheila Onyiah
Program in Human Biology
Stanford University

Prof. Fan Yang, PhD
Department of Orthopedic Surgery
    and Bioengineering
Stanford University School of
    Medicine

Pablo Gurman, MD
Materials Engineering Division
Argonne National Laboratory

Orit Rabinovitz-Harison, MSc
Clinical Trials Director, Tel-Aviv
    Sourasky (Ichilov) Medical
    Center
Tel-Aviv University School of
    Medicine

Prof. Tim B. Hunter, MSc, MD
(Previous) Chair, Department of
    Radiology
University of Arizona College of
    Medicine

Stephen M. Jarrett, DBA
President, Dolphin & Eagle
    Consulting, Inc.

1

Introduction

Yitzhak Rosen, Noel Elman, Emanuel Horowitz

Biomaterials science is a multi-disciplinary field. There are numerous fields involved in assisting in the research and development (R&D) of biomaterials. These fields include, but are not limited to, materials engineering, clinical medicine, mechanical engineering, biomedical engineering, molecular cell biology, histology, bioethics, regulatory affairs, business administration, and commercialization transition.

These fields require an interactive approach, as one can contribute to the others, and vice versa. For example, an unmet clinical need will be an important driving force for the engineering approach. However, the implementation of biomaterials must impact important clinical parameters, which include mortality, morbidity, and quality of life. These parameters need to be used to question the indications for the use of the biomaterial; furthermore, these parameters can be used with additional biological and physiological data to improve the biomaterial or inspire research in the development of more innovative and relevant biomaterials. This is applicable as well for the implementation of biomaterials in biomedical devices and drug delivery systems.

We must take these possibilities into account as best as we can. For example, a biomaterial used in a biomedical device may be implemented in a woman who eventually becomes pregnant. Pregnancy is a hypercoagulable state with possibly a variety of mechanisms in place that may affect the hemocompatibility of that biomaterial and its implementation overall [1]. By understanding the physiological mechanisms of such special states as pregnancy, we may be able to develop better biomaterials that may be applicable to a wider patient population. The integrated approach may be simplified if we continually ask two critical questions:

- While listening to and understanding the patient, what is the patient telling us?
- Will the biomaterial and its implementation truly impact the morbidity, mortality, and quality of life of the patient?

The process involved in the clinical and engineering integration approach is a double-edged sword in terms of its complexity. It is complex, as patients can be quite different from one another. There are numerous diseases that have complex pathophysiological processes, and each patient may react differently to these diseases. Each patient may also react differently to a biomaterial itself as well as to the implementation of a biomaterial in various clinical states. However, there is also a simplicity, which can be viewed as the certain overlap across many patients and disciplines. Listening to and understanding the patient is critical and will assist in elaborating this overlap. Therefore, a critical focus should be our patient.

There are already many books on biomaterials science. This book differs from existing books in that it emphasizes the need for the integrated clinical and engineering approach, an integration that often is lacking. To achieve this objective, the book includes a variety of contributors from many fields, including tissue engineering of musculoskeletal and cardiovascular tissues, neurosurgery, hemocompatibility, regulation, commercialization transition, micro-electro-mechanical systems (MEMS), nanoparticle-based drug delivery, dental implants, and obstetrics/gynecology. Some contributors are engineers, while others are clinicians. Furthermore, the areas of regulation and clinical trials have also been discussed, as these play a pivotal role in biomaterials science. In addition, commercialization transition has been addressed, as it plays an important role in how market needs, as defined by the aforementioned clinical parameters, assist in the research and development of new biomaterials and their implementations. While it is beyond the scope of the book to encompass all fields of biomaterials, the book includes important examples dispersed throughout its chapters that emphasize the need for a clinical and engineering integration approach.

Based on our experiences, without this integration many critical R&D components may be missed. Moreover, R&D resources may become squandered in addressing unnecessary issues. We may miss out on the possibilities of developing biomaterials that may fit a wider patient population needing them. This approach continuously focuses on the patient and always attempts to answer these two critical questions, described herein, from the idea stage all the way to many years thereafter.

In Memory

This book is being dedicated to Professor Moshe Rosen, Ph.D., (RIP),

father of Dr. Yitzhak Rosen. Professor Rosen was a Holocaust Survivor

of a concentration camp, previous chair of the Department of Materials

Science and Engineering at Johns Hopkins University, previous Rector of Ben-Gurion University, loving husband of 45 years to wife Lea and father

to three sons. By being exemplary to Dr. Yitzhak Rosen and many people,

he has truly taught what it means to be a mensch (the Yiddish equivalent

of being a man of noble character having social conscience, honor and

integrity) and to do good deeds for the world at large, for all people [2,3].

Disclaimer: The material in this book, whether related to medicine or any other topic, should be verified as to its accuracy, currency, and preciseness by the reader. It should in no way replace any advice given by a medical professional or any other professional. None of the information provided here should be a substitute for additional reading, advice, experience, or other relevant information in any topic discussed in this book.

References

1. James AH, Grotegut CA, Brancazio LR, Brown H. 2007. Thromboembolism in pregnancy: Recurrence and its prevention. *Semin Perinatol.* June 31(3): 167–75.
2. Leo Rosten (author), Lawrence Bush (editor). 2003. *The New Joys of Yiddish.* Completely Updated. Various Pagings. Three Rivers Press.
3. http://www.merriam-webster.com/dictionary/mensch.

2

Principles of Clinical and Engineering Integration in Hemocompatibility

Yitzhak Rosen, Noel Elman

CONTENTS

Overview

Hemocompatibility, compatibility when coming into contact with blood, is an important component of biocompatibility [1, 2]. It is also a great example where clinical and engineering integration is critical. Many life-saving devices come into contact with blood, whether permanently or temporarily [1–3]. Therefore, in these types of devices, the hemocompatibility component can significantly influence the failure or success of a medical procedure. Any foreign material introduced into the body will impact the behavior of blood in some way; however, the ultimate objective is to minimize the incidence of thrombogenesis [1, 2, 60–62]. It has been suggested that an ultimate design of a biomaterial would be its ability to orchestrate desirable biological effects and then degrade without leaving undesirable metabolites [4]. As far as hemocompatibility is concerned, much effort in biomaterials science has been done towards designing inert materials having a minimized reaction with platelets and coagulation factors [1, 61, 62].

In order to understand hemocompatibility, it is as important to define what is incompatible. A suggested definition of *incompatibility* is a material that induces an unacceptable adverse reaction when placed in contact with blood for a specified time [1, 60–62]. The adverse reactions include the formation of a thrombus, also referred to as a local blood clot, and a possible shedding of this clot, which will undesirably travel elsewhere as an embolus and have devastating effects, such as stroke [1–3]. It should be noted that any foreign material will cause some kind of reaction, whether local and/or systemic, that may or may not be controllable [1, 2].

So why has a whole chapter in this book been dedicated to hemocompatibility? The answers included the following:

1. Many devices, particularly life-saving ones, come into contact with blood. They include catheters for blood access and manipulation, extracorporeal pump oxygenators, hemodialyzers, heart-assist devices, stents, heart valves, and vascular grafts [1–3].

2. The future prospects of permanently implanted artificial organs will have to deal with this important subject [1].

3. The future prospects of biodegradable implants, such as stents, that will come into contact with blood [26, 27, 60].

4. The clinical indications of these devices are being modified by a more comprehensive research and development of improved hemocompatible devices [1, 2, 27].

5. It has been realized that the use of the database of biological knowledge from clinical medicine may result in the modification biomaterials, particularly in the area of surface modification, in order to make them more hemocompatible [1, 9, 30–32].

6. There are synergistic effects from other venues, such as inflammation, that can affect hemocompatibility [1, 5–7]

7. The need for a more comprehensive standard in both design and testing [1].

8. The need for careful examination of the contribution of adjunct therapies, such as oral systemic therapy, to the success and failure of the biomaterial and its implementation in a particular medical technology [2, 3, 8].

9. The need for assessing the degree of contribution, or lack of contribution, of factors such as individual genetic polymorphism and other individual specifications to the success or failure of the biomaterial [5, 6, 8].

10. The need for long-term implanted devices and tissue-engineered products [1, 2, 60].

11. The issue of contact time with blood may be a particularly important factor to consider regarding various biomaterials and their respective medical technologies [1, 2].

Interestingly, there is still a lack of consensus on testing standardization with respect to hemocompatibility. One reason for this is the need for a more comprehensive understanding of the physiological mechanisms leading to materials failure; furthermore, blood interactions have a complex, dynamic, and unpredictable behavior. There are a multitude of biomaterial–blood interactions, many of them not fully understood. Therefore, evaluation of these interactions in order to achieve a complete regulatory consensus cannot be easily performed [1, 2, 60–62].

As with testing, the engineering process of surface modification of biomaterials also lacks consensus. This ultimately has clinical implications in choosing a specific approach for surface modification versus conservative treatments. Moreover, discussions about short- and long-term morbidity and mortality related to hemocompatibilities are taking place, questioning the indications for the minimally invasive implantations of medical devices, such as stents, having direct contact with blood, as well as weighing advantages of stents versus a complete surgical coronary bypass procedure, for example [1, 2, 60–62, 64, 70].

It is beyond the scope of this chapter to encompass the many facets of hemocompatibility. Instead of focusing on a myriad of details that can be found in other references, we have focused on particular principles and considerations that should be taken into account when discussing biomaterials and hemocompatibility, albeit with a strong clinical focus. A key underlying principle is to understand the patient's needs. That is, by listening to the patient, we can ultimately create biomaterials that will have better hemocompatibility with superior indications for their implantation [2, 4]. In this chapter, we have chosen to focus specifically on cardiac stents as a reference point, as they represent an excellent multi-disciplinary example of clinical and engineering venues coming together, with several clinical trials. An important goal of this book is also to stimulate readers to suggest additional questions relevant in the field of biomaterials that integrate clinical and engineering approaches.

Questions That Should Be Addressed

The integration of the clinical and engineering approaches also involves addressing important questions concerning hemocompatibility. Below is a short list that we suggest readers use throughout this chapter. It is recommended, however, to expand on this list when reading this book and to think how the principles of integration can be implemented for each question.

1. What relevant database of knowledge in clinical medicine do we need in order to improve the biomaterial, particularly for surface modification purposes? [8, 27]
2. What are the risk-management issues, that is, benefit versus risk, involved? [1, 2, 7, 8, 27]
3. What systems are involved where clinical and engineering integration is needed to improve the biomaterial (inflammatory, blood, etc.)? [1, 8, 19, 16, 27]
4. How long does the biomaterial need to be in contact with blood? [1]
5. What can we learn by listening to and understanding the patient? [2]
6. What is the patient saying to us about him- or herself, the biomaterials, and their implementation? [1]
7. Can the biomaterial orchestrate desirable biological functions and then degrade into desirable metabolites? [3, 60]
8. How do we address the limiting conditions of keeping the biomaterial and its relevant medical technology in the body? [1, 4–6, 9, 16, 27, 53, 60]
9. What concomitant complicating conditions, also known as "special clinical states," need to be addressed? [3, 10–15, 33–46]
10. How do we deal with the change of influences by the body once the system is implanted? [1, 3]
11. Can the system be modified during its presence in the body when these special states or any other changes arise? [1, 4–6, 9, 16, 26, 27, 30, 53, 54, 56]
12. While integrating the engineering and clinical approaches, how do we advance towards a better standardization testing methodology? [1]
13. What are the short-term and long-term morbidity and mortality issues involved? [1]
14. What can we learn from the end-points of clinical trials to improve the biomaterial? [27, 57–59]

The Patient

The patient's needs represent the most important aspect when addressing hemocompatibility and biomaterials. Ultimately, the patient determines the

validity of the biomaterial. It is therefore critical that when designing biomaterials, we take into account various considerations that focus on the patient
[1, 27, 60]. There are enormous individual differences, yet we can attempt
to better characterize and classify certain similarities among patients that
may explain the successes and failures of a biomaterial once implanted into
a patient [1, 3, 9, 17–23].

Stratifying patients can allow us to achieve several important objectives with respect to the biomaterial being implemented in a particular
medical technology. First, better patient selection, according to relevant
risk factors inherent to the patients, can allow a higher success rate in
a targeted patient population in terms of the hemocompatibility of the
biomaterial. Such factors can include predispositions to thrombogenesis
due to inherent biological factors such as polymorphisms of inflammatory
factors and genetic resistance to anticoagulation adjunct therapy [2, 22–24].
Second, by distinguishing the factors that do or do not contribute to the
success of the biomaterial, we can achieve additional targets for future
surface modification of the biomaterial. This would allow us to enlarge the
targeted patient population. Yet we must be aware that while the attempts
must continue to better stratify patients, individual patient differences can
still occur [2].

Genetic Polymorphism and Individual Variability: Focus on Cardiac Stents

Genetic polymorphism has a critical influence on the development of
thrombosis as well as on the specific treatment response, in that it affects
the efficacy and safety of drugs used in the treatment and prevention of
thrombosis. Genetic polymorphism may impact the systemic and local
response to the surface modification of a biomaterial [2, 6, 22–24]. Cardiac
stents are an example where the impact of genetic polymorphism and
individual variability can be seen [4, 6, 22–24]. The characterization of
inflammation as an important factor of stent restenosis has assisted in
identifying several culprit genes that may impact thrombosis [4, 6]. Much
effort is continuously being allocated to preventing thrombosis by mini-
mizing local inflammation and, the proliferation of particular cells, such
as smooth cells, by the use of drug-eluting stents that carry agents that
prevent smooth-cell proliferation. At the same time, a confluent layer of
endothelial cells is needed within the lumen of the stent to prevent thrombosis [1, 6, 8, 16–21, 27, 50, 60]. In this section, we will discuss the multiple
targets of genetic polymorphisms that have demonstrated predisposition
to thrombosis with respect to cardiac stents.

CardioGene Study

A large study called the CardioGene Study was created under the auspices of the National Heart, Lung and Blood Institute to further understand the factors involved in in-stent restenosis (ISR) in bare mental stents (BMS) for the treatment of coronary artery disease. The overall goal of the study was to understand the genetic determinants of the responses to vascular injury that result in the development of restenosis in some patients but not in others. In this study, global-gene and protein-expression profiling were used to define the molecular phenotypes of patients. Well-defined clinical phenotypes were paired with genomic data to define analyses in order to determine blood gene and protein expression in patients with ISR, investigate the genetic basis of ISR, develop a predictive gene and protein biomarkers database, and identify new targets for treatment. Interestingly, the implications of such a study for biomaterials science can include the following:

- Identifying which patients would less likely benefit from treatment despite a relatively inert biomaterial.
- Identifying new targets to be used for surface modification.
- Providing alternative solutions that emphasize thrombogenic properties of predisposed patients carrying polymorphisms—which may also be helpful for patients without these types of polymorphisms.

Such databases can have enormous potential for improving surface modification of biomaterials in a variety of settings [8].

One potential application of genetic polymorphism testing has been found in the use of drug-eluting stents (DES). DESs, while reducing in-stent restenosis after percutaneous coronary intervention (PCI), have been associated with late stent thrombosis. No accurate method of predicting in-stent restenosis has been found; it should be noted that several risk factors for atherosclerosis do overlap with those for in-stent restenosis. In addition, atherosclerosis candidate genes have been investigated for their possible association with in-stent restenosis [2, 16–24].

Polymorphisms in Inflammation and Proliferation Effects on In-Stent Stenosis

Polymorphisms related to proliferation and inflammation may contribute to in-stent stenosis. Inflammatory activities as well as proliferation of particular cells such as smooth muscle cells can contribute to in-stent stenosis. These effects are related to vascular remodeling after procedures, such as percutaneous coronary stent implantation, that frequently lead to stenosis. One particular enzyme, heme oxygenase 1 (HO-1), is involved in the generation of the endogenous antioxidant bilirubin and carbon monoxide, both of which have anti-inflammatory and antiproliferative effects. Gulesserian et al. showed that the long allele of the HO-1 gene promoter polymorphism, which leads to low HO-1 inducibility, may represent an independent prognostic marker for restenosis after PCI and stent implantation. Interestingly, the effect of this particular allele, with more than 29 repeats, is attenuated in smokers, who have chronic exogenous carbon monoxide exposure [4].

Interleukin (IL)-10 is an important component in the inflammatory response. The Genetic Determinants of Restenosis (GENDER) study by Monraats, which included 3,105 patients treated with percutaneous intervention stent deployment, has indicated that genetic variants in IL-10 may predispose to the risk of restenosis. The primary end-point of this study was target-vessel revascularization. Genotyping of the −2849G/A, −1082G/A, −592C/A, and +4259A/G polymorphisms of the IL-10 gene was assessed along with adjustment for clinical variables. It was demonstrated that three polymorphisms significantly increased the risk of restenosis. The results of this study also indicated that the association of the IL-10 gene with restenosis was independent of flanking genes. Monraats et al. concluded that IL-10 is associated with restenosis; furthermore, Monraats et al. suggest that anti-inflammatory genes also may be involved in developing restenosis. Finally, the authors suggest that a new targeting gene may be used to improve drug-eluting stents [22].

Monraats et al. in another study examined the polymorphisms of genes for caspase-1, interleukin-1-receptor, and protein tyrosine phosphatase nonreceptor type 22, which are important mediators in the inflammatory response. Caspase-1 is also important in *apoptosis*, programmed cell death. Patients with the 5352AA genotype in the caspase-1 gene showed an increased risk of developing restenosis of stents. Monraats et al. suggest that the possibility of screening patients for this genotype may lead to better risk stratification and provide indications for improving individual treatment in addition to providing a new target for drug-eluting stents [6].

Shah et al. identified 46 consecutive cases of PCI with bare-metal stents where the patients subsequently developed symptomatic in-stent restenosis of the target lesion (>/= 75% luminal narrowing) within 6 months. Moreover, 46 matched controls with respect to age, race, vessel-diameter, and gender without in-stent restenosis after PCI with bare-metal stents were also identified. Single-nucleotide polymorphisms from 39 candidate atherosclerosis genes were genotyped for this study. Interestingly, ALOX5AP, a gene within the inflammatory pathway involving chemical inflammatory mediators called leukotrienes and linked to coronary atherosclerosis, has been shown to be associated with in-stent restenosis [9].

Polymorphisms that may contribute to thrombotic events may not always predict an increased rate of these same kinds of events with biomaterials. For example, polymorphisms of receptors involved in platelet adhesion and aggregation-modulating platelet thrombogenicity and found to predispose to premature arterial thromboses in individuals at risk are not necessarily correlated with acute stent thrombosis. Sucker et al., comparing the genotype prevalence of respective polymorphisms in patients with acute coronary stent thrombosis and healthy control subjects, did not find an increased risk of carriers of prothrombotic variants of platelet receptors for this complication [5]. However, being aware of the existence of such variations and delving into the exact causes of in-stent stenosis can ultimately assist in creating enhanced stents with minimized in-stent stenosis [2, 5, 8, 9].

Platelet Receptor Genes

It has been suggested by Rudez et al. that a common variation in the platelet receptor gene P2Y12 may serve as a useful marker for risk stratification for developing restenosis after percutaneous coronary interventions (PCI). Common variations in the P2Y12 gene were assessed by genotyping five haplotype-tagging single-nucleotide polymorphisms (ht-SNPs). These were assessed in 2,062 PCI-treated patients who received a stent. These patients participated in the Genetic Determinants of Restenosis (GENDER) Study. Target vessel revascularization (TVR) was assessed here, too. The study demonstrated that common variation in the P2Y12 gene can predict restenosis in PCI-treated patients [23].

Adjunct Therapy Resistance Stratification

Clopidogrel is a P2Y12 receptor blocker agent used to reduce the risks of acute coronary syndromes and considered an important adjunct therapy for stent deployment together with aspirin, yet clopidogrel-resistance genotypes may occur. It is important to realize that adjunct therapy resistance may be an important contributor to biomaterial failure in selected patients. This should also be taken as a consideration when assessing novel biomaterials and their applications [2, 7, 23, 24, 63].

Common variation in the P2RY12 gene has been demonstrated to be a significant determinant of the inter-individual variability in residual on-clopidogrel platelet reactivity in patients with coronary artery disease. This was corroborated in a study by Rudez et al. of 1,031 consecutive patients with coronary artery disease scheduled for elective percutaneous coronary interventions [23].

Clopidogrel is mentioned here since it plays an important role in adjunct systemic therapy together with aspirin for the success of stent deployment [2, 67, 68]. However, it should be noted that there are individual differences when clopidogrel is used that may influence the failure or success of a stent deployment. Price et al. have shown that platelet reactivity in clopidogrel therapy, as measured by a point-of-care platelet function P2Y12 assay, is associated with thrombotic events after percutaneous coronary intervention (PCI) with drug-eluting stents (DES). Moreover, high post-treatment platelet reactivity measured with a point-of-care platelet function assay has been associated with post-discharge events after PCI with DES, including stent thrombosis. The authors suggest that the investigation of alternative clopidogrel dosing regimens to reduce ischemic events in high-risk patients identified by this assay is warranted [67, 68].

The example of clopidogrel was presented for several reasons. Since clopidogrel is used as an adjunct systemic therapy after stent deployment, its individual variability, which may be assessed by platelet receptor polymorphism, may influence the risk of thrombotic events [8, 24]. Furthermore, this assessment may further assist in deciding whether resistance to adjunct therapy rather than the biomaterial alone may play a role in the risk for thrombosis [2, 24, 67, 68].

Endothelial Cell Trafficking Stratification

Identifying which patients may benefit from the biomaterial and its relevant medical technologies is critical. In fact, careful patient selection with exclusion and inclusion criteria for a particular intervention is often done in clinical medicine. An important reason for such patient selection is to address risk versus benefit [1, 2]. The example below underlines how patient selection according to progenitor endothelial cells capabilities can be influential in the success or failure of implanting a cardiac stent. Endothelial cells, which line the vasculature as a monolayer, play a critical role in the implementation of cardiac stents. They express and excrete a variety of molecules that regulate vascular tone, permeability, inflammation, thrombosis, and fibrinolysis, all of which are important components in hemocompatibility. They are also involved in wound repair. The expression levels of these molecules change according to interactions with the surrounding extracellular matrix and a variety of peripheral cells. They are also a target for pharmacological agents. Interestingly, a failure to re-endothelialize and form a confluent layer on the lumen of the stent is thought to be responsible late (>30 days) thrombosis of cardiac stents.

A clinical study performed by Georges et al. [69] suggests that the characteristics and numbers of circulating endothelial progenitor cells have a potentially important impact on stent restenosis. Patients with angiographically demonstrated in-stent restenosis were compared with patients with a similar clinical presentation that exhibited patent stents. Both groups of patients had similar medication administration that could potentially influence endothelial progenitor numbers. Their characteristics were determined by the colony-forming unit assay, endothelial-cell markers, and adhesiveness. Interestingly, patients with in-stent restenosis and with patent stents displayed a similar number of these cells. However, fibronectin-binding was compromised in patients with in-stent restenosis compared with their controls having patent stents. Furthermore, patients with diffuse in-stent restenosis exhibited reduced numbers of cells in comparison with subjects with focal in-stent lesions. The authors conclude that an intact endothelialization machinery is important for vessel healing after stent placement and as a means of preventing restenosis; moreover, their ability to traffic to damaged vasculature is an important characteristic that could affect stent restenosis. Interestingly, the authors point out a potential, future-risk stratification using such markers and related characteristics of these cells for the likelihood of patients developing in-stent restenosis. Furthermore, this study emphasizes the need for a careful selection of patients for whom such a biomimicry should take place.

These preliminary results can lead to the following:

1. Identification of markers to carefully select patients as candidates for successful stent deployment, as George et al. suggest
2. Identification of cell markers, such as surface ligands, that are needed for adhesion of endothelial cells
3. Immobilization of these markers and/or their relevant counterparts on stents for both patients with their deficiencies as well as for patients with no deficiencies to enhance adhesion

In summary, the work by George et al. corroborates the importance in the success of stent deployment of creating a careful pre-selection of patients by predefined criteria that can be measured by assays [69].

Special Clinical States

There are several clinical states where hemocompatibility may be modified. It is important to be aware of these states, as many patients may be facing them at some point in time. This section will focus on some of the common ones, such as pregnancy, cancer, and autoimmune states. It should be noted, however, that hypercoagulability can be inherent and be acquired in many other ways. Understanding these special clinical states will aid in further optimizing hemocompatibility designs [34–46].

Pregnancy

Pregnancy is considered to be a hypercoagulable state and a risk factor for deep venous thrombosis (DVT). The risk for DVT is further increased when personal or family history of thrombosis or thrombophilia exists. Venous thromboembolism, a phenomenon which includes both deep venous thrombosis (DVT) and pulmonary embolism (PE), complicates an estimated 0.5 to 3.0 pregnancies per 1,000. Thromboembolism is a leading cause of maternal death in the United States, and therefore this risk requires careful evaluation [2, 33, 35, 36, 38–41].

Hypercoagulability of pregnancy is caused by modifications in the plasma levels of many clotting factors. Fibrinogen can be increased up to 3 times the normal value while protein S, a physiological anticoagulant, decreases. Thrombin also increases. Protein C and antithrombin III are not predisposed to change. An impairment in fibrinolysis due to an increase in plasminogen activator inhibitor-1 (PAI-1) and the placenta-synthesized PAI-2 is observed. These changes have been suggested to be a preparation for the prevention of bleeding during labor [33, 35, 36, 38–41].

Other etiologies for hypercoagulability of pregnancy have also been pointed out. Venous stasis can be a culprit, and may occur at the end of the first trimester, from enhanced distensibility of the vessel walls by hormonal effect as well as prolonged bed rest. Acquired etiologies include antiphospholipid antibodies, as in systemic lupus erythematosis, which can exist before pregnancy. Congenital etiologies that can cause hypercoagulability in pregnancy and in the general population include factor V Leiden mutation, prothrombin mutation, protein C and S deficiencies, and antithrombin III deficiency [2, 33–41].

Pregnant women with prosthetic valves have an increased incidence of thromboembolic complications. An important consideration is adequate and effective antithrombotic therapy. Among other important consideration to take into account here is the ability of a therapeutic agent to cross the placenta and cause harm to the fetus. Warfarin, for example, is known to cross the placenta. Since warfarin use in the first trimester of pregnancy is associated with a substantial risk of embryopathy and fetal death, warfarin is typically stopped when a patient is trying to become pregnant or when pregnancy is detected. Typically, heparin, particularly low-molecular-weight heparin, is used alternatively and does not typically cross the placenta. This treatment may be continued until delivery [33–41].

When assessing biomaterials, it is important to take into consideration such hypercoagulable states and their underlying physiological mechanisms, as many patients can have these concomitant conditions. Suggestions would include using models with these coagulation changes to assess these conditions, especially where a specific need for a particular medical device during this condition should arise. Furthermore, altering the coagulation concentrations in order to define a pregnancy-related model may introduce interesting and insightful information as a whole for innovative surface modifications of biomaterials [1, 2, 33–41].

Autoimmune States

There are many autoimmune states in which the body produces antibodies against a variety of antigens. One of the problems that may be faced in these states is hypercoagulability. One particularly noteworthy state is the antiphospholipid syndrome (APS), which is the most common acquired thrombophilia, characterized by venous and arterial thrombosis, recurrent pregnancy loss, and various other clinical manifestations in the presence of antiphospholipid antibodies (aPL) [2, 13, 14]. This syndrome can also perturb the function of endothelial cells, which are important in forming a confluent layer within the lumen of the stent in order to minimize in-stent stenosis. Similar to other autoimmune diseases, the etiology of APS has been suggested to occur from a combination of genetic and environmental factors [2, 13, 14, 42–46].

One important interaction related to thrombosis is that of aPL with endothelial cells (EC). It has been demonstrated that aPL antibodies active endothelial cells in vitro as an enhanced expression of adhesion molecules on human umbilical vein endothelial cells along with enhanced monocyte adherence to ECs in vitro. The adhesion molecules that have been demonstrated to show increased expression include intercellular cell-adhesion molecule-1 (ICAM-1), vascular cell adhesion molecule-1 (VCAM-1), and (E-selectin) [14, 42–45].

The perturbance of ECs in APS has been demonstrated in a clinical study by Cugno et al. This study assessed the plasma levels of soluble adhesion molecules (s-ICAM-1, s-VCAM-1, s-E-selectin), soluble thrombomodulin (sTM), von Willebrand factor (vWF), and tissue plasminogen activator (tPA) using solid-phase assays in 40 selected APS patients as well as 40 healthy subjects matched accordingly by age and sex. Circulating endothelial cells by flow cytometry and brachial artery flow-mediated vasodilation were also evaluated. Their results indicated no noteworthy difference in plasma levels of sTM, s-E-selectin, and s-VCAM-1 between the APS group and controls differ. However, a significant increase in s-ICAM-1 (P = 0.029), t-PA (P = 0.003), and vWF titres (P = 0.002) was observed along with significantly higher levels of circulating mature endothelial cells in patients (P = 0.05), which were decreased when vitamin K antagonists and antiplatelet treatments were administered to the APS patients group. In addition, it was demonstrated that mean brachial artery flow-mediated vasodilation responses were significantly impaired compared with those of healthy subjects (P = 0.0001) [42].

It is evident that the function of ECs can be impaired in APS. Much can be learned about ECs in the APS milieu [14, 42, 43]. Enhanced characterization of ECs in a variety of clinical settings may lead to a better understanding of their role and variability in these settings. This knowledge may be re-applied to attempt to improve surface modification in biomaterials, particularly in cardiac stents, in order to better assist ECs to form a confluent layer within cardiac stents to minimize in-stent thrombosis [1]. That is, more potential targets may be identified for enhanced surface modification of biomaterials [1,8]. That may assist in developing a biomaterial accessible for a larger patient population that would otherwise not be able benefit from biomaterials implanted in their bodies [1].

Since autoimmune states may develop at different ages, it is important to know of their existence and the hypercoagulability potential that may occur in autoimmunity such as APS. For example, a patient with APS implanted with a biomedical device with a specific biomaterial may be more prone to thromboembolic phenomena [1, 43–46]. A variety of modifications in the coagulation system may affect the blood–biomaterial interactions and should be considered. Therefore, an enhanced characterization of the blood–biomaterial interactions in autoimmune models may ultimately lead to the development of an enhanced surface modification of the biomaterial [1].

Cancer

Cancer can lead to an acquired thrombophilic condition associated with a significant risk of thrombosis. Both venous and arterial thromboembolism are common complications for patients with cancer, who also present with a hypercoagulable state. The hypercoagulability, also referred to as the prothrombotic state, of malignancy is due to the ability of tumor cells to activate the coagulation system and cause a variety of associated clinical symptoms [2, 10–12].

There are multifactorial pathogenesis mechanisms for thrombosis in cancer. An important one is attributed to the tumor cells' capacity to interact with and activate the host hemostatic system cells, which can produce and secrete substances that have procoagulant substances and inflammatory cytokines. Tumor cells can allow physical interactions between themselves and a variety of other cells, which can include monocytes, platelets, neutrophils, and vascular cells. The generation of acute-phase reactants, abnormal protein metabolism, hemodynamic compromise, and necrosis can also promote thrombus formation. Anticancer therapies such as surgery, chemotherapy, and hormonal therapy can also assist in inducing procoagulant release, endothelial damage, and stimulation of tissue factor production by host cells [2, 10–12, 15].

One interesting example of hypercoagulability of malignancy was shown with non-small cell lung cancer (NSCLC), which comprises of 75% of all lung cancers. Here, it was shown that human full-length tissue factor (flHTF), the physiological initiator of blood coagulation, is aberrantly expressed in certain solid tumors. Furthermore, flHTF and its soluble isoform, alternatively spliced human tissue factor (asHTF), have been shown to contribute to thrombogenicity of the blood of healthy individuals [15]. It would be interesting to see what the variability of the expression of this factor in blood-biomaterial interaction and assess its role in biomaterial hemocompatibility failure at different timelines of contact with blood [1, 15].

Cancer is quite prevalent in society and thus should be used as a model to assess the thrombogenicity of a biomaterial. As in the case of antiphospholipid antibodies, a more thorough investigation is needed in order to better understand how cancer cells interaction with the various coagulation factors and assess biomaterial–blood interactions [1, 2, 10–13, 15]. Interestingly, existing evidence does not suggest a mortality benefit from oral anticoagulation in patients with cancer, because of the increased risk of bleeding. The potential complications of thrombosis after having a biomaterial implanted are, however, evident [1, 2, 10–12, 15].

Biodegradable and Bioabsorbable Cardiac Stents

It is important to distinguish between biodegradable and bioabsorbable cardiac stents. When using cardiac stents as a reference point, biodegradable stents can refer to polymer-based stents that can degrade and have their by-products assimilate into the body [25, 27, 60]. There are exceptions, where a polymer such as polylactic acid undergoes a degradation of the polymeric chemical backbone, which is controlled mainly by simple hydrolysis and is independent of a biological mediation [1, 25, 27, 28, 60]. Corrodible metallic stents have been considered bioabsorbable, as they directly assimilate into tissues rather than truly degrade [27, 47–49, 51–53, 55, 60].

A variety of biomaterials exist for these stents. Two metals proposed for bioabsorbable stents include Mg-based and Fe-based alloys [50–52, 54–56, 60]. Additional suggested materials involved in clinical evaluation have included poly-L-lactic acid (PLLA), polyglycolic acid (PGA), poly (D, L-lactide/glycolide) copolymer (PDLA), and polycaprolactone (PCL) [51–60].

The potential advantages are enormous in terms of hemocompatibility, assuming that there are minimal initial thrombogenic events and the material can produce its desired mechanical results for the necessary period of time. It is evident that fully biodegradable/bioabsorbable platforms are attracting both clinical and research interest. As mentioned previously, a biomaterial can be designated to orchestrate a necessary event and then degrade into absorbable constituents. The main question is whether these events can be achieved as intended. For example, for stents, it has been questioned how material parameters such as the elastic modulus, yield strength level, and material hardening all influence stent recoil and collapse. Yet biodegradability has shown its success in various animal studies, showing that these stents suggested less neointimal thickening, thrombosis, and inflammation while retaining an adequate radial force [60, 64–66].

A fully biodegradable or bioabsorbable stent, particularly in a drug-eluting stent scenario, would need several important features. A controlled, sustained drug release is required when using drug elution. Sufficient mechanical strength and structural functionality must be maintained in order to prevent negative vessel remodeling, as well as to avoid stent deformity and potential strut fractures. Compatibility with non-invasive coronary angiography is needed in order to maintain follow-ups. No residual stent prosthesis in the area should be present once biodegradability is completed. No potential adverse reactions with the coronary artery should take place. Vasomotion restoration of the artery is necessary [1, 27, 51–62].

Should Cardiac Stents Be Biodegradable/Bioabsorbable?

Overview

The injured vessel, after percutaneous coronary intervention, can necessitate scaffolding. There has not been a consensus about the necessary time for such a scaffolding. Current DESs have demonstrated their capacity in providing scaffolding for injured vessels and limiting in-stent restenosis. Typical permanent polymers used in sirolimus- and paclitaxel-eluting stents include poly(ethylene-co-vinyl acetate), poly(n-butyl-methacrylate), and poly(styrene-b-isobutylene-b-styrene) [17–21, 60, 65].

There have been long-term safety concerns about the permanent nature of the stent material and polymers. Several noteworthy adverse effects that occur with DES include delayed healing, endothelial dysfunction, chronic arterial-wall inflammation, impaired neointimal formation, and late-acquired stent malapposition [9–15, 64]. In addition, particularly serious concerns are late and very late stent thrombosis, which appear long after stent deployment. These can lead to severe clinical outcomes, including death [2, 47, 48, 50–60, 64, 65].

The durable polymers used in DES have been shown to provoke an inflammatory response in animals, such as giant cell infiltration around the stent struts, and progressive granulomatous and eosinophilic reactions. These reactions can increase beyond the first year. Chronic inflammation may decrease efficacy [60, 64–66]. Reports of increased rate of endothelial dysfunction after DES implantation compared with bare-metal stent (BMS) implantation have given impetus to considering biodegradable and bioabsorbable options for cardiac stents [2, 16–21, 57–60].

Moreover, these effects can increase the incidence of very late stent thrombosis, a rare event, after DES implantation [16–21]. In addition, delayed loss of anti-restenotic efficacy has also been observed with the early DES technologies [22, 23]. Chronic arterial-wall inflammation and endothelial dysfunction may be associated with the increased rate of target vessel revascularization at a late stage, which has been found particularly in patients with complex lesions, including those with diabetes [24, 25].

Among the biodegradable polymers implemented, polylactic acid, polyglycolide, and poly(D,L-lactic-co-glycolic acid) are particularly common. These can be completely metabolized as they break into monomers, water, and carbon dioxide. Stents with these biodegradable polymers have antiproliferative agents as eluting agents, which include sirolimus, tacrolimus, biolimus, and paclitaxel [27, 51–60].

TABLE 2.1

Suggested Potential Advantages and Disadvantages of Biodegradable Cardiac Stents. (These have been compared to permanent metallic cardiac stents. Important clinical considerations are mentioned in both the advantages and disadvantages sections.)

Suggested Advantages	Suggested Disadvantages
Possible higher drug-loading capacity [57,70,76]	More data clinical data needed for their use to limit important adverse effects such as late stent thrombosis [60,70,71]
May facilitate enhanced targeted drug delivery, limit smooth muscle cell proliferation and enhance endothelialization on lumen [70]	Not as strong and can result in early recoil post implantation [60]
May reduce need for a protracted dual antiplatelet therapy [81]	May be associated with significant degree of local inflammation [60]
May assist repeated percutaneous revascularization or surgical intervention if metallic stents can be avoided [78]	Relatively slow bioabsorption rate; moreover, still may result in restenosis [60]
Prevention of jailing of side branches and difficulties at ostial lesions [79]	Radiolucent, which can interfere with their positioning [60]
Facilitated magnetic resonance imaging/CT as metallic artifacts seen in metallic stents may be disruptive [80]	Possible difficulties with stent deployment without fluoroscopic visualization assistance [60]
Lack of freeze stent recoil, thereby allowing late favorable positive remodeling [79]	Limited mechanical performance may be abserved [60]
	Limited recoil rate, thereby necessitating thick struts. These can limit their profile and deliver capacities particularly in small vessels [60]
	May need special storage conditions and shorter shelf life [60]

Therefore, there are still not sufficient clinical data to make an assessment for the clinical utility of these biodegradable stents when they are compared to DES. More follow-up data are needed, particularly when assessing late stent thrombosis [2, 60].

Bioabsorbable Stents

When discussing bioabsorbable stents, it is noteworthy to describe metal bioabsorbable stents. Mg- and Fe-based alloys are two common classes. Fe and Mg both possess low toxicity. Metals have superior mechanical properties compared to polymers, as these alloys have similar mechanical properties to 316L, a particularly common alloy for fabricating stents. The alloy 316L has been considered a standard reference for mechanical properties related to new biomaterials for various stent applications [51–60].

An important clinical use for the bioabsorbable Mg-based stent has been in pediatrics. A successful implantation of such a stent into the left pulmonary artery of a preterm baby allowed the reperfusion of the left lung. The reperfusion persisted throughout the 4-month follow-up period during the degradation process of the stent until it was completed. The degradation process was clinically well tolerated by the baby [49, 60].

The bioabsorbable magnesium stent was also evaluated in the human coronary arteries in the PROGRESS-AMS (Clinical Performance and Angiographic Results of Coronary Stent) study [57]. This study was a prospective, multi-center, consecutive, non-randomized study of 63 patients with coronary artery disease that addressed the safety and feasibility of this stent deployment in human coronary arteries. The primary endpoint of this study included major adverse cardiac events (MACE) at 4 months. These were defined as cardiac death, nonfatal myocardial infarction, and ischemia-driven target lesion revascularization. In a sub-study of PROGRESS-AMS, Ghimire et al. assessed the endothelium independent coronary smooth muscle vasomotor function 4 months after implantation of the stent in 5 patients. This group was compared with a control group of permanent metal stents (n = 10) undergoing follow-up angiography and who were free from angiographic restenosis [59].

The bioabsorbable stent was shown to be safe overall, with a high procedural success rate of 99.4%. Moreover, the stent showed its ability to degrade well. No adverse events or distal embolization were observed. In addition, the targeted vessel regained its vasoreactivity properties. A major disadvantage, however, was its association with higher than expected restenosis rates. These results were associated with early recoil and neointima formation [57, 59, 60].

These materials show that ultimately their advantages as well as disadvantages must be correlated with clinical trials and long-term follow-ups. That is, ultimately, their widespread use will be dependent on clinical end-points that include morbidity and mortality parameters being well stratified [2].

Drug-Eluting Balloons (DEB)

While a complete discussion of the topic of drug-eluting balloons is beyond the scope of this chapter, its relevance to hemocompatibility will be discussed. Their salient advantage over DES is their ability to not leave behind an implant. It should be noted, however, DEB cannot overcome the important mechanical limitation of acute recoil, which can be seen with postballoon angioplasty [72].

Several architectures can exist for DEB, yet only one will be discussed here. The DEB may consist of several components, which can include a balloon, elongated members, and reinforcing strands. Reinforcing strands between the inside surface of the balloon and the outside of the elongating member can enhance the balloon when inflated as it is subjected to high pressures. When the balloon is not collapsed, a lumen can exist between the inner surface of the balloon and the outer surface of the elongated member. Within the lumen, flexible members can be disposed. The flexible members can have both an external surface as well as an internal cavity that contains the therapeutic agent critical to the DEB activity. A connecting channel can network the outer surface of the balloon with the internal cavity. The balloon is subjected to high pressures when being inflated [73]. A 60-second drug-elution time can be anticipated [60, 72].

Various materials may be used to construct the DEB. For the strands of the DEB, a polymer blend, metal alloy, laminar or composite construction may be used. Substances for the balloons and members with elongated materials used include polytetrafluoroethylenes, polyethylenes, polypropylenes, polyurethanes, nylons, and polyesters (including polyalkylene terephthalate polymers and copolymers). Therapeutic agents for treating restenosis include sirolimus, tacrolimus, everolimus, cyclosporine, dexamethasone, paclitaxel, actinomycin, geldanamycin, cilostazole, methotrexate, vincristine, and mitomycin [72, 73].

Several important discoveries have given rise to the use of local paclitaxel delivery with through-coated balloons. First, sustained drug release is not required for a long-term antiproliferative effect. Second, a rapid uptake of paclitaxel by vascular smooth muscle cells occurs and can be retained for up to 1 week, thereby allowing a prolonged antiproliferation. Third, paclitaxel has a strong lipophilic nature for its retention to the vessel wall, making it an important therapeutic agent for DEB [72, 74].

Several important advantages for DEB have been indicated by Waksman et al. A homogenous drug transfer to the entire vessel wall along with a rapid release of high concentrations of the drug has been observed. This transfer may be sustained for up to a week in the vessel wall. It should be noted, however, that such delivery has little impact on long-term healing. The absence of a polymer may decrease chronic inflammation and the trigger for late thrombosis. DEBs may assist in avoiding stent placement in bifurcations or small vessels. This may minimize the abnormal flow patterns that may occur with a stent. Local delivery of DEB may also reduce the overdependence on antiplatelet therapy [72].

Several indications have been suggested for the use of DEB. An important indication of the DEB, especially for paclitaxel-eluting balloons, would be for the treatment of in-stent restenosis. Other indications for the use of DEB include tortuous vessels, small vessels, or long diffuse calcified lesions, which can result in stent fracture. In addition, obstructing scaffolding of major side branches or in bifurcated lesions may be another indication of DEB [72, 77].

The efficacy of a paclitaxel-eluting balloon for the treatment of sirolimus-eluting stent (SES) restenosis has been assessed by an important prospective single-blind randomized trial by Habara et al. with a follow-up time of 6 months as the primary end-point. The trial consisting of a total of 50 patients with SES restenosis in which patients were randomly assigned to either a paclitaxel group (n = 25) or a conventional balloon angioplasty group (n = 25). An incidence of recurrent restenosis of 8.7% in the DEB group was shown versus 62.5% for the conventional balloon angioplasty (p = 0.0001). The study showed a target lesion revascularization of 4.3% for the DEB versus 41.7% for the conventional balloon angioplasty (p =0.003). Furthermore, it was shown that a cumulative major adverse cardiac events-free survival was significantly better for the DEB group (96%) compared to the 60% for the balloon angioplasty group (p = 0.005) [75].

Conclusion

In this chapter, we have focused on the patient's needs when addressing biomaterials. Cardiac stents have been addressed extensively in this chapter to emphasize this point. This reflects the understanding of the existence of individual differences between patients and the diverse nature of patient variability, which may be acquired or genetic, or both. An important test of the validity of the biomaterial and its implementation is the analysis of the results of human clinical trials with long-term follow-ups. However, we highlight that it always should be remembered that the final endpoint is the individual patient.

References

1. Ratner BD, Hoffman AS, Schoen FJ, Lemons JE (eds.). 2004. *Biomaterials Science: An Introduction to Materials in Medicine.* 2nd edition, (various pagings). Academic Press.
2. Fauci AS, Braunwald E, Hauser S, Longo D, Jameson JL, Loscalzo J (eds.). 2008. *Harrison's Principles of Internal Medicine,* 17th edition, (various pagings). McGraw-Hill Professional.
3. Fishman G. 2008. Personal communication, National Academies, USA.
4. Gulesserian T, Wenzel C, Endler G, Sunder-Plassmann R, Marsik C, Mannhalter C, Iordanova N, et al. 2005. Clinical restenosis after coronary stent implantation is associated with the heme oxygenase-1 gene promoter polymorphism and the heme oxygenase-1 +99G/C variant. *Clin Chem.* 51(9): 1661–65.
5. Sucker C, Scheffold N, Cyran J, Ghodsizad A, Scharf RE, Zotz RB. 2008. No evidence for involvement of prothrombotic platelet receptor polymorphisms in acute coronary stent thrombosis. *Int J Cardiol.* 123(3): 355–57.
6. Monraats PS, de Vries F, de Jong LW, Pons D, Sewgobind VD, Zwinderman AH, de Maat MP, 't Hart LM, et al. 2006. Inflammation and apoptosis genes and the risk of restenosis after percutaneous coronary intervention. *Pharmacogenet Genomics.* 16(10): 747–754.
7. Tàssies D. 2006. Pharmacogenetics of antithrombotic drugs. *Curr Pharm Des.* 12(19): 2425–35.

8. Ganesh SK, Skelding KA, Mehta L, O'Neill K, Joo J, Zheng G, Goldstein J, et al. 2004. Rationale and study design of the CardioGene Study: Genomics of in-stent restenosis. *Pharmacogenomics.* 5(7): 952–1004.

9. Shah SH, Hauser ER, Crosslin D, Wang L, Haynes C, Connelly J, Nelson S, et al. 2008. ALOX5AP variants are associated with in-stent restenosis after percutaneous coronary intervention. *Atherosclerosis.* (201): 148–151.

10. Falanga A. 2005. Thrombophilia in cancer. *Semin Thromb Hemost.* 31(1): 104–10.

11. Gouin-Thibault I, Achkar A, Samama MM. 2001. The thrombophilic state in cancer patients. *Acta Haematol.* 106(1–2): 33–42.

12. Akl EA, Kamath G, Kim SY, Yosuico V, Barba M, Terrenato I, Sperati F, Schünemann HJ. 2007. Oral anticoagulation for prolonging survival in patients with cancer. *Cochrane Database Syst Rev.* 18(2): CD006466.

13. Sherer Y, Blank M, Shoenfeld Y. 2007. Antiphospholipid syndrome (APS): Where does it come from? *Best Pract Res Clin Rheumatol.* 21(6):1071–78.

14. Pierangeli SS, Harris EN. 2003. Probing antiphospholipid-mediated thrombosis: The interplay between anticardiolipin antibodies and endothelial cells. *Lupus* 12(7): 539–45.

15. Goldin-Lang P, Tran QV, Fichtner I, Eisenreich A, Antoniak S, Schulze K, Coupland SE, Poller W, Schultheiss HP, Rauch U. 2008. Tissue factor expression pattern in human non-small cell lung cancer tissues indicate increased blood thrombogenicity and tumor metastasis. *Oncol Rep.* 20(1): 123–28.

16. Sousa JE, Costa MA, Abizaid AC, Rensing BJ, Abizaid AS, Tanajura LF, Kozuma K, et al. 2001. Sustained suppression of neointimal proliferation by sirolimus-eluting stents: One-year angiographic and intravascular ultrasound follow-up. *Circulation* 104: 2007–2011.

17. Morice MC, Serruys PW, Sousa JE, Fajadet J, Ban Hayashi E, Perin M, Colombo A, et al. 2002. A randomized comparison of a sirolimus-eluting stent with a standard stent for coronary revascularization. *N Engl J Med* 346: 1773–1780.

18. Moses JW, Leon MB, Popma JJ, Fitzgerald PJ, Holmes DR, O'Shaughnessy C, Caputo RP, et al. 2003. Sirolimus-eluting stents versus standard stents in patients with stenosis in a native coronary artery. *N Engl J Med* 349: 1315–1323.

19. Grube E, Silber S, Hauptmann KE, Mueller R, Buellesfeld L, Gerckens U, Russell ME. 2003. TAXUS I: Six- and twelve-month results from a randomized, double-blind trial on a slow-release paclitaxel-eluting stent for de novo coronary lesions. *Circulation* 107: 38–42.

20. Colombo A, Drzewiecki J, Banning A, Grube E, Hauptmann K, Silber S, Dudek D, Fort S, et al. 2003. Randomized study to assess the effectiveness of slow- and moderate-release polymer-based paclitaxel-eluting stents for coronary artery lesions. *Circulation* 108: 788–794.

21. Stone GW, Ellis SG, Cox DA, Hermiller J, O'Shaughnessy C, Mann JT, Turco M, et al. 2004. A polymer-based, paclitaxel-eluting stent in patients with coronary artery disease. *N Engl J Med* 350: 221–231.

22. Monraats PS, Kurreeman FA, Pons D, Sewgobind VD, de Vries FR, Zwinderman AH, de Maat MP, et al. 2007. Interleukin 10: A new risk marker for the development of restenosis after percutaneous coronary intervention. *Genes Immun.* Jan, 8(1): 44–50. Epub 2006 Nov 23.

23. Rudez G, Pons D, Leebeek F, Monraats P, Schrevel M, Zwinderman A, de Winter R, et al. 2008. Platelet receptor P2RY12 haplotypes predict restenosis after percutaneous coronary interventions. *Hum Mutat.* 29(3): 375–80.

24. Rudez G, Bouman HJ, van Werkum JW, Leebeek FW, Kruit A, Ruven HJ, ten Berg JM, de Maat MP, Hackeng CM. 2009. Common variation in the platelet receptor P2RY12 gene is associated with residual on-clopidogrel platelet reactivity in patients undergoing elective percutaneous coronary interventions. *Circ Cardiovasc Genet.* 2(5): 515–21.
25. Ramcharitar S, Serruys PW. 2008. Biodegradable stents. *Minerva Cardioangiol.* 56(2): 205–13.
26. Winger TM, Ludovice PJ, Chaikof EL. 1999. Formation and stability of complex membrane-mimetic monolayers on solid supports. *Langmuir* 15: 3866–74.
27. Waksman R. 2006. Biodegradable stents: They do their job and disappear. *J Invasive Cardiol.* 18(2): 70–74.
28. Grabow N, Martin H, Schmitz KP. 2002. The impact of material characteristics on the mechanical properties of a poly(L-lactide) coronary stent. *Biomed Tech (Berl)* 47 Suppl 1, Pt 1: 503–505.
29. Berg JM, Tymoczko JL, Stryer L. 2002. *Biochemistry,* 5th ed. xxxviii, (various pagings). New York: W.H. Freeman.
30. Defife KM, Hagen KM, Clapper DL, Anderson JM. 1999. Photochemically immobilized polymer coatings: Effects on protein adsorption, cell adhesion, and leukocyte activation. *J Biomater Sci Polym Ed.* 10(10): 1063–74.
31. Woodhouse KA, Klement P, Chen V, Gorbet MB, Keeley FW, Stahl R, Fromstein JD, Bellingham CM. 2004. Investigation of recombinant human elastin polypeptides as non-thrombogenic coatings. *Biomaterials.* 25(19): 4543–53.
32. Jordan SW, Haller CA, Sallach RE, Apkarian RP, Hanson SR, Chaikof EL. 2007. The effect of a recombinant elastin-mimetic coating of an ePTFE prosthesis on acute thrombogenicity in a baboon arteriovenous shunt. *Biomaterials.* 28(6): 1191–97.
33. Dresang LT, Fontaine P, Leeman L, King VJ. 2008. Venous thromboembolism during pregnancy. *Am Fam Physician* 77(12): 1709–16.
34. Snow V, Qaseem A, Barry P, et al. 2007. Management of venous thrombo-embolism: A clinical practice guideline from the American College of Physicians and the American Academy of Family Physicians Panel on Deep Venous Thrombosis/Pulmonary Embolism. *Ann Intern Med.* 146(3): 204–210.
35. Chang J, Elam-Evans LD, Berg CJ, et al. 2003. Pregnancy-related mortality surveillance—United States, 1991–1999. *MMWR Surveill Summ* 52(2): 1–8.
36. de Boer K, ten Cate JW, Sturk A, Borm JJ, Treffers PE. 1989. Enhanced thrombin generation in normal and hypertensive pregnancy. *Am J Obstet Gynecol* Jan, 160(1): 95–100.
37. Wang S, Retzinger GS. 2002. *Hypercoagulability during Pregnancy Lab Lines.* A publication of the Department of Pathology and Laboratory Medicine at the University of Cincinnati. September/October Volume 8, Issue 5.
38. Therapeutic anticoagulation in pregnancy. 2006. Norfolk and Norwich University Hospital (NHS Trust). Reference number CA3017. 9th June [review June 2009].
39. Ageno W, Crotti S, Turpie AG. 2004. The safety of antithrombotic therapy during pregnancy. *Expert Opin Drug Saf* 3(2): 113–18.
40. Bates SM, Greer IA, Hirsh J, Ginsberg JS. 2004. Use of antithrombotic agents during pregnancy: The Seventh ACCP Conference on Antithrombotic and Thrombolytic Therapy. *Chest* 126(3 suppl): 627S–644S.
41. Abadi S, Einarson A, Koren G. 2002. Use of warfarin during pregnancy. *Can Fam Physician* 48: 695–97.

42. Cugno M, Borghi MO, Lonati LM, Ghiadoni L, Gerosa M, Grossi C, De Angelis V, et al. 2010. Patients with antiphospholipid syndrome display endothelial perturbation. *J Autoimmun*. 34(2): 105–10.

43. Pierangeli SS, Colden-Stanfield M, Liu X, Barker JH, Anderson GL, Harris EN. 1999. Antiphospholipid antibodies from antiphospholipid syndrome patients activate endothelial cells in vitro and in vivo. *Circulation* 99(15): 1997–2002.

44. Allen KL, Hamik A, Jain MK, McCrae KR. 2011. Endothelial cell activation by antiphospholipid antibodies is modulated by Kruppel-like transcription factors. *Blood J* 117(23): 6383–6391. [Epub ahead of print].

45. Simantov E, LaSala J, Lo SK, Gharavi AE, Sammaritano LR, Salmon JE, Silverstein RL. 1995. Activation of cultured vascular endothelial cells by antiphospholipid antibodies. *J Clin Invest* 96: 2211–2219.

46. Del Papa N, Guidali L, Spatola L, Bonara P, Borghi MO, Tincani A, Balestrieri G, Meroni PL. 1995. Relationship between anti-phospholipid and anti-endothelial cell antibodies: b2 glycoprotein I mediates the antibody binding to endothelial membranes and induces the expression of adhesion molecules. *Clin Exp Rheumatol* 13: 179–185.

47. Heublein B, Rohde R, Kaese V, Niemeyer M, Hartung W, Haverich A. 2003. Biocorrosion of magnesium alloys: A new principle in cardiovascular implant technology? *Heart* 89: 651–56.

48. Peeters P, Bosiers M, Verbist J, Deloose K, Heublein B. 2005. Preliminary results after application of absorbable metal stents in patients with critical limb ischemia. *J Endovasc Ther* 12: 1–5.

49. Zartner P, Cesnjevar R, Singer H, Weyand M. 2005. First successful implantation of a biodegradable metal stent into the left pulmonary artery of a preterm baby. *Catheter Cardiovasc Interv* 66: 590–94.

50. Stack RE, Califf RM, Phillips HR, et al. 1988. Interventional cardiac catheterization at Duke Medical Center. *Am J Cardiol* 62 (suppl F): 3F–24.

51. Waksman R, Pakala R, Kuchulakanti PK, et al. 2006. Safety and efficacy of bioabsorbable magnesium alloy stents in porcine coronary arteries. *Catheter Cardiovasc Interv* 68: 606–17.

52. Peuster M, Wohlsein P, Brugmann M, et al. 2001. A novel approach to temporary stenting: Degradable cardiovascular stents produced from corrodible metal-results 6–18 months after implantation into New Zealand white rabbits. *Heart* 86: 563–69.

53. Balcon R, Beyar R, Chierchia S, et al. 1997. Recommendations on stent manufacture, implantation and utilization. *Eur Heart J* 18: 1536–47.

54. Waksman R, Pakala R, Baffour R, Seabron R, Hellinga D, Tio FO. 2008. Short-term effects of biocorrodible iron stents in porcine coronary arteries. *J Interv Cardiol* 21: 15–20.

55. Hermawan H, Alamdari H, Mantovani D, Dubé D. 2008. Iron–manganese: New class of degradable metallic biomaterials prepared by powder metallurgy. *Powder Metall* 51: 38–45.

56. Peuster M, Hesse C, Schloo T, Fink C, Beerbaum P, von Schnakenburg C. 2006. Longterm biocompatibility of a corrodible peripheral iron stent in the porcine descending aorta. *Biomaterials* 27: 4955–62.

57. Erbel R, Di Mario C, Bartunek J, et al. 2007. PROGRESS-AMS (Clinical Performance and Angiographic Results of Coronary Stenting with Absorbable Metal Stents) Investigators. Temporary scaffolding of coronary arteries with

bioabsorbable magnesium stents: A prospective, non-randomized multicentre trial. *Lancet* 369: 1869–75.

58. Waksman R, Erbel R, Di Mario C, et al. 2009. PROGRESS-AMS (Clinical Performance Angiographic Results of Coronary Stenting with Absorbable Metal Stents) Investigators. Early- and long-term intravascular ultrasound and angiographic findings after bioabsorbable magnesium stent implantation in human coronary arteries. *JACC Cardiavasc Interv* 2: 312–20.
59. Ghimire G, Spiro J, Kharbanda R, et al. 2009. Initial evidence for the return of coronary vasoreactivity following the absorption of bioabsorbable magnesium alloy coronary stents. *EuroIntervention* 4: 481–84.
60. Waksman R, Pakala R. 2010. Biodegradable and bioabsorbable stents. *Curr Pharm Des* 16(36): 4041–51.
61. Black J. 2005. *Biological Performance of Materials: Fundamentals of Biocompatibility*, 4th edition, (various pagings). CRC Press.
62. Neuman MR (Series editor), SA Guelcher, JO Hollinger. 2006. *An Introduction to Biomaterials*. The Biomedical Engineering Series, (various pagings). CRC Press.
63. Cattaneo M. 2011. Resistance to anti-platelet agents. *Thromb Res.* Feb, 127 Suppl 3: S61–63.
64. Finn AV, Nakazawa G, Kolodgie FD, Virmani R. 2009. Temporal course of neointimal formation after drug-eluting stent placement: Is our understanding of restenosis changing? *JACC Cardiovasc Interv* 2: 300–302.
65. Finn AV, Kolodgie FD, Harnek J, Guerrero LJ, Acampado E, Tefera K, Skorija K, Weber DK, Gold HK, Virmani R. 2005. Differential response of delayed healing and persistent inflammation at sites of overlapping sirolimus- or paclitaxel-eluting stents. *Circulation* 112: 270–78.
66. Rodriguez-Granillo A, Rubilar B, Rodriguez-Granillo G, Rodriguez AE. 2011. Advantages and disadvantages of biodegradable platforms in drug eluting stents. *World J Cardiol* 3(3): 84–92.
67. Price MJ, Barker CM. 2011. Functional testing methods for the antiplatelet effect of P2Y12 receptor antagonists. *Biomark Med.* 5(1): 43–51.
68. Price MJ, Tantry US, Gurbel PA. 2011. The influence of CYP2C19 polymorphisms on the pharmacokinetics, pharmacodynamics, and clinical effectiveness of P2Y(12) inhibitors. *Rev Cardiovasc Med* 12(1): 1–12.
69. George J, Herz I, Goldstein E, Abashidze S, Deutch V, Finkelstein A, Michowitz Y, Miller H, Keren G. 2003. Number and adhesive properties of circulating endothelial progenitor cells in patients with in-stent restenosis. *Arterioscler Thromb Vasc Biol* 23(12): e57–60.
70. Ramcharitar S, Vaina S, Serruys PW. 2007. The next generation of drug eluting stents: What's on the horizon? *Am J Cardiovasc Drugs* 7: 81–93.
71. Camenzind E, Steg PG, Wijns W. 2007. Stent thrombosis late after implantation of first-generation drug-eluting stents: a cause for concern. *Circulation* 115: 1440–1455.
72. Waksman R, Pakala R. 2009. Drug-eluting balloon: The comeback kid? *Circ Cardiovasc Interv* 2(4): 352–58.
73. Holman TJ, Weber J, Shewe S. United States Patent 7,491,188 B2, February 17, 2009.
74. Axel DI, Kunert W, Goggelmann C, Oberhoff M, Herdeg C, Küttner A,Wild DH, et al. 1997. Paclitaxel inhibits arterial smooth muscle cell proliferation and migration in vitro and in vivo using local drug delivery. *Circulation* 96: 636–645.

75. Habara S, Mitsudo K, Kadota K, Goto T, Fujii S, Yamamoto H, Katoh H, Oka N, et al. 2011. Effectiveness of paclitaxel-eluting balloon catheter in patients with sirolimus-eluting stent restenosis. *JACC Cardiovasc Interv* 4(2): 149–54.
76. Erne P, Schier M, Resink TJ. 2006. The road to bioabsorbable stents: Reaching clinical reality? *Cardiovasc Intervent Radiol* 29: 11–6.
77. Baber U, Kini AS, Sharma SK. 2010. Stenting of complex lesions: An overview. *Nat Rev Cardiol*. Sep, 7(9):485–96.
78. Aoki J, Ong AT, Rodriguez Granillo GA, et al. 2005. "Full metal jacket" (stented length > or = 64 mm) using drug-eluting stents for de novo coronary artery lesions. *Am Heart J* 150: 994–99.
79. Hoffmann R, Mintz GS, Popma JJ, et al. 1996. Chronic arterial responses to stent implantation: A serial intravascular ultrasound analysis of Palmaz-Schatz stents in native coronary arteries. *J Am Coll Cardiol* 28: 1134–39.
80. Spuentrup E, Ruebben A, Mahnken A, et al. 2005. Artifact-free coronary magnetic resonance angiography and coronary vessel wall imaging in the presence of a new, metallic, coronary magnetic resonance imaging stent. *Circulation* 111: 1019–26.
81. Pinto Slottow TL, Waksman R. 2007. Overview of the 2006 food and drug administration circulatory system devices panel meeting on drug-eluting stent thrombosis. *Catheter Cardiovasc Interv* 69: 1064–74.

3

Medical Applications of Micro-Electro-Mechanical Systems (MEMS) Technology

Michael A. Huff

CONTENTS

Abstract

The miniaturization of electromechanical systems offers unique opportunities for scientific and technological progress in the medical sciences.
Micromechanical devices and systems are inherently smaller, lighter, and
faster than their macroscopic counterparts, and in many cases are also more
precise. Micromechanical devices are fabricated using many of the same techniques and materials commonly used in integrated circuit (IC) processing
in conjunction with various "micromachining" processes. Therefore, micromechanical devices can be readily integrated with electronics to develop
high-performance closed-loop controlled micro-electro-mechanical systems
(MEMS). While integrated circuit technology has brought unprecedented
computational power ever closer to the point of use, MEMS enables the development of smart systems by providing the required interface between the
available computational power and the physical world through perception
and control capabilities of sensors and actuators. The sensors gather information from the environment; the electronics process this information and
direct the actuators to manipulate the environment for a desired purpose.
MEMS is not just a new technology, it is also a new method for manufacturing
low-cost, miniature, high-performance, functionally sophisticated electromechanical devices and systems for medical applications. MEMS is already having a significant impact in existing biomedical sciences, but what is more
exciting is that it is opening up entirely new areas for treatment, diagnosis,
and scientific study. This chapter will review MEMS technology and why it
is of interest in medical applications, as well as review a few of the current
efforts in employing MEMS technology to important medical applications.

What Is MEMS Technology?

Micro-electro-mechanical systems, or MEMS, is a technology that in its most
general form can be defined as miniaturized electro-mechanical devices and

mechanical structures that are implemented using the techniques of micro-fabrication. The physical dimensions of MEMS devices can vary from over one millimeter on the higher end of the dimensional scale down to below one micron. Likewise, the types of MEMS devices can vary from relatively simple structures in which there are no mechanically moving elements, to extremely complex and sophisticated electromechanical systems with multiple moving elements controlled by microelectronics integrated onto a single substrate. Obviously, the one main criterion of MEMS is that there are at least some elements on the substrate having a mechanical functionality regardless of whether any of the elements move.

While the functional elements of MEMS are miniaturized structures, sensors, actuators, and microelectronics, the most notable (and perhaps most interesting) elements are the microsensors and microactuators. Microsensors and microactuators are appropriately categorized as "transducers," which are defined as devices that convert energy from one form to another. For example, microsensor devices typically convert a mechanical movement into an electrical signal. Conversely, microactuators typically convert electrical signals into mechanical movements. Over the past 25+ years MEMS technologists have created an extremely large number of microsensors for almost every imaginable sensing modality, including pressure, temperature, inertial force, chemical and biological species, magnetic fields, and radiation. Importantly, many of these micromachined sensors have demonstrated performances exceeding those of their macroscale equivalents. For example, the micromachined version of a pressure transducer will usually outperform a pressure sensor made using macroscale manufacturing techniques. Not only is the performance of MEMS devices exceptional, but also their method of production leverages batch fabrication techniques used in the integrated circuit (IC) industry, which translates into low per-device production costs, smaller size and weight, and other benefits as well. Consequently, with MEMS technology it is possible to not only achieve stellar device performance, but to do so at a relatively low cost level. Because of the exceptional performance and cost benefits, the commercial markets for microsensor devices are growing at a rapid rate.

In tandem with developments in microsensors, the MEMS research and development community has also demonstrated a number of microactuators, including microvalves, optical switches, micromirror display arrays, microresonators, micropumps, microflaps on airfoils, and many others. Surprisingly, even though these microactuators are extremely small, they can frequently cause effects at the macroscale level. That is, these tiny actuators can perform mechanical feats far larger than their size would imply. For example, researchers have located small microactuators on the leading edge of airfoils of an aircraft and have been able to steer this aircraft using only these microminiaturized devices [1].

The real potential of MEMS starts to become fulfilled when these miniaturized sensors, actuators, and structures can all be merged onto a common silicon substrate along with integrated circuits. This enables the realization of smart products by augmenting the computational ability of microelectronics with the perception and control capabilities of microsensors and microactuators. The integrated circuits can be thought of as the "brains" of a system, and MEMS augments this decision-making capability with "eyes" and "arms," to allow microsystems to sense and control the environment. Microsensors gather information from the environment through measuring mechanical, thermal, biological, chemical, optical, and magnetic phenomena. The electronics then process the information derived from the sensors and through some decision-making capability direct the microactuators to respond by moving, positioning, regulating, pumping, and filtering, thereby controlling the environment for some desired outcome or purpose. Because MEMS devices are manufactured using batch fabrication techniques similar to ICs, unprecedented levels of functionality, reliability, and sophistication can be placed on a small silicon chip at a relatively low cost. MEMS technology is extremely diverse and fertile, both in its expected application areas and in how the devices are designed and manufactured. Already, MEMS is revolutionizing many product categories by enabling complete systems-on-a-chip to be realized. MEMS is having a huge impact in medical applications, and the number and diversity of MEMS applications in medicine is expected to grow extremely quickly as the technology develops and matures.

Although major progress has been made in recent decades, the history of MEMS technology traces back to the earlier days of the integrated circuit industry. The discovery of the piezoresistive effect in silicon was reported in 1954 by Prof. C. S. Smith, which led to the creation of the entire silicon-based sensor industry. It is still one of the most widely used sensing methodologies in MEMS [2]. Perhaps the first and most far-reaching vision of the promise of miniaturized electromechanical systems was provided in a talk by Prof. Richard Feynman at an annual American Physical Society meeting in 1959 aptly entitled "There's Plenty of Room at the Bottom" [3]. In this talk, Prof. Feynman discussed many possibilities and opportunities of microminiaturized devices, but since the microelectronics industry did not yet exist, he did not foresee how these devices would be made and even questioned whether they would be commercially useful.

One of the first techniques of bulk micromachining, the isotropic etching of silicon, was reported in 1960 [4]. A few years later, in 1967, the still widely used bulk micromachining technique of anisotropic silicon etching was reported [5]. The first published paper in the scientific literature using the term *micromachining* was in 1982 [6]. This paper, entitled "Silicon as a Mechanical Material," published in the *Proceedings of the IEEE* by Kurt Petersen, is probably the most referenced paper in the entire MEMS field and highlighted many of the advantages of silicon as a material for mechanical systems. In the 1980s and 90s, a huge number of papers and patents in the MEMS technology domain occurred, and this growth has steadily continued to the present day. One of the first commercial MEMS products was the micromachined pressure transducer using the piezoresistive properties of silicon discovered by Smith several decades before [7]. Interestingly, this MEMS device technology was also the first to be widely used in medical applications.

The Advantages Offered by MEMS Technology for Medical Applications

There are many important benefits of MEMS technology for medical applications. First, MEMS are made using integrated circuit-like processes, which enable the integration of *multiple and diverse functionalities* onto a single microchip. The ability to integrate miniaturized sensors, miniaturized actuators, and miniaturized structures along with microelectronics has far-reaching implications in countless medical products and applications. For example, implantable and ambulatory medical devices usually require sophisticated and diverse functionality, such as closed-loop control capability for precision drug delivery, but they also must have small size and weight to prevent them from being a burden to the patient or increasing the risk of complications. Moreover, higher levels of functionality also enable higher performance levels and reliability since devices can be made to be self-calibrating, self-maintaining, self-healing, etc.

Second, MEMS borrows many of the production techniques of batch fabrication from the integrated circuit industry, and therefore the per-unit device or microchip cost of complex miniaturized electromechanical systems can be radically reduced—similar to the per-die cost reductions we have experienced in the IC industry. Although the cost of the production equipment and each wafer can be relatively high, the fact that this cost can be spread over many dies in batch fabrication production can drastically lower the per-part cost. With huge pressures to contain medical costs in every industrial country, increasingly sophisticated and complex treatment methodologies will be done on an out-patient and ambulatory basis requiring devices to have multiple and diverse functionality as described above, but without commensurate increases in cost. MEMS is one of the only technologies that can provide this level of technological capability and performance at a low cost level. The lower cost of these miniaturized electromechanical systems also allows them to be easily and massively deployed and more easily maintained and replaced as needed. Consequently, MEMS devices are extremely well suited for disposable medical applications.

Third, integrated circuit fabrication techniques coupled with the tremendous advantages of silicon and many other thin-film materials in mechanical applications allows the reliability of miniaturized electromechanical systems to be exceptionally high. A similar effect was exhibited in the technological transition from discrete electronic components mounted on a printed circuit board to the era of integrated circuits. Discrete electronics soldered onto printed circuit boards do not have nearly the same level of high reliability that is seen with integrated circuits. As miniaturized sensors and actuators are integrated onto a single microchip with electronics, similar improvements in system reliability are being provided by MEMS technology. For example, MEMS inertial sensor technology used in crash airbag sensors for automobiles is providing the extremely high reliability levels demanded by this application. Obviously, most medical applications require exceptional reliability, and MEMS technology is well suited to meet these demands.

Fourth, miniaturization of microsystems enables many benefits, including increased portability, lower power consumption (very important for implantable medical device applications), and the ability to place radically more functionality in a smaller amount of space and without any increase in weight.

Fifth, the ability to make the signal paths smaller allows the overall performance of electromechanical systems to be enormously improved.

The largest growth area in medicine is in alternate care sites, including off-site treatment centers, surgical centers, home care, nursing homes, and ambulatory care. This is being driven by the need to reduce treatment costs as well as improve patient comfort and outcomes. However, most alternate care applications also require the medical devices to be less expensive, smaller, less intrusive, more interoperable, lighter, safer, more user friendly, and more functional than ever before. In short, MEMS is a technology that can simultaneously meet all of these demanding requirements in the medical device market.

How Are MEMS Made?

MEMS fabrication uses many of the same techniques that are used in the integrated circuit domain, such as photolithography, physical vapor deposition, oxidation, diffusion, ion implantation, and LPCVD, and combines these capabilities with highly specialized micromachining processes. The most widely used micromachining processes are discussed below. For information relating to integrated circuit fabrication techniques, readers are referred to [8].

Bulk Micromachining

The oldest micromachining technology is bulk micromachining. This technique involves the selective removal of the substrate material in order to realize miniaturized mechanical structures and components. Bulk micromachining can be accomplished using chemical, physical, or mechanical (as well as combinations of these) methods. However, chemical methods are more widely used in MEMS industry.

A widely used bulk micromachining technique is chemical wet etching, which involves the immersion of a substrate into a solution of reactive chemical that will etch exposed regions of the substrate at measurable rates [9]. Chemical wet etching is popular in MEMS because it can provide a very high etch rate and selectivity. Furthermore, the etch rates and selectivity can be modified by altering the chemical composition of the etch solution, adjusting the etch solution temperature, modifying the dopant concentration of the substrate, and controlling which crystallographic planes of the substrate are exposed to the etchant solution.

The basic mechanism of chemical wet etching involves reactant transport to the surface of the substrate, followed by reaction at the substrate surface between the etchant solution and the substrate material, and then transport of the reaction products from the substrate. If the transport of reactants to the surface of the substrate or transport of reaction products away from the substrate surface are the rate-determining steps, then etching is defined as "diffusion limited" and the etch rate can be increased by stirring the solution. If the surface reaction is the rate-determining step, then etching is "reaction-rate limited" and etch rate is very dependent on etch solution temperature, etch solution composition, and substrate material. In practice, it is usually preferred that the process is reaction-rate limited since this gives more repeatability and higher etch rate.

There are two general types of chemical wet etching in bulk micromachining: isotropic wet etching and anisotropic wet etching [9–12]. In isotropic wet etching, the etch rate is not dependent on the crystallographic orientation of the substrate, and the etching ideally proceeds in all directions at equal rates. The most common isotropic etchant for silicon is a solution of HNO_3, HF, and $HC_2H_3O_2$. The reaction is given by

$$HNO_2 + HNO_3 + H_2O \rightarrow 2HNO_2 + 2OH^- + 2H^+$$

Holes and $(OH)^-$ ions are supplied by HNO_3 when it combines with H_2O and trace concentrations of HNO_2. Note that the reaction is autocatalytic because of the regeneration of HNO_2. Increasing the concentration moves the reaction toward diffusion-limited etch-rate dependence, and etching can be controlled by stirring. Increasing HF concentration or temperature increases the surface reaction rate. In theory, lateral etching under the masking layer etches at the same rate as the etch rate in the normal direction. However, in practice lateral etching is usually much slower without stirring, and consequently isotropic wet etching is almost always performed with vigorous stirring of the etchant solution. Figure 3.1 illustrates the profile of the etch using an isotopic wet etchant with and without stirring the solution.

Any etching process requires a masking material to be used, preferably with a high selectivity relative to the substrate material. Common masking materials for isotropic wet silicon etching include silicon dioxide and silicon

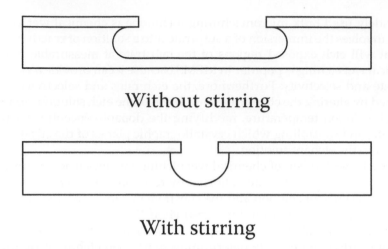

Without stirring

With stirring

FIGURE 3.1
The etch profile, with and without stirring, using an isotropic wet chemical etchant.

nitride. Silicon nitride has a much lower etch rate compared to silicon dioxide and therefore is more frequently used.

The etch rate of some isotropic wet etchant solution mixtures are dependent on the dopant concentration of the substrate material. For example, the commonly used mixture of $HC_2H_3O_2$:HNO_3:HF in the ratio of 8:3:1 will etch highly doped silicon (> 5 x 10^{18} atoms/cm^3) at a rate of 50 to 200 microns/hour, but will etch lightly doped silicon material at a rate 150 times less. Nevertheless, the etch-rate selectivity with respect to dopant concentration is highly dependent on solution mixture.

The much more widely used wet etchants for silicon micromachining are anisotropic wet etchants. Anisotropic wet etching involves the immersion of the substrate into a chemical solution where the etch rate is dependent on the crystallographic orientation of the substrate. The fact that etching varies according to silicon crystal planes is attributed to the different bond configurations and atomic densities that the different planes expose to the etchant solution. Wet anisotropic chemical etching is typically described in terms of etch rates according to the various normal crystallographic places, usually <100>, <110>, and <111>. In general, silicon anisotropic etching etches more slowly along the <111> planes than all the other planes in the lattice, and the difference in etch rate between the various lattice directions can be as high as 1000 to 1. It is thought that the reason for the slower etch rate of the <111> planes is that these planes have the highest density of exposed silicon atoms in the etchant solution, as well as 3 silicon bonds below the plane, thereby leading to some amount of chemical shielding of the surface.

The ability to delineate the different crystal planes of the silicon lattice in anisotropic wet chemical etching provides a higher-resolution etch capability with tighter dimensional control than is possible with isotropic etching. It also provides for two-sided processing to embody self-isolated structures where only one side is exposed to the environment. This assists in the packaging of the device and is very useful for MEMS devices exposed to harsh environments, such as pressure sensors.

Anisotropic etching techniques have been around for over 25 years and are commonly used in the manufacturing of silicon pressure sensors as well as bulk micromachined accelerometers.

Figure 3.2 is an illustration of some of the shapes that are possible using anisotropic wet etching of a <100> oriented silicon substrate, including an inverted pyramidal and a flat bottomed trapezoidal etch pit. Note that the shape of the etch pattern is determined primarily by the slower etching <111> planes. Figure 3.3a and Figure 3.3b are SEM photographs of a silicon substrate after an anisotropic wet etching. Figure 3.3a shows a trapezoidal etch pit that has been subsequently diced across the etch pit, and Figure 3.3b shows the backside of a thin membrane that could be used to make a pressure sensor. It is important to note that the etch profiles shown in these figures are only for a <100> oriented silicon wafer; substrates with other crystallographic orientations will exhibit different shapes. Occasionally, substrates with other orientations are used in MEMS fabrication, but given the cost, lead times, and availability, the vast majority of substrates used in bulk micromachining have <100> orientation.

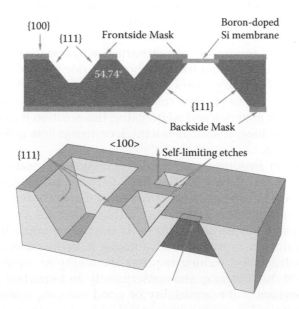

FIGURE 3.2
The shape of the etch profiles of a <100> oriented silicon substrate after immersion in an anisotropic wet etchant solution.

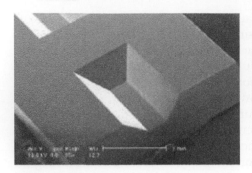
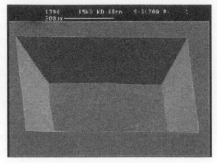

FIGURES 3.3a AND b
SEMs of a <100> oriented silicon substrate after immersion in an anisotropic wet etchant.

There are three basic types of anisotropic wet etchants that are commonly used. The first and by far the most popular anisotropic etchants are aqueous alkaline solutions such as KOH, NH_4OH, NaOH, CsOH, and TMAH [9, 10]. These etchants have high etch rates and a relatively high etch-rate ratio between the <100> and <111> planes. Also, TMAH is sometimes preferred for use on pre-processed microelectronic wafers, since this etchant does not etch aluminum appreciably under certain conditions. The drawbacks of these etchants are that they have a relatively high etch rate of silicon dioxide, which is frequently used as a masking material. There is also the potential for alkali contamination of the wafer using these etchants, although there are cleaning procedures that can be employed to minimize these risks.

Another popular anisotropic wet etchant is ethylene-diamine and pyrocatechol, or EDP. This etchant has a higher etch rate ratio of the <100> and <111> planes and has a larger variety of masking materials that can be used compared to the aqueous alkaline solutions. The drawbacks of EDP are that it is a carcinogenic material, and when using this solution it can be difficult to see the wafer etching. Also, EDP is a thick orange-yellow material and can be hard to clean up.

The last type of anisotropic etchant is hydrazine and water ($N_2H_2:H_2O$). One advantage of this etchant is that it has a very low silicon dioxide etching rate, yet it has a relatively poor etch rate ratio of the <100> and <111> planes. The biggest disadvantage of hydrazine etching is that it is a very hazardous material. Although it was used in the early days of micromachining, given that its disadvantages outweigh its advantages, it is now rarely used.

Useful anisotropic wet etching requires the ability to successfully mask certain areas of the substrate, and consequently an important criterion for selecting an etchant is the availability of good masking materials. Silicon nitride is a commonly used masking material for anisotropic wet etchants since it has a very low etch rate in most etchant solutions. Some care must be exercised in the type of silicon nitride used, since any pinhole defects will result in the attack of the underlying silicon. Also, some low-stress silicon-rich nitrides can etch at much higher rates compared to stoichiometric silicon nitride formulations. Thermally grown SiO_2 is frequently used as a masking material, but some care must be exercised to ensure a sufficiently thick masking layer when using KOH etchants, since the etch rates of oxide can be high. Photoresists are unusable in any anisotropic etchant. Many metals, including Ta, Au, Cr, Ag, and Cu, hold up well in EDP, and Al holds up in TMAH under certain conditions.

In general, the etch rate, etch rate ratios <100>/<111>, and etch selectivities of anisotropic etchants are strongly dependent on the chemical composition and temperature of the etchant solution. The etch rate [R] obeys the Arrhenius law given by

$$[R] = R_o \exp(-E_a/kT) \text{ (micron/hour)}$$

where R_o is a constant, E_a is the activation energy, k is Boltzman's constant, and T is temperature in degrees Kelvin. Both R_o and E_a will vary with the type of etchant, etchant composition, and crystallographic orientation of the material being etched. Fortunately, there is a wealth of published literature characterizing many of the commonly used anisotropic etchants, and readers are referred to [9, 10] for more information.

Frequently, when using bulk micromachining it is desirable to make thin membranes of silicon or control the etch depths very precisely. As with any chemical process, the uniformity of the etching can vary across the substrate, making this difficult. Timed etches whereby the etch depth is determined by multiplying the etch rate by the etch time are difficult to control, and etch depth is very dependent on sample thickness uniformity, etchant species diffusion effects, loading effects, etchant aging, surface preparation, etc. To allow a higher level of precision in anisotropic etching, the MEMS field has developed solutions to this problem, namely etch stops. Etch stops are very useful in controlling the etching process and providing uniform etch depths across the wafer, from wafer to wafer, and from wafer lot to wafer lot. There are two basic types of etch stop methods that are used in micromachining: dopant etch stops and electrochemical etch stops.

Etch stops in silicon are commonly made by the introduction of dopants into the silicon material. The most popular etch stop is heavy p-type doping of silicon with boron (>5 x 10^{19} cm^{-3}) to create an etch stop. The lightly doped region of the wafer will etch at the normal rate and the highly doped region of the silicon will have a very slow etch rate. The dopant is introduced into the silicon using the standard techniques of diffusion or ion implantation followed by an anneal, providing for a controlled depth and reasonable uniformity of the dopants in the substrate. Figure 3.4 is a graph of the normalized etch rate of <100> oriented silicon wafer in KOH at various concentrations as a function of the boron dopant concentration [13, 14]. As can be seen, the etch rate falls off very quickly at dopant concentrations above 10^{19}

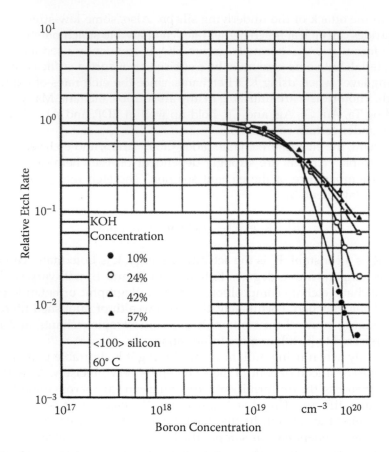

FIGURE 3.4
Etch rate plotted versus boron concentration for <100> oriented silicon wafer at different etchant concentrations.

boron atoms per cm³. One of the problems with boron etch stops is that the surface of the silicon will be so highly doped that it may not be useful. For example, the material at the concentrations required for a good etch stop would not be useful for making a piezoresistive device.

The other etch stop method used in silicon bulk micromachining is the electrochemical etch stop [15–19]. Figure 3.5 is an illustration of a three-terminal electrochemical etch setup. Electrochemical etching of silicon using an anisotropic etchant is useful since it provides very good dimensional control (e.g., diaphragm thickness is reproducible) and can make diaphragms with lightly doped material, which is required for high-quality piezoresistive devices. The disadvantage of electrochemical etch stops is that they require special fixturing to each wafer in order to make electrical contacts, and an electronic control system is needed to control and apply the correct voltage potential to the wafer during the etch.

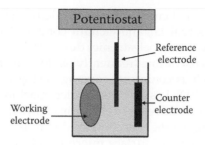

FIGURE 3.5
The setup for a three-terminal electrochemical etch of silicon.

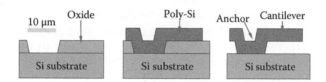

FIGURE 3.6
A surface micromachining process.

Surface Micromachining

Surface micromachining is another very popular technology for the fabrication of MEMS devices. There is a very large number of variations of how surface micromachining is performed, depending on the materials and etchant combinations that are used [20, 21, 22]. However, the common theme involves a sequence of steps starting with the deposition of some thin-film material to act as a temporary mechanical layer onto which the actual device layers are built, followed by the deposition and patterning of the thin-film device layer of material referred to as the structural layer, and then followed by the removal of the temporary layer to release the mechanical structure layer from the constraint of the underlying layer, thereby allowing the structural layer to move. A diagram of a surface micromachining process is given in Figure 3.6, showing an oxide layer being deposited and patterned. This oxide layer is temporary and is commonly referred to as the sacrificial layer. Next, a thin film layer of polysilicon is deposited and patterned, this being the structural mechanical layer. Finally, the temporary sacrificial layer is removed and the polysilicon layer is now free to move as a cantilever.

Some of the reasons surface micromachining is so popular is that it provides for precise dimensional control in the vertical direction. This is because the structural and sacrificial layer thicknesses are defined by deposited film thicknesses, which can be accurately controlled. Also, surface micromachining provides for precise dimensional control in the horizontal direction, since the structural layer tolerance is defined by the fidelity of the photolithography and etch processes used. Other benefits of surface micromachining are that a large variety of structural, sacrificial, and etchant combinations can be used, and some are compatible with microelectronics devices to enable integrated MEMS devices. Surface micromachining frequently exploits the deposition characteristics of thin films such as conformal coverage using low-pressure chemical vapor deposition (LPCVD). Lastly, surface micromachining uses single-sided wafer processing and is relatively simple. This allows higher integration density and lower resultant per die cost compared to bulk micromachining.

One of the disadvantages of surface micromachining is that the mechanical properties of LPCVD structural thin-films are usually unknown and must be measured. Also it is common for these types of films to have a high state of residual stress, frequently necessitating a high-temperature anneal to reduce residual stress in the structural layer. Also, the reproducibility of the mechanical properties in these films can be difficult to achieve. Additionally, the release of the structural layer can be difficult owing to a stiction effect whereby the structural layer is pulled down and stuck to the underlying substrate by capillary forces during release. Stiction can also occur in use, and an anti-stiction coating material may be needed.

The most commonly used surface micromachining process and material combination is a phosphosilicate glass (PSG) sacrificial layer, a doped polysilicon structural layer, and the use of hydrofluoric acid as the etchant to

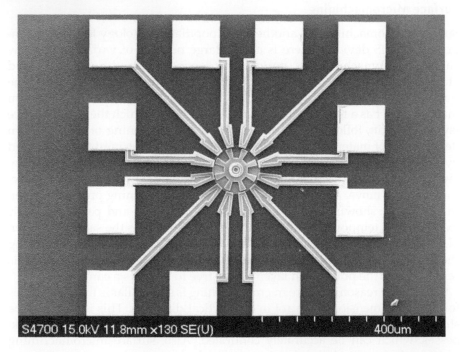

S4700 15.0kV 11.8mm ×130 SE(U)   400um

FIGURE 3.7
SEM of a polysilicon micromotor fabricated using a surface micromachining process.

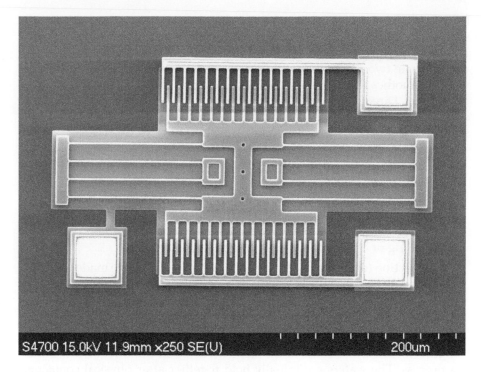

S4700 15.0kV 11.9mm ×250 SE(U)   200um

FIGURE 3.8
SEM of a polysilicon resonator structure fabricated using a surface micromachining process.

remove the PSG sacrificial layer and release the device. This type of surface micromachining process is employed in the widely used MEMSCap MUMPS process technology [22]. Figure 3.7 and Figure 3.8 are SEMs of two surface micromachined polysilicon MEMS devices.

Another variation of the surface micromachining process is to use a metal structural layer, a polymer layer as the sacrificial layer, and an O_2 plasma as the etchant. The advantage of this process is that the temperature of the sacrificial and the structural layer depositions are sufficiently low so as not to degrade any microelectronics in the underlying silicon substrate, for processes that integrate MEMS with electronics. Also, since the sacrificial layer is removed without immersion in a liquid, problems associated with stiction during release are avoided. A process similar to this is used to produce the Texas Instruments digital light processor (DLP) device used in projection systems [22].

Wafer Bonding

Wafer bonding is a micromachining method that is analogous to welding in the macroscale world and involves the joining of two (or more) wafers together to create a multi-wafer stack. There are three basic types of wafer bonding, including direct or fusion bonding; field-assisted or anodic bonding; and bonding using an intermediate layer. In general, all bonding methods require substrates that are very flat, smooth, and clean, in order for the wafer bonding to be successful and free of voids. Direct or fusion bonding is typically used to mate two silicon wafers together, or alternatively, to mate one silicon wafer to another silicon wafer that has a thin film on the surface such as an oxide layer. Direct wafer bonding can be performed on other combinations, such as bare silicon to a silicon wafer with a thin film of silicon nitride on the surface as well. The basic wafer bonding process has five steps [23]:

1. Hydration and cleaning of the wafer surfaces (RCA clean, piranha immersion, etc.)
2. Physical contacting and pressing together of wafers (must be done quickly and in a clean environment since the substrates will have a surface charge after cleaning that will attract particles and prevent bonding or cause voids)
3. Inspection (usually infrared inspection if the wafers are made of silicon) of pre-anneal bond quality
4. Elevated temperature anneal,
5. Infrared inspection of post-anneal bond quality

Figure 3.9 illustrates some of the steps involved in the direct wafer bonding process. The wafers are initially held together after physical contact as a result of the hydrogen bonds created by hydration of the surfaces. The two wafers to be bonded can be pre-processed and then aligned during the bonding procedure so as to register features on the top wafer to features on the bottom wafer. After the elevated temperature anneal, the bond strength between the two wafers can be similar to that of single crystal silicon. There is relatively low residual stress in the bonded layers after anneal. It is imperative that the wafers are flat, smooth, and clean prior to bonding and that the

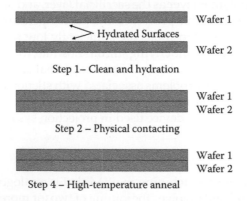

FIGURE 3.9
Direct or fusion wafer bonding.

bonding process is performed in a very clean environment. The hydration of the surfaces makes them highly charged and attractive to particles in the environment. Any particles attached to the wafer surfaces prior to bonding will result in voids between the two wafers or the inability of the wafers to bond. Although the anneal can be performed at lower temperatures, the bond strengths tend to strengthen as the anneal temperature is increased.

As mentioned, wafer bonding is analogous to welding in the macroscale world. Wafer bonding is used to attach a thick layer of single crystal silicon onto another wafer. This can be extremely useful when it is desirable to have a thick layer of material for applications requiring appreciable mass or in applications where the material properties of single crystal silicon are advantageous over those of thin-film LPCVD materials. Direct wafer bonding is also used to fabricate silicon-on-insulator (SOI) wafers having device layers several microns or more in thickness.

Another popular wafer bonding technique is anodic bonding, which is illustrated in Figure 3.10. In anodic bonding a silicon wafer is bonded to a Pyrex® 7740 wafer using an electric field and elevated temperature [24–26]. The two wafers can be pre-processed prior to bonding and can be aligned during the bonding procedure. The mechanism by which anodic bonding works is based on the fact that Pyrex 7740 has a high concentration of Na^+ ions; a positive voltage applied to the silicon wafer drives the Na^+ ions from the Pyrex glass surface, thereby creating a negative charge at the glass surface. The elevated temperature during the bonding process allows the Na^+ ions to migrate in the glass with relative ease. When the Na^+ ions reach the interface, a high field results between silicon and glass, and this combined with the elevated temperatures fuses the two wafers together. As with direct wafer bonding, it is imperative that the wafers are flat, smooth, and clean and that the anodic bonding process is performed in a very clean environment. Any particles on the wafer surfaces will result in voids between the two wafers. One advantage of this process is that Pyrex 7740 has a thermal expansion coefficient nearly equal to that of silicon, and therefore there is a low value of residual stress in the layers. Anodic bonding is a widely used technique for MEMS packaging.

In addition to direct and anodic bonding there are other wafer bonding techniques that are used in MEMS fabrication. One method is eutectic bonding, which involves the bonding of a silicon substrate to another silicon

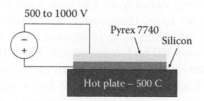

FIGURE 3.10
The setup for anodic wafer bonding.

substrate at an elevated temperature using an intermediate layer of gold on the surface of one of the wafers [27]. Eutectic bonding works because the diffusion of gold into silicon is extremely rapid at elevated temperatures. In fact this is a preferred method of wafer bonding at relatively low temperatures.

Another wafer-bonding technique used in MEMS is glass frit bonding [28]. In this process, glass is spun or screen-printed onto a substrate surface. Next, this wafer is physically contacted to another wafer, and then the composite is annealed in order to flow the glass intermediate layer and thus bond the two wafers together.

Last, various polymers can be used as intermediate layers to bond wafers, including epoxy resins, photoresists, polyimides, and silicones [29]. This technique is commonly used during various fabrication steps in MEMS such as when the device wafer becomes too fragile to handle without mechanical support.

High-Aspect-Ratio MEMS Fabrication Technologies

Deep Reactive Ion Etching of Silicon

Deep reactive-ion etching, or DRIE, is a relatively newer fabrication technology that has been widely adopted by the MEMS community [30, 31]. This technology enables very high-aspect-ratio dry plasma etches to be performed into silicon substrates. The sidewalls of the etched holes are nearly vertical, and the depth of the etch can be hundreds or even thousands of microns into the silicon substrate.

Figure 3.11 illustrates how deep reactive ion etching is accomplished. The etch is a dry plasma etch and uses a high-density plasma to alternatively etch the silicon and deposit an etch resistant polymer layer on the sidewalls. The etching of the silicon is performed using a SF_6 chemistry, whereas the deposition of the etch-resistant polymer layer on the sidewalls uses a C_4F_8 chemistry. Mass flow controllers alternate back and forth between these two chemistries during the etch. The protective polymer layer is deposited on the sidewalls as well as on the bottom of the etch pit, but the anisotropy of the etch removes the polymer at the bottom of the etch pit faster than the polymer is removed from the sidewalls. The resultant etched sidewalls are not optically smooth. If the sidewalls are magnified under SEM inspection, a characteristic washboard or scalloping pattern is seen in the sidewalls. The etch rates on most commercial DRIE systems varies from 1 to over 5 microns per minute. DRIE systems are single wafer tools. Photoresist can be used as a masking layer for DRIE etching. The selectivity with photoresist and oxide varies, but typical values are 75 to 1 and 150 to 1, respectively. For a through-wafer etch, a relatively thick photoresist masking layer will be required. The aspect ratio of the etch can be as high as 30 to 1 or more, but a value of 15 to 1 is more typical. The process recipe depends on the amount of exposed silicon owing to loading effects in the system, with larger exposed areas etching at a much faster rate compared to smaller exposed areas. Consequently, the etch must

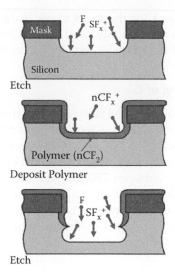

FIGURE 3.11
How deep reactive ion etching works.

FIGURE 3.12
SEM of the cross section of a silicon wafer demonstrating high-aspect-ratio and deep trenches that can be fabricated using DRIE technology.

frequently be characterized for the exact mask feature and depth to obtain desirable results. DRIE is a commonly used process technology for the implementation of MEMS and microfluidic devices for medical applications.

Deep Reactive Ion Etching of Glass

Glass substrates can also be etched deep into the material with high-aspect ratios. This is a much newer process technology and it has been gaining in popularity in MEMS fabrication [32]. Figure 3.13 shows a structure fabricated into fused silica using this technology. The typical etch rates for high-aspect-ratio glass etching range between 250 and 500 nm per minute. Depending on the depth of the photoresist, metal or a polysilicon can be used as a mask. This deep, high-aspect-ratio etching technology can be used to make MEMS and microfluidic devices from glass materials, which obviously has considerable utility in medical applications.

LIGA

Another popular high-aspect-ratio micromachining technology is called LIGA, which is a German acronym for "Lithographie, Galvanoformung, Abformung" [33]. This is primarily a non-silicon based technology and requires the use of synchrotron generated X-ray radiation. The basic process, outlined in Figure 3.14, starts with the cast of an X-ray radiation sensitive Polymethylmethacrylate (PMMA) onto a suitable substrate. A special X-ray mask is used for the selective exposure of the PMMA layer using X-rays. The PMMA is then developed and will be defined with extremely smooth and

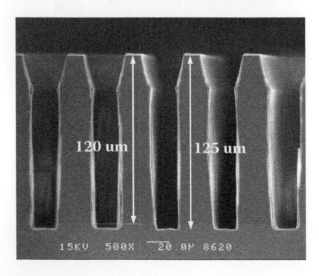

FIGURE 3.13
SEM of deep, high-aspect-ratio trenches etched into fused silica using a plasma etch technology.

nearly perfectly vertical sidewalls. Also, the penetration depth of the X-ray radiation into the PMMA layer is quite deep and allows exposure through very thick PMMA layers, up to and exceeding 1 mm. After the development, the patterned PMMA acts as a polymer mold and is placed into an electroplating bath and nickel is plated into the open areas of the PMMA. The PMMA is then removed, thereby leaving the metallic microstructure.

Because LIGA requires a special mask and a synchrotron (X-ray) radiation source for the exposure, the cost of this process is relatively expensive. A variation of the process that reduces the cost of the micromachined parts made with this process is to reuse the fabricated metal part (step 5) as a tool insert to imprint the shape of the tool into a polymer layer (step 3) (see

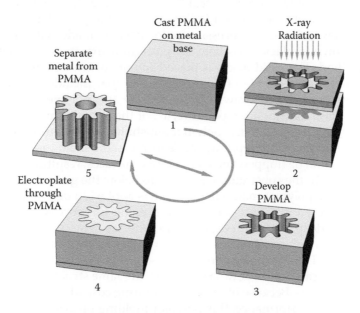

FIGURE 3.14
The steps involved in the LIGA process to fabricate high-aspect-ratio MEMS devices.

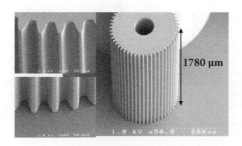

FIGURE 3.15
SEM of a tall high-aspect-ratio gear made using LIGA technology.

hot embossing section), followed by electroplating of metal into the polymer mold (step 4) and removal of the polymer mold (step 5) [34]. Obviously this sequence of steps avoids the need for a synchrotron radiation source each time a part is made and thereby significantly lowers the cost of the process. The dimensional control of this process is quite good, and the tool insert can be used many times before it is worn out.

Hot Embossing

Hot embossing is a process used for replicating deep high-aspect-ratio structures in polymer materials by fabricating the metal tool insert using LIGA or comparable technology and then embossing the tool insert pattern into a polymer substrate, which is then used as the part. Figure 3.16 illustrates the hot embossing process. A mold insert is made with the inverse pattern (shown in black cross-hatched pattern) using an appropriate fabrication method. The mold insert is placed into a hot embossing system (see Figure 3.17 for an example of a hot embosser) that includes a chamber in which a vacuum can be drawn. The substrate and polymer are heated to above the glass transition temperature, T_g, of the polymer material, and the mold insert is pressed into the polymer substrate. The vacuum is critical for the polymer to faithfully replicate the features in the mold insert, since otherwise air would be trapped between the two surfaces, resulting in distorted features. Finally, the substrates are cooled to below the glass-transition temperature of the polymer material and force is applied to de-emboss the substrates. As shown in Figure 3.18, hot embossing can successfully replicate complicated, deep, and high-aspect-ratio features. This process can make imprints into a polymer hundreds of microns deep with very good dimensional control. The advantage of this process is that the cost of the individual polymer parts can be very low compared to the same structures made using other technologies. Because of the overwhelming cost advantages combined with very good performance, this polymer molding process is very popular for producing microfluidic components for medical applications [35].

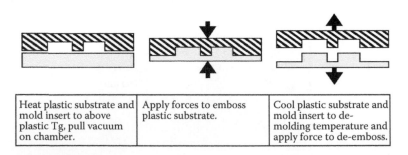

| Heat plastic substrate and mold insert to above plastic Tg, pull vacuum on chamber. | Apply forces to emboss plastic substrate. | Cool plastic substrate and mold insert to de-molding temperature and apply force to de-emboss. |

FIGURE 3.16
The hot embossing process to create microdevices. (Courtesy of the MEMS and Nanotechnology Exchange at CNRI)

FIGURE 3.17
A hot embossing platform during use. (Courtesy of the MEMS and Nanotechnology Exchange at CNRI)

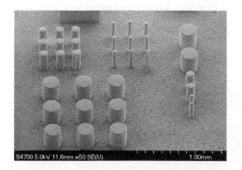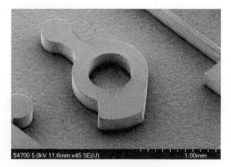

FIGURE 3.18
SEM of a variety of small test structures made in a plastic substrate using hot embossing technology at the MEMS and Nanotechnology Exchange. The height of the plastic microstructures is nearly 300 um, and the smallest features have a diameter of about 25 um. (Courtesy of the MEMS and Nanotechnology Exchange at CNRI)

Other Micromachining Technologies

In addition to bulk micromachining, surface micromachining, wafer bonding, and high-aspect-ratio micromachining technologies, there are a number of other techniques used to fabricate MEMS devices. We shall review a few of the more popular methods in this section, but readers are referred to [9, 10] for an exhaustive catalog of MEMS processes.

XeF₂ Dry Phase Etching

Xenon difluoride (XeF_2) in a vapor state is an isotropic etchant for silicon [36]. This etchant is highly selective with respect to other materials commonly used in microelectronics fabrication, including LPCVD silicon nitride, thermal SiO_2, and aluminum. Since this etchant is a completely dry release process, it does not suffer from the stiction problems of wet release processes. Figure 3.19 is a cross-sectional SEM of a cantilever beam made by isotropic etching of the silicon substrate using XeF_2 to partially undercut the metal material of the cantilever.

This etchant is popular with micromachining of microstructures in pre-processed CMOS wafers where openings in the passivation layers on the surface of the substrate are made to expose the silicon for etching.

Electro-Discharge Micromachining

Electro-discharge micromachining, or micro-EDM, is a process used to machine a conductive material using electrical breakdown discharges to remove material [9]. A working electrode is made from a metal material onto which high-voltage pulses are applied. The working electrode is brought into close proximity to the material to be machined, which is immersed in a dielectric fluid. The minimal sizes of features that can be made with micro-EDM are dependent on the size of the working electrode and how it is fixtured, but holes as small as tens of microns have been made using this method. One issue with micro-EDM is that it is a slow serial process.

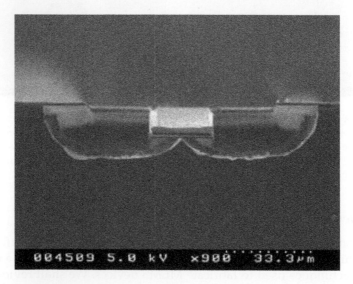

FIGURE 3.19
SEM of a beam undercut using XeF_2 isotropic etching of silicon.

Laser Micromachining

Lasers can generate an intense amount of energy in very short pulses of light and direct that energy onto a selected region of material for micromachining [9]. Among the many types of lasers now in use for micromachining are CO_2, YAG, and excimer. Each has its own unique properties and capabilities suited to particular applications. Factors that determine the type of laser to use for a particular application include laser wavelength, energy, power, and temporal and spatial modes; material type; feature sizes and tolerances; processing speed; and cost. The action of CO_2 and Nd:YAG lasers is essentially a thermal process, in which focusing optics are used to direct a predetermined energy/power density to a well-defined location on the work piece to melt or vaporize the material. Another mechanism, which is nonthermal and referred to as photoablation, is the exposure of organic materials to ultraviolet radiation generated from excimer, harmonic YAG, or other UV sources. Similar to microEDM, laser micromachining can produce features on the order of tens of microns, but it is a serial process and therefore slow. Figure 3.20 is a photo of some very small holes made in a medical catheter using laser micromachining.

Focused Ion Beam Micromachining

Another versatile tool for performing micromachining is the focused ion beam (FIB) [9]. The accelerating voltages are adjustable from few keV to several hundred keV. The spot sizes can be focused down to below 25 nm, making it capable of producing extremely small structures. The user can input a 3-D CAD solid model of desired etching topology; the computer-controlled stage with sub-micron positional accuracy allows very precise registration of sample. In addition to material removal, the FIB can also be used to perform ion-induced deposition, lithography, implantation doping, mask repair, device repair, and device diagnostics. Many of these tools can also be outfitted with a secondary column for mass analysis of particles removed from the substrate using uSIMS.

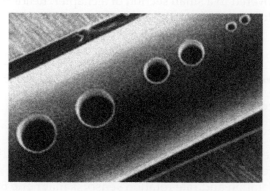

FIGURE 3.20
Very small holes made in a medical catheter using laser micromachining.

MEMS Process Sequence Development and Process Integration

The making of MEMS devices involves bringing together a multiplicity of processing steps into what is called a *process sequence* [22]. The number of processing steps involved in fabricating a MEMS device can vary from less than ten for a simple process sequence to a few hundred for a more complicated one. A common thread in MEMS device implementation is that most MEMS process sequences are customized to the device that is being fabricated. That is, the process sequence for any one MEMS device is likely to be vastly different from the process sequence used to implement any other MEMS device. This is quite a different situation from what is found in microelectronics, where there are typically only three components made, namely transistors, resistors, and capacitors, and a small number of process sequences commonly used for fabrication, including CMOS, Bipolar, and BiCMOS. In contrast, there are a vastly larger number of MEMS devices, each having its own unique and custom process sequence. This means that the development of the process sequence for any MEMS device is typically very challenging, and most successful MEMS development projects engage MEMS fabrication experts early and extensively. Process integration is defined as understanding, characterizing, and optimizing to the greatest extent possible the interrelationship of the individual processing steps in a process sequence. Given the customization of MEMS process sequences, it should not be surprising that process integration is of critical importance in any MEMS development effort. Specifically, skilled MEMS fabrication technologists having relevant practical experience are needed to successfully develop a process sequence for a MEMS device. Moreover, the skills of these technologists must be directly relevant to MEMS device process development, that is, device development skills learned in other related fields such as microelectronics are usually not sufficient and do not directly transfer well to MEMS process sequence development. The learning of MEMS process development and process integration requires many years of education and practical experience and cannot be adequately covered in a small section of a chapter. Readers are referred to [22] for more information about this subject.

MEMS Device Technologies

Microsensor Technology

Many types of sensors exist that use micromachined devices, including pressure; acoustic; temperature (including infrared focal plane arrays); inertial (including acceleration and rate rotation sensors); magnetic field (Hall, magnetoresistive, and magnetotransistors); force (including tactile); strain; optical; radiation; and chemical and biological [9, 12, 37, 38]. As discussed above, sensors are transducers that convert one form of energy, such as mechanical force, to another form of energy, usually an electrical signal. There are several basic physical principals by which sensors function, including resistive, magnetic, photoconductive piezoresistive, piezoelectric, thermocouples and thermopiles, diodes, and capacitive. All of these sensing principles have been successfully demonstrated in MEMS sensor devices. There are far too many variations of MEMS sensors to review even a fraction of them reported in the literature, and therefore we will limit ourselves to reviewing only a few of the transduction principles used in the implementation of microsensors. Readers are referred to [9, 11, 12, 37, 38] for more information.

A piezoresistive material is one in which the resistance is influenced by applied mechanical strains [9, 11, 12, 37, 38]. This phenomenon is most prominent in semiconductors where the strain induces changes in the electronic band structure of the material, thereby making the carrier scattering rates dependent on direction of transport. This effect can be used to make a variety of sensors by placing the piezoresistive elements at a position where the strain is maximized. A useful quality factor of piezoresistive materials is the gauge factor (GF), which is given by the normalized change in resistance divided by the strain. A higher gauge factor implies a more sensitive piezoresistor-to-mechanical deformation. Silicon as a material can have a very high gauge factor, approaching 200 in some special configurations. In comparison, metal resistors typically have gauge factors of around 2. Piezoresistors are used primarily as strain measurement sensors where the resistor (strain-sensing element) is placed on a compliant surface or structure such as illustrated in Figure 3.21 for pressure sensors and accelerometers. Note that micromachining allows the substrate to be selectively removed, which significantly reduces the stiffness in the sensing region of the device. This allows the stain to be maximized in a localized region of the sensor where the piezoresistors are located, thereby also maximizing the sensitivity of the sensor device. Typically, the piezoresistors are placed into a Wheatstone bridge circuit configuration, although other configurations can be used as well. The fabrication of two types of pressure sensors using the piezoresistive effect to transduce the mechanical strain into an electrical signal is described below.

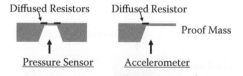

FIGURE 3.21
Two types of MEMS silicon sensors made using the piezoresistive effect in semiconductors. Piezoresistors are formed by diffusing a suitable type and concentration of dopants into the substrate material. The substrate is partially removed to reduce the mechanical stiffness where the piezoresistors are located. This has the effect of increasing the strain under loading at these locations and, more importantly, of increasing the sensitivity of the sensors.

Capacitive sensing is very commonly utilized in MEMS sensors owing to its inherent simplicity [9, 11, 12, 37, 38]. In general, the capacitance of a two terminal device is given by

$$C = (e_o e_r A)/d$$

where e_o is the dielectric constant of free space, e_r is the relative dielectric constant of any material between the electrodes, A is the area of the capacitor, and d is the separation of the electrodes [39]. In general, capacitors can be made to function as sensors in five ways, as shown in Figure 3.22: (a) varying the distance between electrodes, (b) varying the position of the center electrode relative to two outer electrodes giving a differential measurement, (c) varying the overlap between electrodes, (d) varying the differential overlap between electrodes and (e) varying the position of the dielectric in the space between electrodes. As an example of how a capacitive structure can be used to implement an accelerometer, see Figure 3.23, where a pair of electrodes is separated by a distance that can change as the compliant cantilever is deflected under acceleration loading in the direction normal to the device.

Another widely used physical phenomenon for making sensors in MEMS is the piezoelectric property of some materials [9–12, 37, 38]. Piezoelectricity occurs when a mechanical strain produces an electrical polarization (i.e., voltage potential) in the material. Likewise, an applied electrical field will also induce a mechanical strain in the material. The first effect is primarily used for sensors, while the second is usually employed for actuators. Silicon and germanium are centrosymmetric crystals and

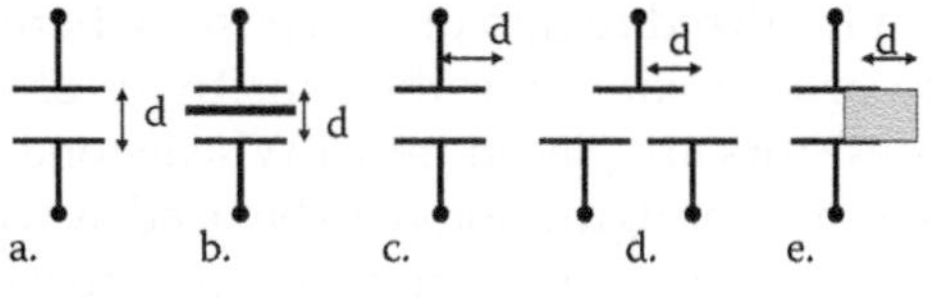

a.   b.   c.   d.   e.

FIGURE 3.22
Various configurations for using a capacitor as a sensing element.

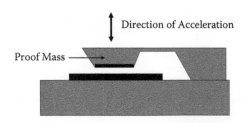

↑ Direction of Acceleration

Proof Mass

FIGURE 3.23
MEMS capacitive-based accelerometer.

therefore display no piezoelectric effect (unless strain is induced). Other materials that lack a center of symmetry, such as quartz, lead zirconate titanate (PZT), and zinc oxide (ZnO), are commonly used as piezoelectric materials in MEMS sensors. The latter two can be deposited in thin-film form on silicon or other material substrates.

Microactuator Technology

A variety of basic principals are used to implement MEMS actuators, including electrostatic, piezoelectric, magnetic, magnetostrictive, bimetallic, and shape memory alloy. Each of these has its respective advantages and disadvantages, and therefore careful consideration of the specific application requirements must be part of the selection process. We shall review a few of the more popular methods for MEMS microactuators here. Readers are referred to [9–12, 37, 38] for a more comprehensive review.

Electrostatic actuation is based on the mutual attraction of two oppositely charged plates. The force F generated between plates under the application of a voltage potential V is given by

$$F = 1/2 \; (e_o \, e_r \, A) \; (V/d)^2$$

where e_o is the permittivity of free space, e_r is the relative permittivity of a dielectric, A is the area of the plates, V is the applied voltage potential on the plates, and d is the separation between the plates [41]. There are some inherent advantages of electrostatic microactuators that make them attractive and popular for MEMS devices, including being easy to fabricate and integrate with electronics, having very low power consumption during operation, and being able to exhibit very high mechanical bandwidths. Some of the disadvantages of electrostatic actuators are that the force is non-linear with displacement and applied voltage, the resultant force is relatively small, and the operating voltages can be relatively high.

Another popular method of implementing microactuators for MEMS is based on the bimetallic effect. The bimetallic effect uses the differing thermal expansion coefficients of two different materials to realize a thermal-based microactuator. When these two materials are made into a composite structure and heated, a thermally induced stress is generated in the structure if it is sufficiently compliant. The thermal strain is given by

$$\alpha = (a_{filma} - a_{filmb}) \; (T_{ele} - T_{amb})$$

where a_{filma} and a_{filmb} are the thermal expansion coefficients of the top and bottom films, respectively, and T_{ele} and T_{amb} are the temperature of the bimetallic element and ambient, respectively [9, 12]. Some of the attributes of a bimetallic microactuator for MEMS are high power consumption for heating, low mechanical bandwidth, relatively complex design and fabrication, relatively large deflections (at least compared to electrostatic actuation), a linear deflection versus power relationship, and sensitivity to environmental conditions.

A simple bimetallic microactuator can be made by depositing a thin-film of aluminum onto a thin compliant silicon cantilever (Figure 3.24) while passing current through the aluminum layer. As the aluminum layer heats up through Joule heating, the cantilever will bend because of the different expansion coefficients of the two materials, with the aluminum metal having a larger expansion under heating than the silicon semiconductor material. Obviously, the choice of materials used in a bimetallic microactuator will depend on the application.

Yet another popular material for implementing microactuators in MEMS is shape-memory alloys [40–42]. A shape-memory alloy is a material that undergoes a martensite to austenite phase change upon heating. During this phase change, the material will return to its unstrained shape (i.e., the material has memory). Figure 3.25 is an illustration of the shape memory effect. Shape memory alloys can be sputter deposited in thin film form on silicon wafers. Heating of the film is usually achieved by Joule heating. The shape memory effect is a reversible effect and can be repeated many times. The attributes of shape memory in MEMS actuators include the following:

Aluminum thin film

Silicon

FIGURE 3.24
A bimetallic microactuator made from a thin film aluminum layer on a silicon cantilever. When heated, the aluminum has a higher thermal expansion coefficient than silicon and causes the cantilever to deflect downward.

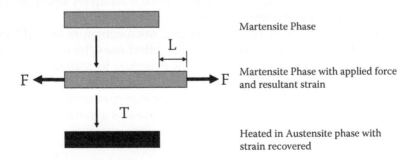

Martensite Phase

L

F ⟵   ⟶ F

Martensite Phase with applied force and resultant strain

T

Heated in Austensite phase with strain recovered

FIGURE 3.25
The shape memory effect. The SMA sample in the martensite phase (top) is at room temperature and unstrained. In the middle, the sample is strained at room temperature. At bottom, the SMA sample is heated and the sample undergoes a phase change from the martensite to the austenite phase, and the strain is completely recovered. This cycle can be repeated over and over.

they exhibit very high energy densities; they have enormous recoverable strain levels (strains of over 8% have been demonstrated); since they are thermally activated, the power consumption can be high and the mechanical bandwidth low; the processing is somewhat complicated; and the material can fatigue if repeatedly cycled at high strain levels.

MEMS Design Tools

MEMS design is frequently more demanding than the design of integrated circuits. In the integrated circuits domain, a process technology and the associated design rules usually already exist. The designers merely need to build these into their computer-aided design tool, which usually only considers the electrical effects, and come up with a design. In MEMS technology, the situation is much more complicated. As we have already mentioned, a customized process sequence must frequently be developed for each device type as part of the product development effort, and design rules are not known until a process sequence is finalized. Furthermore, the material properties are usually not fully known and are highly dependent on the process sequence and conditions, which also are not known beforehand. Also, many MEMS devices have several physical phenomena (electrical, mechanical, thermal, chemical, etc.) occurring simultaneously that leads to many strongly coupled fields, thereby making the design process more challenging. Importantly, the IC designer usually does not need to know much about the fabrication, whereas in MEMS design, the designer must be an expert in MEMS fabrication [22, 43].

Fortunately, there are some design tool capabilities now available to the MEMS community that are suitable for process, physical, device, and systems modeling [44]. The process-modeling tools are essentially the same as those used by the integrated circuit industry and enable the designer to create process models and mask artwork. Numerical techniques are available to simulate the processing steps. Although these tools are quite good for predicting electrical behavior, they are not very good for predicting mechanical material properties. One important feature of MEMS process design tools is the ability to create representative 3-D renderings of the devices. The physical-level design tools are used to model the behavior of components in the real 3-D continuum using partial differential equations. These tools can be analytical or numeric; the numeric techniques include finite-element, boundary-element, and finite-difference methods. These tools are based mostly on finite-element tools that are modified versions of FEM tools used in macroscale design. The device level models are macro-models or reduced-order models that capture the physical behavior of a component (over a limited range) and are compatible with the system level models. Care must be exercised to ensure that the dynamic range of the model is not overextended. System level models are high-level block diagrams and lumped-parameter models that describe the system as a coupled set of ordinary differential equations.

Backend MEMS Processes: Device Release and Die Separation

Backend processes for MEMS are usually more challenging then those found in microelectronics. One reason is that many MEMS products have movable elements that are made from silicon or other thin film materials, and these elements can be easily broken during handling, assembly, packaging, and testing. The backend processes for MEMS frequently involve the releasing of the movable elements and die separation. In some MEMS process sequences, particularly surface micromachining, a sacrificial material layer is removed to "release" the moveable element on the MEMS device. The release of the movable elements prior to die separation means that the fragile MEMS devices must survive the cutting and separation processes, which frequently require special fixturing. Likewise, if the die separation is performed first, a means to perform the release of the MEMS device in a batch mode rather than individually will require a special fixturing.

Another issue is the release process itself. Many times the release is performed using an immersion into a strong chemical solution, such as hydrofluoric acid, for the removal of a sacrificial oxide layer. The wafer must then be rinsed and dried and the retracting liquid can cause the movable element to be brought into contact with the underlying substrate and remain stuck. This is commonly referred to as stiction and a widely used solution to this problem is CO_2 super-critical point drying [45]. Additionally, sometimes a vapor hydrofluoric (HF) acid is used instead of immersion to avoid stiction during release.

While super-critical drying or vapor HF etch may solve the problem of stiction during release, it is not uncommon for compliant MEMS devices to come into contact with the underlying substrate during use as well. This can also result in stiction, which can cause the MEMS device to become inoperable. Self-assembled monolayers and environmental control of the MEMS device in the package are frequently used to overcome the stiction effects during use [46].

MEMS assembly, testing and packaging are other demanding areas of the technology. There are very few MEMS wafer level functional test systems available; the test equipment that is available is very application specific. MEMS testers must usually apply suitable stimulus to device under test (pressure, inertia, etc.). Consequently, a test system for a pressure sensor will invariably be very different from a tester for an inertial sensor.

Assembly in MEMS is demanding because many MEMS devices are very fragile, so standard pick and place equipment may not be appropriate. Also, many standard die attach adhesives are often not appropriate, and the effect of stress on them from the use of an adhesive and as a function of time must be considered.

MEMS packages often must allow controlled access of the die to the environment (e.g., pressure sensors require a port allowing the pressure to be applied to the strain-sensitive membrane) and simultaneously protect die from all other environmental effects. Obviously, considerations for material biocompatibility must be taken into account in the packaging for MEMS used in medical applications. This will limit the choices of materials and techniques used for packaging. Furthermore, there are no MEMS packaging standards, and existing packaging solutions tend to be highly proprietary. Even for mature MEMS markets, packaging solutions are frequently fragmented by the use of different package types to accommodate different customer needs. Consequently, the cost of the packaged MEMS device is frequently dominated by the cost of the assembly, testing, and packaging of the component. It is not uncommon for this cost to be 50 to 90% of the total packaged device cost.

The Materials Used in MEMS

There is an enormous diversity of materials used to fabricate MEMS devices, including semiconductors, metals, glasses, ceramics, and polymers. Furthermore, since the functionality of MEMS devices is not only electrical, but mechanical, chemical, thermal, etc., the choice of a material or a set of materials for a given application frequently depends on its electrical as well as non-electrical material properties. Silicon is the most popular material used in MEMS. This is partly owing to the established infrastructure and extensive knowledge base of silicon as a material and how to fabricate with it. Silicon is an extremely good mechanical material that has a yield strength nearly equal to that of stainless steel, and its strength-to-weight ratio is one of the highest of all engineering materials. Nevertheless, if silicon is strained beyond its limit, it will catastrophically fail. This is unlike most metals that plastically deform if loaded beyond their yield points. Another important fact about silicon is that it is an anisotropic material and therefore the material properties will vary significantly depending on the orientation of the crystal axes. Although silicon is the most commonly used substrate material, various types of glass substrates, such as fused silica and Pyrex are also commonly used in MEMS technology. Depending on the application, metal, polymer, ceramic, or other semiconductor substrates are also sometimes employed.

A variety of thin films are also used in MEMS fabrication, with the more popular ones being polysilicon, silicon nitride, deposited glasses, aluminum, platinum, and gold. These thin films frequently are deposited using either vapor chemical methods, such as low-pressure chemical vapor deposition (LPCVD) and plasma-enhanced chemical vapor deposition (PECVD), or by means of physical vapor deposition, including sputtering and evaporation. These deposited thin-film materials can have a large residual (or built-in) stress. The residual stresses in thin films are dependent on material type, temperature of deposition, and method of deposition, and can vary from highly compressive to highly tensile.

It is extremely desirable that the mechanical properties of the materials used in a MEMS device are accurately determined before initiating design. Unfortunately, this can be difficult to achieve in practice since the material properties, especially many of the mechanical material properties, are highly dependent on the exact processing conditions and sequence used in the fabrication of the MEMS device. Therefore, since most every MEMS device has its own customized process sequence, the development of a MEMS device and the fabrication sequence must be done in conjunction with the materials properties measurements. Once they are experimentally determined, the measured material property data can then be fed back into models in order to improve or optimize the design.

The measurement of properties of thin film materials can be challenging. For example, the film cannot be removed from the substrate and have a load-deflection measurement done on it without changing the stress state of the film. Fortunately, the MEMS community has devised a number of different test structures that can be used to measure the most important of the material properties (Figure 3.26). Readers are referred to [11] for more information.

Biocompatibility of MEMS Materials

The choice of materials used in any MEMS device will need to satisfy the specific requirements of the intended application, which are often very challenging. However, in the medical domain, the challenges are often considerably more difficult since they usually entail the more stringent and rigorous requirements of biocompatibility and biostability. These requirements include the following: the materials do not cause or induce toxicity in any surrounding tissues; the materials do not react with any other materials, tissues, serums, or solutions that they are in contact with to form toxic byproducts; and the materials do not cause significant negative chemical, electro-chemical, or mechanical effects on the surrounding tissue. Moreover, the surrounding environment of the MEMS device must not compromise the performance of the device. For example, a MEMS drug delivery device should not have its operational performance affected by the tissue it is implanted into during its operational lifetime. Also, if the intended application of a MEMS device

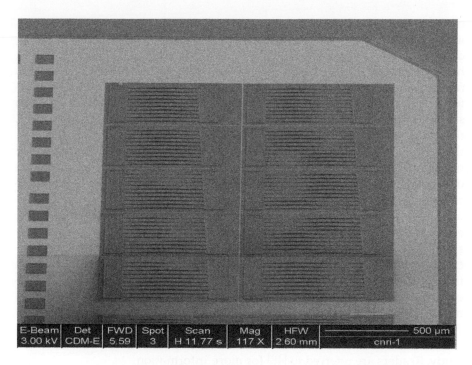

FIGURE 3.26

SEM of an array of thin film polysilicon beams that are clamped at both ends and that have a compressive residual stress. As can be seen, the beams are buckled by the compressive residual stress in the polysilicon layer. Fortunately, the residual stress can be reduced or eliminated by a high temperature anneal. For example, it is very common to perform a high-temperature anneal on polysilicon thin films that have been deposited using LPCVD to reduce the high compressive residual stress in these layers in a surface micromachining process.

involved implantation, the device must withstand long-term exposure to the physiological environment into which it is placed. This includes enduring the influence of the surrounding tissues on the device's function over its lifetime of operation.

Despite these very demanding requirements, MEMS devices have been used in several medical applications, primarily as sensor devices. In most of these applications the devices are not used in implanted applications, and the devices are packaged in such a way that the sensor is not in direct contact with tissue. Instead the device is packaged within an enclosure that is made of approved and commonly used suitable materials, usually synthetic polymers such as polyethylene and polyurethanes. Importantly, the medical industry tries to use already-approved and "off-the-shelf" materials since the cost of the experimental work for new materials for new applications is frequently prohibitive.

Nevertheless, recently several research groups have been examining the biocompatibility of MEMS devices and materials, particularly for implantable medical applications. In one recent published study [47], the biocompatibility and biofouling of the materials used in a MEMS-based implantable drug delivery device was studied to determine the vivo inflammatory and wound healing response of the materials used in the device, including gold, silicon nitride, silicon dioxide, silicon, and a formation of SU-8. These tests were performed using the cage implant system, with the materials being placed into stainless steel cages and implanted into a rodent model. At 4, 7, 14, and 21 days, the leukocyte concentrations were measured. The researchers reported that the inflammatory responses elicited by the MEMS materials examined were not significantly higher than those of the empty cage control tests over the duration period of the experiments. They also examined fouling of the materials by inspecting samples that had been implanted for 4, 7, 14, and 21 days into rodents. The extracted sample materials were inspected using a scanning electron microscope to measure the amount of surface cell attachment on the samples, including macrophages and foreign body giant cells, two of the principal ways of determining biofouling. Significantly, the researchers reported that the materials gold, silicon nitride, silicon dioxide, the formation of SU-8 examined, and silicon were all found to be biocompatible, with gold, silicon nitride, silicon dioxide and SU-8 all exhibiting reduced biofouling.

As demonstrated by this research and others, the issue of biocompatibility of MEMS materials is a difficult subject that requires careful and extensive study. Readers are referred to [48] for more information.

Applications of MEMS in Medicine

There is a wide variety of applications for MEMS in medicine. The first and by far the most successful application of MEMS in medicine (at least in terms of number of devices and market size) are MEMS pressure sensors, which have been in use for several decades [49, 50]. The market for these pressure sensors is extremely diverse and highly fragmented, with a few large markets and many smaller ones. Nevertheless, the contribution to patient care for all of these applications has been enormous. More recently, MEMS inertial sensors, specifically accelerometers and rate sensors, are being used as activity sensors. Perhaps the foremost application of inertial sensors in medicine is in cardiac pacemakers, in which they are used to help determine the optimum pacing rate for patients based on their activity level. MEMS devices are also starting to be employed in drug delivery devices, for both ambulatory and implantable applications. MEMS electrodes are also being used in neuro-signal detection and neuro-stimulation applications. A variety of biological and chemical MEMS sensors for invasive and non-invasive uses are beginning to be marketed. Lab-on-a-chip and miniaturized biochemical analytical instruments are being marketed as well.

Given the fragmentation of the applications and markets for these devices combined with the tremendous growth of new applications of MEMS in medicine, an exhaustive list of all current applications is nearly impossible and would be quickly dated. Therefore, we shall review several examples of the MEMS devices in most widespread use today. The theme to the success of MEMS technology in these medical applications can be directly attributed to several unique benefits of MEMS technology, including low cost, high performance, high reliability and stability, small size, and increased functionality (on-chip temperature compensation and calibration).

MEMS Pressure Sensors

The pressure sensor was the first MEMS device of significant technological and economic importance. These types of sensor devices continue to represent a large and important part of the MEMS component industry with annual sales of well over $1B for an enormous diversity of applications, including automotive, aerospace, industrial control, medical, and environmental control. Essentially these devices are composed of a thin diaphragm of material that is usually made of silicon that deflects under the application of a differential pressure loading across the diaphragm. A number of schemes can be used for transduction of the diaphragm deflection into an electrical signal, including piezoresistive, optical, capacitive, and piezoelectric. Nevertheless, it is most common to employ the piezoresistive effect of silicon in these sensors. The piezoresistive effect demonstrated in semiconductors whereby a relatively large change in resistance occurs when the semiconductor is subjected to a strain was first reported by Smith in 1954 [2]. Basically, the piezoresistors are positioned on the diaphragm at locations where the strain is the largest as the diaphragm is deflected. This transduction mechanism is attractive owing to the large size of the piezoresistive effect in semiconductors, thereby enabling high sensitivity levels to be obtained, the relative simplicity of implementation and readout circuitry required, and low cost.

These types of sensors have now been on the market for several decades. Given the length of time that these devices have been produced for the commercial marketplace, it should be no surprise that there has been a significant progression of technology development associated with silicon-based pressure sensors over the years. In fact, the historical development of pressure sensors in many ways reflects the broader development of silicon micromachining and MEMS technology with the first use of isotropic and anisotropic wet chemical etching, eutectic bonding, anodic bonding, and direct silicon fusion bonding. Sze and Brysek [12, 51] both have an excellent review of the history of MEMS pressure sensors.

The first application of a MEMS device for medical use was the pressure sensor. Currently, these type of devices represent the largest market for MEMS in the medical commercial sector with millions of sensors used each year, including tens of millions of disposable pressure sensors that are employed to monitor the blood pressure of the patient through the patient's intravenous (IV) tubing; more than a million intrauterine pressure sensors that monitor pressure around an infant's head during delivery; more than half a million disposable angioplasty devices that are used to monitor pressure in balloon catheters; and a smaller but still significant number of pressure sensors to measure pressure across the membrane in kidney dialysis systems accounting for tens of thousands of devices per year. Additionally, there are several smaller MEMS medical pressure sensor markets, which include MEMS pressure sensors for measuring pressure in the stomach or other organs during endoscopic procedures; MEMS pressure sensors to monitor for obstructions in the fluid lines of drug infusion pumps; and MEMS pressure sensors used in sphygmomanometers and other noninvasive blood pressure monitors.

By far the largest market for MEMS pressure sensors in the medical sector is the disposable sensors used to monitor blood pressure in IV lines of patients in intensive care. These devices were first introduced in the early 1980s and captured the majority of the market very quickly. These devices replaced other technologies that cost over $500 and that had a very substantial recurring cost since they had to be sterilized and recalibrated after each use. MEMS disposable pressure sensors are delivered pre-calibrated in a sterilized package from the factory at a cost of around $10. These devices are connected to an artery in the patient through a saline-filled IV line. The sensor device is isolated from the saline solution by a non-toxic, non-allergenic polymer gel, which is located between the deflecting silicon membrane of the device and the solution. This protects the on-chip sensor circuitry from the solution and the patient from stray currents from the sensor. We discuss the fabrication of one of the most prominent disposable pressure sensor devices sold in the market in more detail below.

A similar MEMS pressure sensor device and package is used to measure intrauterine pressure during birth. The device is housed in a catheter that is placed between the baby's head and the uterine wall. During delivery, the baby's blood pressure is monitored for problems during the mother's contractions.

Some of the other notable examples of MEMS pressure sensor applications include the following:

- MEMS pressure sensors are used in hospitals and ambulances as monitors of a patient's vital signs, specifically the patient's blood pressure (see above) and respiration. The MEMS pressure sensors in respiratory monitoring are used in ventilators to monitor the patient's breathing.
- MEMS pressure sensors are used for eye surgery. They are employed to measure and control the vacuum level used to remove fluid from the eye, which is cleaned of debris and replaced back into the eye during surgery.

- Special hospital beds for burn victims that employ inflatable mattresses use MEMS pressure sensors to regulate the pressure inside a series of individual inflatable chambers in the mattress. Sections of the mattress can be inflated as needed to reduce pain as well as improve patient healing.
- Physician's office and hospital blood analyzers employ MEMS pressure sensors as barometric pressure correction for the analysis of concentrations of O_2, CO_2, calcium, potassium, and glucose in a patient's blood.
- MEMS pressure sensors are used in inhalers to monitor the patient's breathing cycle and release the medication at the proper time in the breathing cycle for optimal effect.
- MEMS pressure sensors are used in kidney dialysis to monitor the inlet and outlet pressures of blood and the dialysis solution and to regulate the flow rates during the procedure.
- MEMS pressure sensors are used in drug infusion pumps of many types to monitor the flow rate and detect for obstructions and blockages that indicate that the drug is not being properly delivered to the patient.
- Many types of medical drilling equipment also use MEMS pressure sensors to monitor blood and/or other internal fluids during the drilling process.

The physically smallest MEMS pressure sensor is the GE NovaSensor intracardial catheter-tip blood pressure sensor, which is used for diagnostics during cardiac catheterization. The size of this device measures only 150 microns by 400 microns by 900 microns [52]. The implementation of this device as well as a few other types of pressure sensors for medical applications is described in more detail below.

Integrated MEMS Pressure Sensor for Disposable Medical Applications

The integrated pressure sensor (IPS) process technology was originally developed and put into production by Motorola (now Freescale Semiconductor) in 1991 and represents one of the most successful and long-standing high-volume MEMS products in the medical market. This sensor employs the piezoresistive effect to measure the deflection of a thin silicon membrane and combines bipolar microelectronics for signal conditioning and calibration on the same silicon substrate as the sensor device, thereby making it a fully integrated MEMS product (actually the first fully integrated high volume MEMS product). The transduction approach taken to measure membrane deflection under pressure loading is somewhat unusual since it uses a single piezoresistor element to measure strain, as opposed to the conventional approach of using multiple, distributed piezoresistors (e.g.,

Wheatstone bridge). Freescale has used two types of piezoresistive transducers, the "X-ducer™" and the "Picture Frame." The original "X-ducer"
design resembles an "X" located at the edge of the pressure-deflecting membrane (Figure 3.27a). The X-ducer design had the benefit of reducing the offset distribution that is an undesirable attribute of some Wheatstone bridges.
The process technology has been improved over the years, tracking recent
developments in micromachining and design, and Freescale now uses what
is called a "picture frame" piezoresistor configuration that allows approximately 40% greater output signal than the X-ducer design (Figure 3.27b).
Other process improvements include using an electrochemical etch stop to
precisely control the pressure-sensing membrane thickness and reduction of
the area consumed by the sensor [53].

The fabrication process sequence begins with a single crystal p-type silicon
substrate (Figure 3.28) [54]. A diffusion to create an n+ region is performed,
and this will be used to form the buried layer for the bipolar transistor
devices. A 15-micron-thick layer of n-type silicon is epitaxially grown on
the surface of the wafer, and this is followed by a deep diffusion to create
p+ regions to electrically isolate the bipolar transistors (Figures 3.28a, b).

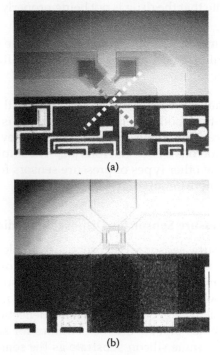

(a)

(b)

FIGURE 3.27 (See color insert.)
(a) Optical photograph of top surface of Freescale Semiconductor pressure sensor that employs
the original "X-ducer." An "X" has been added to show the transducer layout. (b) The newer
"picture frame" piezoresistor configuration. (Reprinted with permission, copyright Freescale
Semiconductor, Inc.)

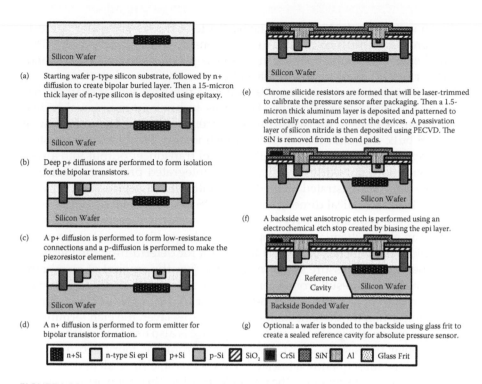

(a) Starting wafer p-type silicon substrate, followed by n+ diffusion to create bipolar buried layer. Then a 15-micron thick layer of n-type silicon is deposited using epitaxy.

(b) Deep p+ diffusions are performed to form isolation for the bipolar transistors.

(c) A p+ diffusion is performed to form low-resistance connections and a p-diffusion is performed to make the piezoresistor element.

(d) A n+ diffusion is performed to form emitter for bipolar transistor formation.

(e) Chrome silicide resistors are formed that will be laser-trimmed to calibrate the pressure sensor after packaging. Then a 1.5-micron thick aluminum layer is deposited and patterned to electrically contact and connect the devices. A passivation layer of silicon nitride is then deposited using PECVD. The SiN is removed from the bond pads.

(f) A backside wet anisotropic etch is performed using an electrochemical etch stop created by biasing the epi layer.

(g) Optional: a wafer is bonded to the backside using glass frit to create a sealed reference cavity for absolute pressure sensor.

n+Si ☐ n-type Si epi ■ p+Si ☐ p-Si ▨ SiO₂ ■ CrSi ▦ SiN ▥ Al ▧ Glass Frit

FIGURE 3.28
Process technology for Freescale Manifold Air Pressure (MAP) Sensor.

Subsequently, a diffusion is performed to create a p+ region that will be used to make low-resistance connections. This is followed by performing another diffusion to form p– regions that will be used to make the piezoresistors of the sensor device (Figure 3.28c). Yet another diffusion is performed to create an n+ region to make the bipolar transistor emitter and to provide ohmic contact to the n-type epi (Figure 3.28d).

A layer of chrome silicide (CrSi) is deposited, patterned, and etched to form resistors. These CrSi resistors on the top surface of the substrate are laser trimmed after packaging to calibrate the sensor offset and adjust the sensor span. A 1.5-micron-thick layer of aluminum is deposited, patterned, and etched to make electrical contact and connection to the transistors and sensor elements of the device. A layer of PECVD silicon nitride is then deposited to passivate and protect the top surface of the substrate. The SiN is removed only from the electrical pads by a photolithography and etch process (Figure 3.28e). Then, the wafer is immersed in a wet anisotropic etch solution, and electrical contact is made to the n-type silicon layer that was epitaxially grown to allow an electrochemical etch stop. As the etchant solution reaches the junction between the p-type silicon substrate and the n-type silicon epi layer, the etching process terminates, thereby allowing for precise control of the sensor membrane thickness (Figure 3.28f). As an optional step, a silicon wafer may be bonded to the backside of the sensor wafer to create sealed pressure reference cavities for each sensor device (Figure 3.28g). This would be used for implementing an absolute pressure sensor. Wafer bonding is performed using a glass frit layer between the two substrates.

From a fabrication standpoint, the Freescale pressure sensor example has several very interesting attributes. First, this technology is unusual in that it integrates a bipolar transistor microelectronics process technology with a MEMS bulk micromachining process technology. The bipolar transistors form amplifiers that convert the millivolt-level transducer output into a volt-level device output. Nearly all other MEMS-integrated process technologies that have been demonstrated merge CMOS with MEMS devices. Second, control over the mechanical dimensions of MEMS devices made by bulk micromachining are typically not very precise; however, the Freescale technology employs an electrochemical etch stop that enables precise control of the membrane thickness, which is extremely important for determining the mechanical stiffness of the pressure sensor membrane (i.e., the amount of membrane deflection that is to result from a certain level of pressure loading). This etch stop layer is an epitaxially grown n-type layer that is reverse-biased during the etching; when the etchant solution reaches this layer, the etch terminates. This same epitaxial layer in the process sequence is also used in the fabrication of the bipolar transistors. Third, the microelectronics is fabricated first, and the MEMS are subsequently fabricated. This is possible since all of the MEMS processing steps subsequent to the microelectronics fabrication are performed at relatively low processing temperatures. Fourth, the process uses <100> oriented silicon wafers to enable the micromachining steps to be done. Normally, <111> oriented wafers are used for the fabrication of bipolar transistor devices, and therefore Freescale needed to develop a special bipolar transistor process for this substrate orientation. Fifth, the device wafer can be bonded to another wafer to form a sealed reference cavity for implementing an absolute pressure transducer, or it can be left "as is" to implement a differential pressure transducer without any major changes to the process sequence. Last, the chrome silicide resistors enable the circuits and sensors to be easily trimmed for calibration, thereby allowing any offsets to be inexpensively eliminated.

The major disadvantages of this process are that it employs wet anisotropic bulk micromachining to implement the pressure sensor, and therefore a large amount of die area is consumed by the sidewalls of the exposed crystallographic planes in the silicon substrate. This is costly compared to the area that would be used to implement a surface-micromachined sensor having the same membrane dimensions. Also, the wet-etch process must expose only the back of the wafer to etchant; this requires specialized etch fixturing. Finally, bipolar transistors consume large amounts of power compared to CMOS electronics, and thus the Freescale integrated MEMS pressure sensor has higher power consumption levels than some other technologies. The bipolar circuitry cannot be used to form complex digital circuits, so this fabrication process is limited to analog devices.

Extremely Small-Sized Catheter-Tip Disposable MEMS Blood Pressure Sensor

While the previous device example has been an outstanding success in the medical market, its success is more directly tied to the much larger automotive market. The next device example is of a non-integrated pressure sensor that was entirely developed for the medical market. This device is notable because of its extremely small size, allowing it to be used on cardiac catheters. The unique fabrication technology to fabricate the pressure sensor devices allows the small size to be achieved; this was developed by GE NovaSensor, as first reported, in 1988 [52]. It can be used to make both low- and high-range pressure sensors with some slight variations in the process sequence, but we shall review only the low-pressure sensor process technology. Readers can refer to [52] for information on the fabrication of the high-pressure sensor device.

The process begins with the use of a standard thickness silicon wafer with <100> orientation (Figure 3.29a). A thin film layer is grown or deposited onto the surface of the wafer followed by a photolithography and etch to expose square openings in the masking layer. The wafer is then etched using an anisotropic wet chemical etchant to form pyramidal pits in the surface of the wafer (Figure 3.29b). This resultant shape is due to the fact that the thickness of the wafer is nominally 525 microns and the masking openings are 250 microns, and therefore the anisotropic etchant self-terminates on the exposed <111> crystallographic planes, resulting in an etch pit depth of approximately 175 microns. The masking layer is removed, and then a second silicon wafer that is p-type with an n-type epitaxial layer is directly bonded to the wafer having the previously etched pyramids on the surface (Figure 3.29c).

The thickness of the epitaxial silicon layer will set the thickness of the diaphragm of the resultant pressure sensor. Subsequently, the bulk of the bonded top wafer is removed using a controlled-etch process, thereby leaving only the bonded epitaxial layer on the bottom wafer (Figure 3.29d). Piezoresistors are then made in selected regions on the silicon diaphragm using ion implantation (Figure 3.29e). Metal is deposited, patterned, and etched to form electrical connections to the piezoresistors on the diaphragm. Next, the composite wafer is lapped and polished back to reduce its thickness to about 140 microns, thereby opening up the cavity from the backside of the diaphragm (Figure 3.29f). An image of three of these tiny pressure sensors located on the head of a pin is shown in Figure 3.30.

Wireless MEMS Pressure Sensors

A more recent advancement in MEMS pressure-sensor technology enables sensors to be interrogated remotely; that is, the pressure reading from the sensor is wirelessly transmitted to an external reader. These devices do not require wiring in order to operate. This is most important when the pressure

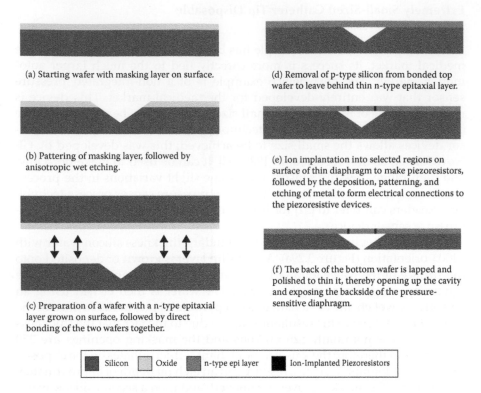

(a) Starting wafer with masking layer on surface.

(b) Pattering of masking layer, followed by anisotropic wet etching.

(c) Preparation of a wafer with a n-type epitaxial layer grown on surface, followed by direct bonding of the two wafers together.

(d) Removal of p-type silicon from bonded top wafer to leave behind thin n-type epitaxial layer.

(e) Ion implantation into selected regions on surface of thin diaphragm to make piezoresistors, followed by the deposition, patterning, and etching of metal to form electrical connections to the piezoresistive devices.

(f) The back of the bottom wafer is lapped and polished to thin it, thereby opening up the cavity and exposing the backside of the pressure-sensitive diaphragm.

| ■ Silicon | □ Oxide | ▨ n-type epi layer | ■ Ion-Implanted Piezoresistors |

FIGURE 3.29
Cross-section of the process technology for the NovaSensor pressure sensor.

sensor is to be implanted into the patient to monitor a body or device function for extended periods of time.

One example of this technology is being developed for monitoring abdominal aortic aneurysms (AAA). An aneurysm is a region of an artery that has a weakened wall and expands as a result of the patient's blood pressure. If left untreated, the artery can rupture and result in death. AAA is the 13th leading cause of death in the United States and the 3rd leading cause of sudden death in males over 60 years in age. The prevalence is approximately 2.7 million patients in the United States alone. A common treatment is to place a stent-graft in the artery [55]. However, there is no way to determine whether the graft is leaking once it is installed.

To allow for pressure monitoring in this artery, a new wireless pressure sensor technology was developed that essentially merges a capacitive-based pressure sensor with an inductor to form a resonant tank circuit. The resonant frequency of the circuit is proportional to the square root of the product of the capacitance and the inductance. The capacitor value varies with pressure and thereby changes the resonant frequency. This configuration is extremely simple and is installed using a delivery catheter.

FIGURE 3.30
Three ultra-small silicon pressure transducers located on the head of a pin.

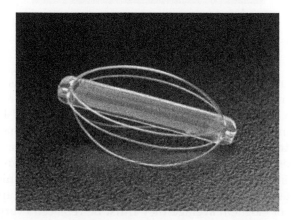

FIGURE 3.31
Photograph of the CardioMEMS implantable wireless pressure transducer device.

Importantly, the device requires no power source since the external reader merely scans the frequency spectrum and remotely detects when a resonant frequency is obtained. This technology has been approved by the FDA and is being marketed by CardioMEMS [55]. A photograph of the sensor device is shown in Figure 3.31. The external reader used to wirelessly interrogate the implanted wireless pressure sensor in shown in Figure 3.32.

There are many other applications for wireless pressure sensors. One example is the monitoring of the pressure inside a patient's cranium in hydrocephalus. Hydrocephalus is one of the most common birth defects, occurring in one

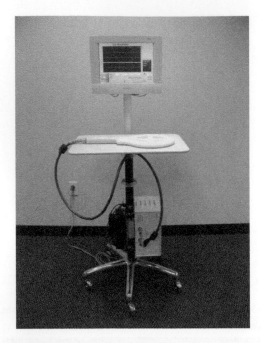

FIGURE 3.32
External reader used to interrogate the CardioMEMS implantable wireless pressure transducer device.

out of one thousand births. The primary characteristic of this disease is the excessive accumulation of cerebrospinal fluid (CSF) in the inter-cranial cavity. This causes an abnormal dilation of the brain ventricles and harmful pressure on the brain tissues, which can lead to severe brain damage, blindness, or death. The treatment is to place a ventriculoperitoneal shunt mechanism, essentially a one-way check valve, so that when the CSF pressure in the intra-cranial cavity increases, the fluid is allowed to drain away. This shunt valve must be maintained over the patient's lifetime, since these shunt valves can become sticky or clog. On average they must be replaced every six years or so, resulting in about 36,000 shunt-related procedures per year. Therefore, an implantable pressure sensor to monitor the pressure inside the cranial cavity is highly beneficial. This application also requires a wireless interrogation mechanism, since any lead wires would be very problematic. Therefore, researchers are working on resonant frequency devices similar in device function to that of the wireless sensor described above. A photograph of a recently developed wireless intracranial pressure sensor is shown in Figure 3.33 [56].

MEMS Inertial Sensors

Similar to the development of MEMS pressure sensors, mostly the automotive market drove most of the development of MEMS inertial sensors, primarily

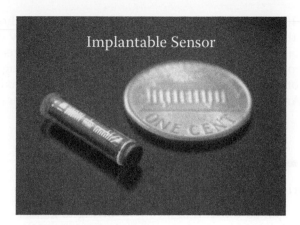

FIGURE 3.33
Implantable wirelessly interrogated pressure sensor designed to measure the intracranial cerebrospinal fluid pressure of hydrocephalus patients.

for crash airbag deployment sensors. These inertial sensors have proven to have excellent performance and reliability as well as a low cost. This technology has allowed the proliferation of a multitude of airbag sensors in recent-model vehicles and is attributed to having saved many lives in auto accidents.

MEMS inertial sensor technology also has many applications in medicine. These devices typically employ either piezoresistive, piezoelectric, or capacitive transduction mechanisms. These devices incorporate a proof mass that is suspended by one or more tethers that deflect when the proof mass experiences an inertial force (see discussion above). The movement of the proof mass is detected and converted into an electrical signal by the transduction mechanism. Ideally, the inertial sensor gives an output signal that is directly related to the position of the proof mass. These devices can be used to measure vibration frequency and amplitude, as well as in DC applications for angular measurement.

One important application of MEMS inertial sensors is in cardiac pacemakers, which are implanted in the patient's chest to regulate heartbeat. The benefit of the inertial sensor for this application is that it can measure the activity level of the patient in his or her daily routine, and the signal from the sensor can be used to pace the heart at the appropriate level. That is, when the patient engages in physical activity, the sensor detects this increased activity level and outputs this information to a microprocessor in the pacemaker to increase the rate of the heart so that the patient can maintain a higher activity level—similar to what the heart would do normally. The Model 40366 variable-capacitance silicon accelerometer from Endevco Corporation is one MEMS accelerometer used in this application. It is packaged in a surface mount configuration with a small footprint [57].

Another application of MEMS inertial sensors is in patient-activity monitoring during sleep or daily motion. These are increasingly being used on elderly patients to make sure that they are well, performing their daily exercise, etc. If three devices are used and configured orthogonally, the movement of the patient in three directions can be accurately detected and monitored. Some of the devices used for these applications include the STMicroelectronics LIS244AL dual-axis accelerometer having a full-scale output range of +/− 2 g, the STMicroelectronics LIS302 three-axis accelerometer, and the Analog Devices ADXL311 two-axis accelerometer [58].

MEMS Clinical Diagnostics

MEMS are well suited for clinical medicine owing to its fast response time, small size, capability for multiple functionality, accuracy, low cost, and ability to dramatically reduce sample size and reagent usage. Moreover, MEMS technology allows medical diagnostics to be performed on a handheld unit at the patient's bedside, so-called point-of-care, or POC, thereby allowing quicker test results and lower overall testing costs. A variety of clinical tests have been performed using MEMS technology, including enzyme-linked immunosorbent assay (ELISA) [59], polymerase chain reaction (PCR) [60, 61], capillary electrophoresis [62], and electroporation [63]. It is beyond the scope of this section to cover all of the different MEMS clinical diagnostic systems reported in the literature; therefore, readers are referred to [64] for a review on this subject area. Nevertheless, to illustrate the capability of MEMS in clinical medicine a successful demonstration of a MEMS-based diagnostic system that has been on the market for over a decade is described below.

A portable clinical blood analyzer from i-STAT, a division of Abbott Laboratories, uses a silicon substrate as a MEMS-based disposable chemical sensor cartridge for POC blood-chemistry analysis. It is a handheld device with the electronics in the reader; the serum manipulation and chemical reaction takes place within the disposable cartridge. The sensor electrodes, gels, and membranes are made using MEMS fabrication processes on the disposable cartridge. The microfluidic channels are of molded plastic. The handheld instrument contains all of the electronics that interface with the disposable MEMS cartridge, a screen to display test results, a key-pad for controlling the unit, and actuators to operate the microfluidic components (Figure 3.34). This instrument automatically performs calibration, steers the blood sample through the microfluidic channels on the test cartridge, takes measurements on the serum sample, and displays the results on the screen of the handheld unit [65]. This instrument can save hours compared to traditional clinical laboratory analysis. The tests that can be performed include sodium, potassium, chloride, urea, glucose, hematocrit, ionized calcium, PO_2, pH, PCO_2, creatinine, lactate, celite, and kaolin ACT [66].

Other POC MEMS or microfluidic devices approved for use by the FDA and on the market include Abaxis's PicoloXpress™, Agilent's 2100 Bioanalyzer™,

FIGURE 3.34
The i-STAT clinical diagnostic POC handheld unit.

Biosite's Triage™, Cepheid's Xpert™, MicroParts' GmbH Bilichek™, and the Veridex CellSearch™ [64].

MEMS Medical MicroProbes

A number of researchers have been developing microfabricated neural inter-faces for a variety of medical applications [67]. These are usually miniaturized probes that can be used to electrically sense or stimulate the surrounding tis-sue. Within this category of devices are also miniaturized probes that can be used to sense or deliver chemicals in the tissue. The goal has been to develop probes that can be relatively permanent and have high fidelity and a large bandwidth. Perhaps the first reported work on using MEMS to implement microprobes was in 1970 [68]. Since that time, a large amount of progress has been reported in the literature [69–71].

One application of these probes is for cortical implants. Direct stimula-tion of the visual cortex of the brain was demonstrated in the late 1960s using an array of surface electrodes [72]. The advantage of this approach is that direct interfacing with the brain bypasses the peripheral sensory organs, which may be damaged by injury or disease [73, 74]. Figure 3.35 illustrates one configuration for a cortical implant for visual prosthesis. A key element in this design is the array of microprobes that is inserted

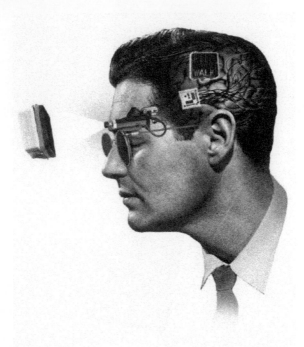

FIGURE 3.35 (See color insert.)
Illustration of one implementation of a cortical implant for visual prosthesis.

into the brain tissue. The microprobes stimulate the surrounding tissues via injected electrical currents to provoke a series of luminous points to appear (an array of pixels) in the field of vision of the blind person (See Figure 3.36). The visual prosthesis system is composed of two distinct subsystems: the implanted microprobe array and an external electronic controller [73].

The implanted microprobe array is inserted into the visual cortex and connected to the controller by a wire to receive the visual and control data [75, 76]. Energy to power the microprobe array is also supplied through the connecting wiring. More sophisticated systems connect the microprobe array to the controller wirelessly using a radio frequency communication linkage. The implantable microprobe array contains all of the circuits necessary to generate the electrical stimuli and to oversee the interface between the microprobes and the surrounding biological tissue. A battery-operated camera is positioned externally to capture and process the image data as well as to generate a signal to electrically stimulate the implanted microprobes.

Visual prosthetics based on cortical implants have been tested on several patients and have demonstrated proof of principle. Patients with these implants have been able to obtain some limited visual functionality, such

1024 sites and 64 data channels on 400µm centers

FIGURE 3.36 (See color insert.)
Photograph of MEMS microelectrode array for visual prosthesis.

as the ability to recognize levels of brightness, shades of colors, and geo-metrical shapes [77]. However, presently these devices are far from having a performance level remotely equivalent to the replacement of a living eye. It is important to note that these types of electrode array architectures are likely to be well suited for implanting in other sensory or motor regions of the cerebral cortex for other medical conditions, including deafness, epilepsy, paralysis, and Parkinson's disease [76].

Future Trends

The future of MEMS includes higher levels of integration, more func-tionality for each device, and smaller dimensional scales. As fabrication capabilities continue to improve, it is expected that it will be possible in the future to easily integrate a multiplicity of sensor and actuator types on a single slab of silicon with state-of-the-art electronics. Moreover, the integra-tion of other types of technologies, such as photonics and nanotechnology, will become increasingly prevalent. This will enable enormous amounts of functionality to be squeezed into a very tiny amount of space and at a very low relative cost. The size scale of many MEMS devices is continu-ously being reduced by advancements in fabrication technologies. This fact, coupled with the benefits derived from making things smaller as a result of the scaling laws (as well as economic forces) will provide a relentless driv-ing force for continuous size reductions. The current trend toward reducing the length of hospital stays is putting a greater emphasis on outpatient and home care. Many of the monitoring products originally developed for hos-pitals are being made less expensive and less complicated for use in home care environments. The market for lower-cost MEMS devices is accordingly expanding.

Conclusions

MEMS is revolutionizing the design of mechanical and electromechanical systems through miniaturization, batch fabrication, and integration with electronics. MEMS technology is not about a specific application or device, nor a single fabrication process. Rather, this technology provides new and unique capabilities for the development of smart products for many applications, including medicine and biology. MEMS has a promising future in the medical arena since it is one of the few technologies that can meet many of the stringent requirements for health care environments in the modern era, along with enormous functionality in a small package, high performance and reliability, and low cost.

References

1. Huang, P.H., Folk, C., Silva, C., Christensen, B., Chen, Y.F., Lee, G.B., Chen, Minjdar, et al. 2001. *Applications of MEMS Devices to Delta Wing Aircraft: From Concept Development to Transonic Flight Test*, AIAA Paper No. 2001-0124. Reno, Nevada, January 8–11.
2. Smith, C.S. 1954. Piezoresistance effect in germanium and silicon. *Physical Review* 94: 42–49.
3. Feynman, R.P. 1961. There's plenty of room at the bottom. In *Miniaturization*, H.D. Gilbert, ed., pp. 282–296. New York: Reinhold Publishing.
4. Robbins, H., and Schwartz, B. 1960. Chemical etching of silicon, II: The system HF, HNO_3, H_2O. *Electrochemical Society* 107: 108.
5. Holmes, P.J. 1967. Practical applications of chemical etching. In *The Electrochemistry of Semiconductors*, P.J. Holmes, ed., p. 329. New York: Academic Press.
6. Petersen, K.E. 1982. Silicon as a mechanical material. *Proceedings of the IEEE* 70: 420–457.
7. Maluf, N., and Williams, K. 2004. *An Introduction to Microelectromechanical Systems Engineering*, 2nd ed., p. 89. Artech House Inc.
8. Van Zant, P. 2000. *Microchip Fabrication*, 4th edition. McGraw-Hill.
9. Madou, M. 1997. *Fundamentals of Microfabrication*. Boca Raton, FL: CRC Press.
10. Ghodssi, R., Lin, P., eds. 2010. *MEMS Materials and Processing Handbook*. Springer Press.
11. Gad-el-Hak, M., ed. 1995, 2002. *The MEMS Handbook*. Boca Raton: CRC Press.
12. Sze, S. 1994. *Semiconductor Sensors*. New York: Wiley.
13. Seidel, H., Csepregi, L., Heuberger, A., and Baumgartel, H. 1990. Anisotropic etching of crystalline silicon in alkaline solutions: II, influence of dopants. *J. Electrochem. Soc.* 137: 3626.
14. Raley, N.F., Sugiyami Y., van Duzer, T. 1984. <100> silicon etch-rate dependence on boron concentration in ethylenediamine-pyrocatechol-water solutions. *J. Electrochem. Soc.* 131: 61.

15. Zwicker, W.K., and Kurtz, S.K. 1973. Anisotropic etching of silicon using electrochemical displacement reactions. In *Semiconductor Silicon*, Huff, H.R. and Burgess, R.R., eds. p. 315.

16. Jackson, T.N., Tischler, M.A., Wise, K.D. 1981. An electrochemical p-n junction etch stop for the formation of silicon microstructures. *IEEE Electron Device Letters*, EDLM-2: 44.

17. Glembocki, O.J., Stanlbush, R.E., and Tomkiewicz. 1985. Bias-dependent etching of silicon in aqueous KOH. *J. Electrochem. Soc.* 132: 145.

18. Kloech, B., Collins, S.D., de Rooij, N.F., and Smith, R.L. 1989. Study of electrochemical etch-stop for high precision thickness control of silicon membranes. *IEEE Trans. Electron Devices*, ED-36: 663.

19. McNeil, V.M., Wang, S.S., Ng, K.Y., and Schmidt, M.A. 1990. An investigation of the electrochemical etching of <100> silicon in CsOH and KOH. Tech. Dig. IEEE Solid-State Sensor and Actuator Workshop, Hilton Head, S.C., p. 92.

20. Howe, R.T., and Muller, R.S. 1983. Polycrystalline and amorphous silicon micromechanical beams: Annealing and mechanical properties. *Sensors and Actuators* 4: 447.

21. Fan, L.S., Tai, Y.C., and Muller, R.S. 1991. IC-processed electrostatic micromotors. Presented at the Int. Electron Devices Meeting (IEDM), p. 666.

22. Huff, M.A. et al. 2010. Process integration. Chapter 14 in *MEMS Materials and Processing Handbook*, eds. R. Ghodssi and P. Lin. Springer Press.

23. Huff, M.A., Schmidt, M.A. 1992. Fabrication, packaging, and testing of a wafer-bonded microvalve. IEEE Solid-State Sensor and Actuator Meeting, Hilton Head, SC, June 22–25.

24. Wallis, G., and Pomerantz, D.L. 1969. Field assisted glass-metal sealing. *J. Appl. Phys.* 40: 3946.

25. Johansson, S., Gustafsson, K., and Schweitz, J.A. 1988. Strength evaluation of the field assisted bond seals between silicon and Pyrex glass. *Sens. Mater.* 3:143.

26. Johansson, S., Gustafsson, K., and Schweitz, J.A. 1988. Influence of bond area ration on the strength on FAB seals between silicon microstructures and glass. *Sens. Mater.* 4: 209.

27. Tiensuu, A.L., Schweitz, J.A., and Johansson, S. 1995. In situ investigation of precise high strength micro assembly using Au-Si eutectic bonding. 8th International Conference on Solid-State Sensors and Actuators, Transducers 95, Stockholm, Sweden, June, p. 236.

28. Editorial. 1981. *Sealing Glass*. Corning Technical Publication, Corning Glass Works.

29. den Besten, C., van Hal, R.E.G., Munoz, J., and Bergveld, P. 1992. Polymer bonding of micromachined silicon structures. *Proceedings of the IEEE Micro Electro Mechanical Systems*, MEMS 92, Travemunde, Germany, p. 104.

30. Larmar, F., and Schilp, P.1994. Method of Anisotropically Etching of Silicon, German Patent DE 4,241,045.

31. Bhardwaj, J., and Ashraf, H. 1995. Advanced silicon etching using high density plasmas. *Proc. SPIE*, Micromachining and Microfabrication Process Technology Symp., Austin, TX, Oct. 23–24, vol. 2639, p. 224.

32. Unpublished internal etch data from the MEMS and Nanotechnology Exchange, CNRI, Reston, Virginia.

33. Ehrfeld, W., Bley, P., Gotz, F., Hagmann, P., Maner, A., Mohr, J., Moser, H.O., Munchmeyer, D., Schelb, W., Schmidt, D., and Becker, E.W. 1987. Fabrication of microstructures using the LIGA process. *Proc. IEEE* Micro Robots and Teleoperators Workshop, Hyannis, MA, Nov.
34. Menz, W., Bacher, W., Harmening, M., and Michel, A. 1991. The LIGA technique: A novel concept for microstructures and the combination with Si-technologies by injection molding. IEEE Workshop on Micro Electro Mechanical Systems, *MEMS 91*, p. 69.
35. Unpublished internal hot embossing data from the MEMS and Nanotechnology Exchange, CNRI, Reston, Virginia.
36. Chu, P.B., Chen, J.T., Yeh, R., Lin, G., Hunag, C.P., Warneke, B.A., and Pister, K.S.J. 1997. Controlled pulse-etching with xenon difluoride. *Proc. Inter. Conf. Solid-State Sensors and Actuators, Transducers 97*, June, p. 665.
37. Kovacs, G.T.A. 1998. *Micromachined Transducers Sourcebook*. New York: McGraw-Hill.
38. Elwenspoek, M., and Wiegerink, R. 2001. *Mechanical Microsensors*. Berlin, Germany: Springer.
39. H.A. Haus and J.R. Melcher. 1989. *Electromagnetic Fields and Energy*. Prentice-Hall.
40. Kahn, H., Benard, W., Huff, M., and Heuer, A. 1996. Titanium-nickel shape-memory thin-film actuators for micromachined valves. *Materials Research Society Symposium Proceedings*, 444, presented at Fall MRS meeting, December, Boston, MA.
41. Benard, W. and Huff, M. 1997. Thin-film titanium-nickel shape memory alloy microfluidic devices. Spring Electrochemical Society Meeting, June, Montreal, Canada.
42. Benard, W., Kahn, H., Heuer, A., Huff, M. 1997. A titanium-nickel shape-memory alloy actuated micropump. IEEE International Conference on Solid-State Sensors and Actuators, Transducers 97, Chicago, IL, June.
43. Senturia, S.D. 2001. *Microsystem Design*, p. 720. New York: Springer.
44. See: http://www.coventor.com/.
45. Mulhern, G.T., Soane, D.S., and Howe, R.T. 1993. Supercritical carbon dioxide drying of microstructures. *Proc. 7th Int. Conf. Solid-State Sensors and Actuators, Transducers 93*, Yokohama, Japan, p. 296, June.
46. Ashurst, R., Carraro, C., Chinn, J.D., Fuentes, V., Kobrin, B., Maboudian, R., Nowak, R., Yi, R. 2004. Improved vapor-phase deposition technique for anti-stiction monolayers. *Proceedings of SPIE: The International Society for Optical Engineering* 5342(1), 204–211.
47. Voskerician, G., et al. 2003. Biocompatibility and biofouling of MEMS drug delivery devices. *Biomaterials* 24a: 959–1967.
48. Greco, R.S. 1994. *Implantation Biology*. CRC Press.
49. Joseph, H., Swafford, B., Terry, S. MEMS in the medical world. *Sensors Magazine*. See http://archives.sensorsmag.com/articles/0497/medical/.
50. Panescu, D. 2006. MEMS in medicine and biology. *IEEE Engineering in Medicine and Biology Magazine*, September/October.
51. Brysek, J., Petersen, K., Mallon, J.R., Christel, L., and Pourahmadi, F. 1990. *Silicon Sensors and Microstructures*. Fremont, CA: NovaSensor Company Books.
52. Petersen, K., Barth, P., Poydock, J., Brown, J., Mallon, J., and Brysek, J. 1988. Silicon fusion bonding for pressure sensors. IEEE Solid-State Sensor and Actuator Workshop, Hilton Head South Carolina, June 6–9, p. 144.

53. G. Bitko, A. McNeil, and R. Frank. 2000. Improving the MEMS pressure sensor. *Sensors Magazine* July, 17(7).
54. Information provided by Andrew McNeil of Freescale Semiconductor, August, 2009.
55. See http://www.cardiomems.com/.
56. See http://www.mems-issys.com/implantable_intracranial.shtml.
57. See http://www.endevco.com/product/prodpdf/40366.pdf.
58. See http://www.eetimes.com/electronics-news/4075662/ MEMS-sensors-good-medicine.
59. Lachame, F., et al. 2008. On-chip nanoliter immunoassay by geometrical magnetic trapping of nanoparticle chain. *Anal. Chem.* 80: 2905–2910; S.S. Saliterman. 2006. *Fundamental of BioMEMS and Medical Microdevices.* New York: Wiley.
60. Kopp, M. U., de Mello, A. J., Manz, A. 1998. *Science* 280: 1046–1048.
61. Lin, Y.C., Huang, M.Y., Young, K.C., Chang, T.T., and Wu, C.T. 2000. *Sens. Actuators B* 71: 2–8.
62. Effenhauser, C.S., Bruin, G.J.M., Paulus, A. 1997. *Electrophoresis* 18: 2203–2213.
63. Huang, Y., and Rubinsky, B. 1999. *Biomedical Microdevices* 2: 145–50.
64. Rosen Y., and Gurman, P. 2006. MEMS and microfluidics for diagnostics devices. *Current Pharmaceutical Biotechnology* 2010: 11; and S.S. Saliterman. *Fundamentals of BioMEMS and Medical Microdevices.* New York: Wiley.
65. M.J. Madou. 2002. *Fundamentals of Microfabrication: The Science of Miniaturization*, 2nd edition. CRC Press.
66. See http://www.abbottpointofcare.com.
67. Michael R. Neuman. Biopotential electrodes. 1995. Chapter 50 in *The Biomedical Engineering Handbook*, Joseph D. Bronzino, ed. CRC Press and IEEE Press.
68. Wise, K.D., Angell, J.B., Starr, A. 1970. An integrated-circuit approach to extracellular microelectrodes. *IEEE Trans. Biomed. Eng.*, BME 17: 238–247.
69. Najafi, K. 1997. Micromachined systems for neurophysiological applications. In *Handbook of Microlithography, Micromachining, and Microfabrication.* SPIE-International Society for Optical Engine, Chicago, pp. 517–569.
70. Donoghue, J.P. 2002. Connecting cortex to machines: Recent advances in brain interfaces. *Nat. Neurosci.* 1085–1088.
71. Wise, K.D. 2005. Silicon microsystems for neuroscience and neural prostheses. *IEEE Eng. Med. Biol. Mag.* 24: 22–29.
72. Brindley, G. S., Lewin, W. S. 1968 The sensations produced by electrical stimulation of the visual cortex. *J. Physiol.* 196: 479–493.
73. Wise, K. D. and Najafi, K. 1991. Microfabrication techniques for integrated sensors and microsystems. *Science* 254: 1335–1342.
74. Normann, R. A., Maynard, E. M., Guillory, K. S., Warren, D. J. 1996. Cortical implants for the blind. *IEEE Spectrum* 54–59.
75. Dobelle, W.H. 1994. Artificial vision for the blind: The summit may be closer than you think. *ASAIO Journal* 40(4) October/December: 919–922.
76. Wise, K.D., et al. 2008. Microelectrodes, microelectronics, and implantable neural microsystems. *Proceedings of the IEEE* 96(7) July.
77. Dagnelie, G., and Schuchard, R.A. 2007. The state of visual prosthetics: Hype or promise? Guest Editorial, *JRRD* 44(3).

Other Information

Readers are referred to three very popular additional sources of information concerning MEMS technology. The first source is the web site MEMS and Nanotechnology Clearinghouse, which is located at http://www.memsnet.org and is an informational portal about MEMS technology that includes events, news announcements, directories of MEMS organizations, and a MEMS material database. The second source is the MEMS and Nanotechnology Exchange located at http://www.mems-exchange.org. This web site represents a large MEMS foundry network and offers MEMS design, fabrication, packaging, product development, and related services, as well as considerable information about MEMS and Nanotechnologies. Lastly, readers are referred to several electronic discussion groups concerning MEMS technology that have very active participation from thousands of MEMS developers and researchers from around the world. These sites can be accessed through http://www.memsnet.org/about/groups.

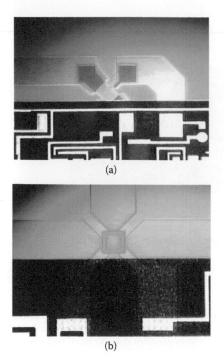

(a)

(b)

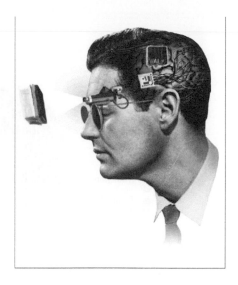

FIGURE 3.35
Illustration of one implementation of a cortical implant for visual prosthesis.

FIGURE 3.27
(a) Optical photograph of top surface of Freescale Semiconductor pressure sensor that employs the original "X-ducer." An "X" has been added to show the transducer layout. (b) The newer "picture frame" piezoresistor configuration. (Reprinted with permission, copyright Freescale Semiconductor, Inc.)

1024 sites and 64 data channels on 400μm centers

FIGURE 3.36
Photograph of MEMS microelectrode array for visual prosthesis.

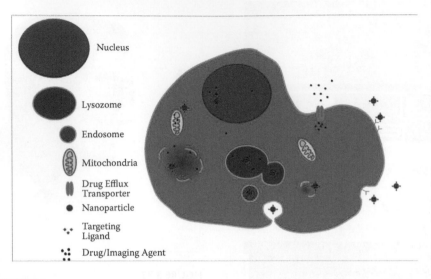

FIGURE 4.2

Specific cellular or intracellular targets may increase nanoparticle treatment efficacy. To avoid the degradation of the payload, nanoparticles may need to actively escape endosomal or lysosomal compartments. Delivery of the payload to specific organelles may be mediated by the nanocarrier.

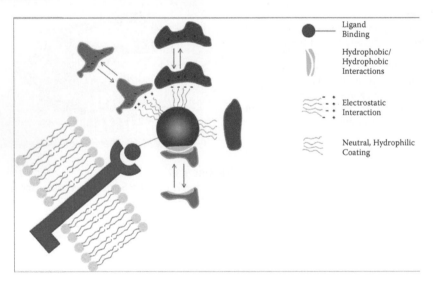

FIGURE 4.3

Once within the body, nanoparticle fate will be determined by interaction with proteins. The physiochemical properties and surface engineering of a nanoparticle will determine the particle/protein interaction. Electrostatic interactions between the nanoparticle surface charge and the positive and negative domains of proteins, along with non-polar interactions between hydrophobic regions, will determine protein adsorption. Hydrophilic coatings, such as PEG, can confer "stealth" properties. Various ligands attached to the nanoparticle surface can target cell-specific moeities and mediate cellular processes.

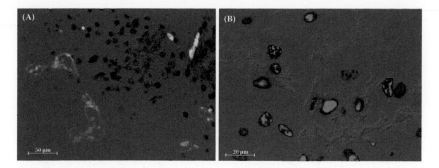

FIGURE 8.1
In-vitro construction of engineered vascularized cardiac muscle using a multi-cellular strategy of hES-CMs, endothelial cells, and embryonic fibroblasts seeded within a porous polymer scaffold of PLLA/PLGA. (A) The endothelial cells (vWF, green) within the scaffold self-organized to lumen vessel structures located in close proximity to the hES-CMs (troponin I red). (B) Higher magnification reveals that the hES-CMs matured to a certain degree, presenting developed cytoplasm and sarcomeric pattering (troponin I, red). Nuclei are stained with DAPI (blue).

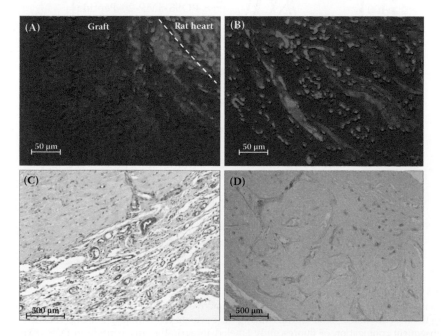

FIGURE 8.2
Transplantation of the engineered human vascularized cardiac tissue demonstrating localization of the hES-CMs (troponin I, red) in the graft area next to the myocardium (A), and structural maturation of the CMs in the graft area (B). The graft area was occupied with intense vascularization, as detected by staining with aSMA antibody (host and human derived vessels, brown, C) and with human specific endothelial CD31 antibody (human implanted vessels, brown, D). Nuclei were stained with DAPI (A, B in blue) or with hematoxylin (C, D in blue).

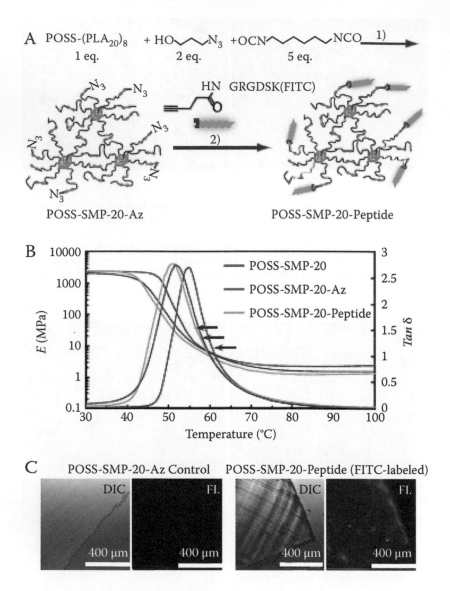

FIGURE 9.2
Chemical modification of POSS-SMP with a bioactive peptide: (A) Synthetic scheme illustrating the introduction of azido groups during the covalent cross-linking of POSS-(PLA20)8 and subsequent conjugation of fluorescently labeled integrin-binding peptide to POSS-SMP via "click" chemistry. (B) Storage modulus (E')-temperature curves and loss angle (Tan δ)-temperature curves (denoted by black arrows) of POSS-SMP-20, POSS-SMP-20-Az, and POSS-SMP-20-Peptide. (C) Differential interference contrast (DIC) and fluorescent (Fl) micrographs confirming the covalent conjugation of the fluorescently labeled peptide via click chemistry. (From *PNAS* 107:7652–7657, 2010. With permission from the Publisher.)

4

Nanoparticles to Cross Biological Barriers

Thomas Moore, Elizabeth Graham, Brandon Mattix, Frank Alexis

CONTENTS

KEY TERMS: Drug delivery, pulmonary delivery, oral delivery, nasal delivery, transdermal delivery, circulation time, biodistribution, excretion, blood–brain barrier

Abstract

Polymeric nanoparticle delivery systems have the potential to significantly impact the treatment of cancer. Nanoparticle drug delivery systems offer the ability to design a delivery vehicle that maximizes the therapeutic index of a drug encapsulated into nanoparticles (NPs), target it to cancerous tissue, and release it in a controlled manner for optimal dosing. This chapter describes the barriers associated with various delivery routes, specifically tissue, cellular, and molecular. Nanoparticle delivery approaches across tissue barriers such as pulmonary, nasal, oral, and transdermal are discussed along with systemic biodistribution and interactions of nanoparticles with their biological environment. A discussion of design considerations to optimize the delivery system is included to provide a guide for the engineering of a delivery system for specific applications.

Introduction

Nanotechnology [1] has a very broad definition depending on scale, and nanomedicine is a multidisciplinary field using nanotechnology for medical applications such as therapy and diagnosis [2–8]. While the National Nanotechnology Initiative (NNI) defines nanotechnology as materials that have an upper size limit of 100 nm (www.nanogov.com), many nanocarriers ranging from 5 to 150 nm are now in various phases of clinical trials [7–9]. Nanomedicine can provide an opportunity for improved drug development because early clinical phases often cast light on the side effects limiting the drug's therapeutic dose. In addition, a sharp increase over the past few years in the number of patent applications and high-impact factor publications in this area highlight the level of interest in nanomedicine by both academic and industry investigators [10]. In 2009, the NNI budget of $1.5 billion was allocated to a number of funding agencies, reflecting a steady growth in addition to the $8 billion invested since 2001. To develop a common direction in the fight against cancer, the National Cancer Institute (NCI) showed a clear commitment to nanomedicine with the recent allocation of $144 million. This initiative was established with clear objectives to form academic and commercial partnerships, establish outstanding training programs, leverage additional funds, and reduce the risk of investment in new products (www.nano.cancer.gov).

Current therapies and diagnostic systems are still lacking in precision, efficacy, sensitivity, and safety. A large percentage of drug candidates (50–80%) developed by the pharmaceutical industry fail during clinical trials. In addition, the efficacy of many drugs that are being approved for therapy are still limited for certain diseases, leaving ample room for improvements using new technologies. Generally, therapeutic drugs have to be administered in high-dose regimens because of short half-life, thereby increasing the systemic toxic side effects. The physico-chemical properties of drugs, such as solubility, charge, molecular size, and stability, affect their systemic pharmacokinetics. In general, nanocarriers deliver drugs in the optimum dosage, increase drug's bioavailability, enhance drug's efficacy, improve patient compliance, and allow the use of drugs with a broad range of physico-chemical properties. Nanotechnologies could thus provide new approaches for delivering drugs (small molecules, proteins, nucleic acids) at the optimal dose in a controlled release to specific tissues, cells, and even cellular organelles.

Nanotechnologies could also re-integrate drugs that previously failed in clinical trials and displace certain classes of drugs by achieving suitable pharmacokinetic and toxicological properties [11, 12]. TNP-470 is an example of a drug used for cancer therapy that had encouraging efficacy, but was stopped in clinical trials because of its neurotoxic side effects. However, it has reemerged as a successful nanodelivery formulation [13–15]. A polymer-drug conjugate formulation reduced neurotoxic side effects by preventing the compound from crossing the blood–brain barrier [16, 17].

Polymer, antibody, and small-molecule drug conjugates have been leading the nanocarriers market, and many molecules are still in various phases of clinical translation [18–21]. Early efforts in nanomedicine were focused on understanding and engineering the properties of already available therapeutic and diagnostic modalities, rather than development of novel therapies. The last few years saw tremendous progress in the use of many nanoparticles, such as liposomes, nanocrystals, micelles, nanoemulsions, polymeric nanoparticles, hybrid nanoparticles, being utilized for therapy and diagnosis [22]. One example of a recently successful nanomedicine formulation is Abraxane, an albumin-bound paclitaxel formulation that allows higher dosage of the drug without additional side effects. Today, commercial nanomedicine is at a nascent stage of development, but its potential is growing largely owing to recent success of certain nanomedicine formulations, improved understanding of the benefits of non-toxic nanomaterials, and the development of novel materials and drug delivery systems.

Clinical Impact of NPDDS

The physico-chemical properties of nanoparticles can be engineered at the molecular level [23, 24]; their shape, size, and charge can be controlled, and the surface density of the targeting ligand can be optimized for specific applications [25–37]. The fine-tuning of nanocarrier properties controls drug distribution throughout the body for optimum efficacy. One can develop a nanocarrier that will accumulate in certain tissues owing to its physico-chemical properties, as well as select a therapeutic load based on the distribution properties of the nanocarriers. Tuning of the physico-chemical properties of nanoparticles to achieve site-specific accumulation is an attractive approach that takes advantage of physiological defects and cellular interactions with nanomaterials. For example, albumin-bound paclitaxel nanoparticle formulations (Abraxane, 130 nm) were tested for in vivo efficacy using various tumor-cell-line xenograft models (lung, breast, ovarian, prostate, and colon) in mice [38]. MX-1 breast cancer and SK-OV-3 tumor xenograft models showed almost complete reduction of tumor size and the highest survival time differences when compared to Taxol formulations. This result is believed to be due to a more efficient transport across the endothelial barrier. Furthermore, Abraxane was approved by the Food and Drug Administration (FDA) in 2008 for breast cancer therapy. A comparative preclinical study of the albumin-bound paclitaxel (Abraxane) and paclitaxel formulated in Cremophor (Taxol) showed ~50% greater clearance (CL) and volume of distribution (Vz), allowing a tolerated dose of 280 mg/m^2. The higher volume of distribution correlated with the albumin-drug complex is believed to be associated with the higher intratumoral concentration of Abraxane and the higher allowable dosages that displayed comparable side effects to the Taxol formulation [12]. Similar results have also been shown with a PLA-PEG micellar formulation (Genexol-PM) [39]. After intravenous administration of Genexol-PM in mice models, the biodistribution of paclitaxel showed 2- to 3-fold higher concentrations in multiple tissues, including tumor tissue, with a peak ~400 ug/g approximately 2 h post-injection when compared to Taxol. The in vivo antitumor efficacy of Genexol-PM, as measured by reduction in tumor volume of SK-OV-3 human ovarian cancer and MX-1 human breast cancer xenograft mice, was substantially greater than that of Taxol [40]. There have been numerous nanomedicine technologies that have been developed using approved materials and new materials designed for specific applications [41–44]. Thus, the scale of complexity has drastically increased to design safer and more effective nanomaterials for therapy and diagnosis.

This early success is leading to tremendous expectations to develop novel technology platforms for therapy and diagnosis using multifunctional systems for improved sensitivity of detection, and improved specificity and material properties [45–50]. It is anticipated that the next generation of multifunctional NPs will provide molecular information such as cellular protein expression and molecular changes as a result of treatments. This information could be used to evaluate the therapeutic efficacy of drugs and extent of tumor invasion and metastasis for therapy assessment [51–55]. For example, using intravital microscopy, R. K. Jain et al. [51] have shown substantial pathologic differences between cancerous and healthy tissues. Tumors displayed leaky, dilated blood vessels, abnormal basement membranes and ECM, decreased pH, and impeded blood and lymphatic flow. An understanding of angiogenesis mechanisms, intercellular tumor interactions, stromal microenvironment, biomechanics, extracellular matrix components, cell mobility, and in vivo cell signals could impact the design of more efficient therapeutic drug delivery systems.

Crossing Epithelial Barriers

Although numerous agents need to be administered through oral [56–59], nasal, or pulmonary drug delivery [60–63] to increase patient compliance, these routes of administration represent significant barriers. Instead of spending large amounts of resources to modify a drug to improve its tissue adsorption, nanocarriers can be engineered to transport any type of drug across biological barriers, increase adsorption, and reduce drug degradation in the site microenvironment [64–71]. Molecules can be transported through an epithelial membrane by either paracellular transport or transcellular transport. Paracellular transport is a one-step passive phenomenon that occurs when molecules move between cells by crossing tight junctions [72]. Transcellular transport can be either an active or passive two-step process and occurs when the molecule crosses both the apical and basolateral membrane. In most epithelia, tight junctions have a relatively high permeability, as this surface is involved in vectorial transport for the absorption of nutrients or other necessary molecules [72].

Pulmonary Delivery

Pulmonary drug delivery is an efficient route of administration for rapid systemic delivery and quick onset of pharmacological activity. It avoids first-pass hepatic clearance, has reduced enzymatic degradation, and presents a high surface area for drug absorption [73, 74]. The lower respiratory tract is composed of a large tube, the trachea, which branches into two bronchiole tubes, and then a series of diverging conduits that decrease in size and terminate in the alveoli [75, 76]. The alveoli are small sacs that provide a large surface area, approximately 100 m^2 [74, 75, 77], for gas exchange. The cellular makeup of the alveolar sacs consists of thin, simple squamous epithelial cells (type I cells), larger epithelial pneumocytes (type II cells), and alveoli-specific macrophages [78]. Type II cells secrete pulmonary surfactant that aids in reducing lung surface tension and inhibits protein degradation [79, 80]. Type II cells are also the progenitor of type I cells, which represent the most prevalent cell type and serve as the site of gas exchange. The alveolar macrophages protect the lungs by continually migrating throughout the alveoli, engulfing microorganisms and particulate matter. A mucus layer also covers the airway and bronchioles [81]. This layer is cited as being approximately 15 μm in the airway and 55 μm in the bronchioles. This mucus is generally biphasic with a solution-type layer and a gel layer. The solution layer is moved upward by what is known as the mucociliary escalator, which acts to clear particulates from the airway. Mucus in the nasal cavity is turned over approximately every 20 minutes, whereas mucus turnover in the pulmonary airway is thought to be more on the scale of every 4–6 hours [81]. The physiological characteristics of the pulmonary system, such as high alveolar surface area coupled with thin squamous cells, provide rapid access to a high volume of blood. However, drug delivery via the pulmonary route must overcome several barriers in order to be effective, including premature deposition in the upper airways [82], entrapment by the mucus membrane and clearance by the mucociliary escalator [81, 83], and alveolar macrophage clearance [74].

Micro- and nanoparticle formulations to bypass barriers in pulmonary drug delivery have been investigated. In particular, pulmonary delivery of nanocarriers may be employed to treat lung-specific pathologies, or can be directed towards systemic drug delivery. Lung-specific ailments may include asthma, cystic fibrosis, fungal infection, lung cancer, and tuberculosis [75, 84, 85]. The diffusion of nano-sized molecules in the lungs (drugs, nanoparticles, peptides) appears to be governed by factors such as size, surface charge [86], and hydrophobicity [79]. With regards to aerosol delivery, sub-1-μm particles are liable to be exhaled because of their small size and virtually nonexistent inertial impact, yet nanoparticles <100 nm are able to penetrate deeply in the lungs and settle in the alveoli [82, 87, 88]. Thus <100 nm vectors deposit in the alveoli and are able to be absorbed through the epithelial cells and enter the bloodstream [77, 82, 88]. Moreover, nanoparticles <100 nm may avoid clearance via macrophage phagocytosis [82] and are able to diffuse through the mucus layer in the lungs [82]. In vivo studies conducted by Kwon et al. [89] show that inhaled 50-nm fluorescently labeled magnetic nanoparticles were able to enter systemic circulation and accumulate in tissues such as the liver, testis, spleen, lung, and brain. Yacobi et al. [86] investigated the transcellular transport across rat alveolar epithelial cell monolayers using polystyrene nanoparticles with varying sizes (20 nm and 100 nm) and surface charges. Positively charged nanoparticles were shown to exhibit a higher flux than negatively charged particles, and 20-nm particles were shown to cross the cell monolayer more rapidly than 100-nm particles. The influence of hydrophobicity on pulmonary absorption in vivo of small, 60–700 Da, molecules was reviewed by Patton et al. [79]. A range of molecules were deposited in vivo at the bifurcation of the trachea via injection. This model has shown to be representative of lung, specifically alveolar, absorption as it avoids absorption in the nasopharynx and oropharynx. Hydrophobic molecules were shown to rapidly absorb through the lung within a matter of minutes, while hydrophilic molecules were reported to take an average of 60 minutes to absorb 50% of the administered dose. Furthermore, Dames et al. [90] investigated the feasibility of magnetically targeting the lungs using superparamagnetic iron oxide nanoparticles (SPIONs). Polyethylenimine-coated SPIONs with a hydrodynamic diameter of 80 nm were delivered via aerosol in vivo and preferentially directed to one lung using a magnetic field. This type of technology may have implications as a local targeting mechanism.

With inhaled therapies, it has been shown that particles must be <5 μm in diameter to deposit in the deeper regions of the lungs, and the optimal size for alveolar deposition is 1–3 μm. Larger particles deposit via impaction or sedimentation in the bronchial pathways, and particles >10 μm are trapped in the oropharynx by impaction [82, 88]. These particles will in turn be cleared by the mucociliary escalator. Currently, the future of particle-borne systemic pulmonary drug delivery seems to be up in the air. In 2007, Pfizer pulled Exubera, an inhalable <5-μm insulin-carrying particle [91], off the market after discovering an increased risk of cancer in former smokers [92]. Though these were not nano-scale formulations, the findings have put a damper on the pulmonary delivery of insulin, and consequently on the investigation of particle-borne pulmonary delivery. Research will therefore need to show that nanoparticle-mediated pulmonary drug delivery is more effective than conventional treatment, and is safe across a wide patient demographic. That being said, the pulmonary route is an efficient desirable candidate for nanocarrier-mediated drug delivery because it offers rapid absorption and extensive bioavailablility.

Transmucosal Delivery

The transmucosal administration route is efficient owing to the inherent absorptive properties of mucosal surfaces and the rich blood supply that translates to rapid transport into systemic circulation [70]. Mucus, an adhesive gel composed of a densely woven network of natural mucin polymers interspersed with a variety of glycoproteins, creates an effective barrier to diffusion across mucosal surfaces [81]. Furthermore, mucus is constantly secreted and turned over, an action that serves to clear particulate matter. Delivery across the mucosal layers will require nanocarriers to traverse this thick web and evade adhesion to the sticky mucin fibers. The thickness of mucus layers is known to vary between different types of tissues [81]. In addition to the physical variability of mucosal membranes, the pH can vary greatly depending on the physiological locations of the mucosal surface. Studies have demonstrated that lung and nasal pH are generally neutral, while a cross-sectional pH gradient exists in gastric mucus. This gastric pH can change from ~1–2 at the luminal surface to a pH ~7 at the epithelial surface [81]. These variations between mucosal membranes must be considered when designing a nanoparticle for transmucosal delivery.

Two specific transmucosal routes of interest are the nasal and oral routes. The nasal route has three possible modes of entry: the nasal epithelium [93], the bronchial epithelium [94], and possibly a direct link to the brain via the olfactory nerve. The oral route has two possible modes of entry: through the oral mucosa, such as buccal, gingival, sublingual, or palatal routes [70], and through the gastrointestinal epithelia. Drug delivery via transmucosal routes occurs via three primary modes: paracellular uptake, endocytosis by enterocytes, and endocytosis by membranous microfold cells (M cells) [95]. Since cells lining mucosal membranes are arranged in monolayers, particles can translocate between tight junctions of neighboring cells, thereby inducing paracellular translocation.

The nasal epithelium has moderate permeability, low enzymatic activity, and the ability to avoid first-pass metabolism, and it displays rapid onset of pharmacological activity, making it an attractive route for drug delivery. Moreover, the ease of intranasal delivery may improve patient compliance owing to the possibility of self-administration. However, drugs delivered via the intranasal route must avoid many obstacles as well. Primary concerns of intranasal delivery include a limited volume of drug that can be delivered into the cavity (25–200 µL), irritation of the nasal mucosa, a molecular weight cutoff around 1 kDa, and large interspecies variability. Furthermore, because of the high viscosity of the lining mucosa, the diffusion rate of compounds across the mucosal membrane must be greater than the mucociliary clearance rate [96]. Vila et al. [97] have clearly shown the advantages of nanoparticles (~200 nm) vs. microparticles (1.5 um) in crossing the nasal mucosa. The results, illustrated using fluorescent microscopy, show that nanoparticles with a stealth surface are transported across the nasal mucosa barrier. However, no quantifiable data is provided, and the transport mechanism is unknown, therefore limiting the assessment of efficacy. Research has been done into direct nose to brain delivery of drugs through the olfactory nerve and trigeminal nerves, the olfactory nerve delivering drugs to the rostral brain areas and the trigeminal nerve to caudal brain areas. Wang et al. [98] have been able to improve the absorption of estradiol into the cerebrospinal fluid (CSF) through encapsulation in 260-nm chitosan nanoparticles, and have shown significantly higher amounts of estradiol in the CSF through intranasal administration over intravenous administration. This suggests direct transport of the drug to the brain via the olfactory nerve. Another in vivo study investigated poly(lactide)-poly(ethylene glycol) (PLA-PEG) nanoparticles functionalized with wheat germ agglutinin (WGA), a lectin [99]. Nanoparticles ~90 nm were loaded with the fluorescent molecule 6-coumarin. After intranasal administration, significantly higher concentrations of the WGA-functionalized nanoparticle were observed in the olfactory bulb, olfactory tract, cerebrum, and cerebellum compared to unfunctionalized dye-loaded particles. This study suggests that direct nose to brain delivery is possible when employing nanocarriers. However, intracellular transport via the olfactory nerve could prove difficult because of the variability in nerve diameter, which ranges from 100 to 700 nm [100].

The orotransmucosal route can occur through the buccal, sublingual, palatal, or gingival mucosa and is attractive owing to the high rate of blood flow, the avoidance of destruction by gastric acid, and the avoidance of first-pass metabolism [70]. The routes of orotransmucosal delivery vary in permeability, with sublingual being the most permeable, with buccal, and palatal and gingival being the least permeable. Permeability is based on the relative thickness of the epithelia and the amount of keratinization. Thinner epithelia with less keratinization are the most permeable. Drug absorption through the oral mucosa is a diffusion-driven process, thus buccal and sublingual tissues are the principle focus of orotransmucosal drug delivery because of their relatively high permeabilities. The sublingual mucosa is most often used for drug delivery of acute disorders because of its high permeability and high blood flow. Challenges of the orotransmucosal route include the hydrophobic and hydrophilic barriers of the oral mucosa that must be overcome, involuntary swallowing, which can result in drug loss, and an enzymatic barrier that causes rapid degradation of proteins and peptides. These challenges could be overcome with engineered nanocarriers designed for optimal absorption through both the hydrophilic and hydrophobic barriers as well as enhanced mucoadhesion. Encapsulation may protect drugs from enzymatic degradation.

Orogastric drug delivery is an effective route of administration owing to high patient compliance. However, there are several challenges, such as low oral bioavailability of the drug from degradation in the stomach, inactivation and digestion of the drug by proteolytic enzymes, and slow diffusion across intestinal epithelium because of the high molecular weight and low lipophilicity of most drugs [71]. As mentioned previously, the mucosal lining thickness varies within the gastrointestinal tract, and pH variability will also affect nanoparticle efficiency [81]. Multiple nanocarrier approaches have been investigated for orogastric drug delivery, including floating particles in the stomach, bioadhesive particles, and cell-specific targeting particles. M cells are a subtype of epithelial cells found in Peyer's patches in tissue such as the intestinal mucosa [102]. M cells sample the lumen of the small intestine and initiate immune sensitivity in lymphocytes. Various pathogens utilize M cells as a means of crossing the intestinal tract, and nanoparticles may also take advantage of this route [103–105]. Fluorescent confocal microscopy studies of PLGA-PVA nanoparticles functionalized with *Ulex europaceous* agglutinin 1 (UEA-1), a lectin, showed capability as a transmucosal carrier [103]. Particles approximately 400 nm in diameter were shown to localize in the Peyer's patches of mice after oral administration. In another study, ~200-nm nanoparticles were synthesized with varying formulations of poly(lactide-co-glycolide) (PLGA), PLGA-PEG, or poly(caprolactone)-poly(ethylene glycol) (PCL-PEG) [104]. These were functionalized with an RGD-binding domain, and in vitro studies showed preferential translocation across Caco-2 and Raji co-cultures. Furthermore, orally administered nanoparticles were localized in Peyer's patches of mice. In regards to floating particles, drugs are administered orally and can be encapsulated within a functional carrier to improve physical interactions with the biological environment of the stomach. Sato et al. [106] investigated riboflavin-encapsulated microballoons, studying their buoyancy and pharmacokinetics. By encapsulating riboflavin within a solid polymer microballoon, they demonstrated improved excretion half-life of the drug from the increased buoyancy. Moreover, improved release kinetics of the drug were also observed. The use of bioadhesive nanoparticles has also been investigated for oral drug delivery. Chitosan nanoparticles have been shown to promote paracellular permeability of intestinal epithelium using an Ussing chamber technique [101]. The group fabricated chitosan-coated poly(isobutylcyanoacrylate) core-shell nanoparticles and introduced them to the mucosal side of the intestinal epithelium on the "donor" side of the chamber. Nanoparticle diffusion across the biological barrier from the donor chamber to the acceptor chamber was tracked using ^{14}C mannitol as a tracer. Results from the study indicated the chitosan-coated nanoparticles had increased bioadhesion to the mucosal membrane, which then allowed for more rapid diffusion via paracellular transport across the intestinal epithelium. Furthermore, it has been shown that negatively charged particles derived from very hydrophilic polymers enhanced bioadhesive properties and are readily absorbed by M cells and absorptive enterocytes lining the oral canal [95]. Nanocarrier technologies will need to protect the drug molecule from acidic degradation, efficiently avoid mucus adhesion, and facilitate the delivery across the gastrointestinal tract. This can be achieved by tailoring the size, surface charge, and biofunctionality of the nanocarrier. Moreover, targeting specific cell types for internalization or modifying the physiochemical properties may aid in orogastric delivery efficiency.

Transdermal Delivery

Skin acts as a biological barrier, serving to offer thermal insulation, prevent water loss, and protect the body's internal organs from the external environment and foreign substances. Proponents of the transdermal delivery route seek to alleviate the pain associated with needle-mediated delivery methods, decrease the production of dangerous waste from needles, and improve patient compliance [107]. Transdermal delivery can avoid first-pass hepatic clearance while also avoiding pitfalls of oral delivery such as gastrointestinal tract degradation [108]. Sustained delivery can also be achieved, prolonging drug concentrations in the blood at therapeutic levels. Transdermal therapy evidently presents advantages over traditional delivery routes such as hypodermic injection, intravenous delivery, and oral administration, but there are definite barriers that transdermal delivery must overcome. Skin is generally less than 2 mm thick and composed of several layers: the epidermis, dermis, and hypodermis [109]. The outermost layer, the epidermis, is further divided into the stratum corneum, stratum granulosum, stratum spinosum, and stratum basale. Transdermal delivery endeavors to deliver drugs into either systemic circulation or locally in the skin. The primary barrier to transdermal delivery being the stratum corneum, the outermost 10–20 µm of the epidermis [109, 110]. The stratum corneum is composed of corneocytes, dead squamous cells filled with keratin filaments enclosed by an envelope of cross-linked proteins and surrounded by a semi-continuous matrix of lipids [111]. Therefore, to overcome these barriers, therapeutics used in transdermal delivery have generally been lipophilic molecules with a size less than 500 Da [112].

There are several approaches to increase drug flux through the stratum corneum for transdermal delivery, and these can be classified as physical or chemical approaches. Physical methods to increase skin permeability include iontophoresis, sonoporation, electroporation, and microneedle delivery. Iontophoresis is a method in which charged drugs are mobilized through the skin via an electromotive force [113]. By modulating the electrical current, drug dosage can be moderately controlled [114]. Sonoporation is a method that uses ultrasonic sound waves to enhance the permeability of the skin [115]. Sound waves are able to disturb the lipid structure within the stratum corneum, thus increasing the permeability of lipophilic drug

TABLE 4.1

Advantages and Disadvantages of Delivery Route

Delivery Route	Advantages	Disadvantages	Reference
Intravenous	• Rapid bioavailability • Conventional method of sustained drug delivery	• Pain associated with injection site • Precipitation of solubilized drug due to phase separation of drug formulation • Inflammation of vein walls • Renal and hepatic clearance	[222]
Pulmonary	• High available surface area for particle absorption • Rapid bioavailability • Rapid onset of pharmacological activity • Low enzymatic activity • Reduced first-pass hepatic clearance	• Pulmonary macrophage clearance • Difficulty controlling particle deposition in the airway • Links discovered between microparticle-borne pulmonary delivery and occurrence of cancer in former smokers	[73, 74, 77]
Nasal	• Porous endothelial basement membrane provides favorable for particle diffusion • Rapid onset of pharmacological activity • Low enzymatic degradation • Reduced first-pass hepatic clearance	• Possible loss of olfaction in parts of the nasal passage • Relatively small surface area for absorption	[93]
Orogastric	• Non-invasive • Improved patient compliance • No technical equipment required • Self-administration and increased convenience of "home-based" therapy c	• Enzymatic degradation • Low bioavailability due to degradation in the stomach • Low diffusivity across the intestinal epithelium	[66]
Orotransmucosal	• Non-invasive • Improved patient compliance • Delivers to highly vascularized tissue • Bypass gastrointestinal tract metabolism and first-pass hepatic clearance • No technical equipment required	• Hydrophobic/hydrophilic barriers of the oral mucosa. Enzymatic degradation • wAbsorption variable	[70]
Transdermal	• Minimized pain relative to needle-mediated delivery • Improved patient compliance • Reduced first-pass hepatic clearance	• Impermeability of the skin as a biological barrier • May require chemical/physical methods to improve diffusivity through the epidermis	[107, 108]

molecules. Electroporation is a method in which a sequence of electrical pulses can be used to disrupt the lipid structures in the skin. This has been shown to create microscopic pores within the disrupted stratum corneum, thus allowing the transport of drugs such as methotrexate through the skin [116]. Microneedles, fabricated from metal, silicon, or polymer [117–119], have been used to create a relatively painless mechanism for penetrating the skin and delivering therapeutics. An advantage of using these microneedles is that with a smaller needle comes less compression on the tissue as the needle is inserted. This decreased compression can lead to less stimulation of pain receptors as well as a decrease in discomfort associated with the insertion of a larger-diameter needles.

Chemicals enhancers and bio-molecules are another means of increasing the permeability of the skin [120, 121]. Chemical enhancers work by partially solubilizing the lipid structure of the stratum corneum and have been investigated via high-throughput methods developed by Mitragotri et al. [120, 122]. Chemical enhancers make up a broad spectrum of chemical compounds, such as surfactants and fatty acids. These formulations can include single-component systems, for instance ethanol [121], or multi-component systems, such as sodium laureth sulfate with phenyl piperazine. Aside from chemical enhancers, biochemical methods to penetrate the skin have been investigated. Chen et al. [123] used a short, synthetic peptide to increase the transdermal delivery of insulin in vivo. It appears the peptide TD-1 (ACSSSPSKHCG) creates a temporary inlet by which insulin is able to penetrate. Time-lapse studies of insulin blood concentration after topical administration following TD-1 treatment showed a marked decline after 15 minutes. These results suggest that while coadministration of TD-1 with insulin was not requisite for improved transdermal delivery, the effects of TD-1 were limited over time. Further investigation revealed that the peptide increased FITC-labeled insulin concentrations deep within hair follicles, a region populated with vasculature. The connection between hair follicle penetration and improved systemic delivery has not been completely substantiated. Magainin is a natural peptide that has also been shown to increase skin permeability by disrupting the lipid structure within the stratum corneum [124]. This peptide has previously been reported to increase the permeability of bacterial membranes [125] by tightly binding lipids, creating a tension that forms pores. When used in conjunction with the chemical enhancer N-lauroyl sarcosine in an ethanol solution, magainin enhanced transdermal permeability [124]. While these methods offer a means to increase skin permeability, nanoparticle formulations may offer a vehicle for the transdermal delivery of drugs. Nanoparticle drug delivery systems could have an impact on the delivery of large molecules or non-lipophilic drugs that are not as viable for transcutaneous delivery.

Nanoparticles serve as vehicles for delivering a wide range of drugs, and physico-chemical properties, including size, surface charge, and physical morphology of nanoparticles, can be controlled for optimal transdermal delivery. The aim of nanoparticle treatments will be to enhance the permeation of drugs through the skin, shield drugs from metabolism occurring during transport, and prolong drug-residence time in systemic circulation. Many nano-sized formulations are currently being investigated for transdermal delivery, in a diverse range of treatments. Liposomes [109] have been used to deliver a variety of therapeutics [126–128]. Dubey et al. [126] developed melatonin-loaded elastic liposomes approximately 126 nm in diameter for the transcutaneous treatment of jet lag. Liposomal formulations loaded with 1% melatonin were incubated with human cadaver skin at 32°C for time periods up to 24 hours in a custom-built Franz diffusion cell. Melatonin levels in a compartment across the cadaver skin were measured using high-performance liquid chromatography (HPLC). This study showed improved melatonin flux across skin with elastic liposomes as transdermal vector in comparison to free drug. Moreover, topical liposomal formulations loaded with ibuprofen have shown equivalent effectiveness compared to orally administered ibuprofen in clinical trials [129]. In addition to the apparent benefits of transdermal delivery in the medical field, nanoparticle technologies have also found a niche in other markets. The cosmetics industry looks to nanoparticles for improved skin hydration, delivery of acne medication, lubricant properties to increase comfort, and improved aesthetic value of their products [130]. Isotretinoin, a retinoic acid derivative used to treat acne, has been loaded into solid lipid nanoparticles 30–60 nm in diameter. This formulation was able to reduce systemic distribution of isotretinoin and elevate concentrations in the skin [131]. Lipid nanoparticles have been investigated for their ability to form a monolayer on the skin and retain the skin's moisture. Polymeric and lipid nanoparticles may also act as solid lubricants, thus improving patient comfort. Nanoparticle formulations can be synthesized in ways that reduce irritation responses and may also be used to serve as controlled release of active ingredients.

Mechanical stimuli have been used in conjunction or in series with liposomal vectors to further enhance skin permeability and drug uptake. The physical methods to improve skin permeability are generally done prior to application of topical liposomal solutions. Thus, the skin permeability is temporarily improved via micro pores formed by this treatment, and the liposomal vectors can diffuse across the skin with less difficulty. Badkar et al. [134] delivered 274 nm liposomes encapsulating colchicine by first increasing skin permeability with electroporation and then mediating liposomal transdermal delivery via iontophoresis. Microneedles represent another physical method employed to improve transdermal drug delivery and have been used in conjunction with liposomes to transdermally deliver docetaxel [135]. McAllister et al. [118] have delivered 25-nm and 50-nm latex nanoparticles across human cadaver and mouse epidermis by increasing skin permeability with polymeric, metallic, and silicon microneedles. Furthermore, iontophoresis has been combined with 110-nm enkaphalin-loaded, charged liposomes to successfully transport drugs across the skin barrier [132]. Micellar formulations and core-shell nanoparticles have also been investigated for transdermal delivery [108, 136, 137]. Nanoparticle formulations can be combined with other delivery mechanisms such as microneedles [138], chemical enhancers [121], or mechanical stimuli [139] in order to optimize skin permeation. It is believed that the controlling factors in transcutaneous transport of nanoparticles are size, surface charge, and lipophilicity.

The transdermal route is a promising avenue for drug delivery. Along with current advances in transdermal delivery methods, nanoparticles may serve as an excellent vehicle for carrying drugs. In this regard, nanoparticles will need to be either localized in the skin for cutaneous delivery, or be delivered transdermally into systemic circulation. Transdermal delivery is effective owing to its ability to deliver doses over long periods of time, avoid first-pass clearance in the liver, and avoid metabolic degradation in the GI tract.

Long-Circulating NPs

The inherent material properties of nanoparticles can be used for specific applications as a non-ligand-mediated targeted systems. Anderson et al. [43] developed new materials defined as "lipidoid" for siRNA delivery using high-throughput in vitro screening. Their work determined the best formulation to deliver siRNA in the liver at single nanomolar concentrations based on serum stability and toxicity using macrophages, HeLa, and HepG2 cells. In vivo non-human primate results showed significant knockdown at a very low, 2.5 mg/kg dose. In contrast, long-circulating NPs will require reduced mononuclear phagocyte system (MPS) uptake to increase their circulation time and accumulation in the site to be treated. Nanocarriers can be engineered to reduce their clearance from systemic circulation [140–142]. Surface functionalization can be tuned to increase residence time in the blood, reduce nonspecific body distribution, and, in some cases, target specific tissues or cell surface antigens with targeting ligands such as peptides, apatmers, antibodies, and small molecules. For instance, it is well established that hydrophilic polymers, most notably poly(ethylene glycol) (PEG), can be grafted, conjugated, or adsorbed to the surface of nanoparticles to form a corona, which provides steric stabilization and confers "stealth" properties that reduce rapid clearance, such as the prevention of protein adsorption. Thus, over the past 20 years, numerous approaches to improve nanoparticle blood residence and accumulation in specific tissues for therapy and diagnosis have been developed [140, 143, 144]. The resultant rapid clearance is due to interaction with blood cells and proteins, phagocytosis by the mononuclear phagocyte system (MPS) in the liver, and filtration by the spleen [142, 145]. Non-PEGylated liposome drug-encapsulated formulations accumulated more in the liver than PEGylated formulations. This resulted in a lower drug concentration in tumor xenograft models [146].

In addition, it was shown that nanoparticles larger than 50 nm did not differentially accumulate in the liver [147]. However, nanoparticles smaller than 50 nm substantially accumulated in the liver (~60% of injected dose), and these results were confirmed by a reduced amount of nanoparticles circulating in the blood. Splenic filtration size limit was found to be ~100 nm. It is also well accepted that the surface charge and surface functional groups of nanoparticles can affect their interactions with their microenvironments [7, 148–150]. Cedervall et al. [148] found that protein adsorption kinetics and characteristics depend on particle size and surface hydrophobicity. The results suggest that a layer of albumin is adsorbed on ~100% of the 200-nm nanoparticles' surface, while there is ~60% surface coverage for smaller nanoparticles (70 nm). In contrast, nanoparticles with a hydrophilic surface showed 10–20% surface coverage of protein. The higher curvature of smaller nanoparticles reduces the adsorption of larger proteins. Interestingly, the results showed competitive binding between high affinity proteins and lower affinity proteins, whereby an adsorption exchange occurred and high affinity proteins displaced other adsorbed proteins despite being at lower concentrations. The same group [149] showed, using polystyrene beads, that protein adsorption was significantly dependent on the size and charge of nanoparticles. Mass spectroscopy analysis has identified proteins with differing functions, such as immunoglobulin, lipoproteins, complement pathway proteins, and coagulation factor proteins. The fraction of the proteins adsorbed on the surface of the nanoparticles will clearly affect their in vivo behavior, such as complement activation, biodistribution, blood circulation half-life, and aggregation.

Vascular endothelium will prevent circulating NP diffusion into tissues and thus promote liver and spleen filtration. Typically, most endothelia are continuous with tight junctions between the cells and an underlying basement membrane, which prevents most nanoparticles from exiting the circulation via the paracellular route. In the liver, the discontinuous endothelium is fenestrated, thus allowing nanoparticles from 50 to 100 nm to pass across the endothelium to the underlying parenchymal cells. In other tissues, such as the spleen, the endothelium lacks a basement membrane and has larger fenestrations, which allows nanoparticles to exit circulation. Under certain physiological and pathological conditions, the endothelium can become leaky, consequently allowing the accumulation of nanoparticles determined by size exclusion and distribution of the leaky vasculature. This phenomenon is enhanced in some cancers by the reduced lymphatic drainage, leading to the accumulation of nanoparticles in tissues through a process termed the *enhanced permeability and retention* (EPR) *effect* [151–154]. Therefore, nanoparticles can take advantages of specific physiological defects of the endothelium to accumulate at specific sites and increase the local drug concentration.

In addition to the paracellular route of exit from the circulation, NPs can be engineered to cross physical barriers in a process known as *transcytosis*, the intracellular transport across a cell [69, 155, 156]. This allows the passage of macromolecules through the endothelium/epithelium, and is thought to be size dependent, whereby larger macromolecules pass through less easily than smaller molecules [157–160]. Schnitzer et al. [159, 160] have developed an antibody that binds specifically to lung caveola for the delivery of therapeutics across the endothelium barrier. The luminal expression of caveola on the lung endothelium enabled the transport from blood circulation to reach tissue accumulation of ~90% in less than one hour. Thus, the strategy to target caveola differentially expressed in specific tissues offers exciting possibilities for imaging and therapy.

Tissue Diffusion: Effect of Size, Charge, and Shape

Once out of the circulation, nanoparticles will accumulate in the peripheral extracellular matrix of the leaky tissues. The tumor microenvironment imparts special challenges that nanoparticles must overcome to successfully deliver therapeutic agents. As a result of the fenestrated, abnormal vasculature and reduced lymphatic development, tumors are characterized by an elevated interstitial fluid pressure, hypoxic solid tumor centers, and an acidic tumor microenvironment [51, 161–163]. The disorganized, heterogeneous, and tortuous tumor vasculature causes reduced blood flow [161], and the high interstitial fluid pressure creates a hydrostatic barrier opposing the convective transport of drugs and nanoparticles into solid tumors.

Next, nanocarriers must negotiate the labyrinth of ECM components to penetrate tumor tissue and reach their intended target. Diffusion of nanoparticles can be diminished by their interaction with the interstitial matrix and the tortuosity of the interstitium [161, 164]. In general, shape, size, and surface properties of the nanocarriers will contribute to their ability to penetrate tissues. Studies have demonstrated that channels exist within the extracellular matrix for these particles to diffuse [165], and further evidence exists for similar channels in the brain [165–167]. Once outside the circulation, there is evidence of how well nanomaterials can diffuse through tissues [163, 165, 168–170]. Previous studies on the diffusion of macromolecules through tumor interstitium provide insight into the parameters effecting intratumor diffusion. Studies conducted on the diffusion of inulin, bovine serum albumin, and dextran through ex vivo murine fibrosarcoma and polymeric ECM models showed the molecular weight cutoff to be >40,000 Da. Krol et al. [171] investigated the available volume fraction (K_{AV}), the interstitial space within the ECM that can contain therapeutic agents, by quantifying the diffusion of fluorescently labeled macromolecules in the ECM of rat fibrosarcoma and also gels composed of 2% gelatin, 1% chondroitin sulfate, and PBS. Indications of tissue diffusivity come from in vitro studies using 3D models [172–177] or in vivo intradermally or subcutaneously injected nanoparticles [169]. Particles with diameters larger than 60 nm tend to have difficulty diffusing through the collagen matrix and tend to concentrate in the vascular periphery [161]. Multicellular spheroid (MCS) cultures were used to model the parameters controlling nanoparticle penetration into solid tumors [178]. Polystyrene beads of varying sizes, 20–200 nm, were tested against spherical cultures of human cervical carcinoma (SiHa), and it was discovered that tumor penetration through the extracellular matrix was limited to particles <100 nm. Apparently larger particles were unable to maneuver through the collagen network. Subsequently, collagenase was immobilized onto the nanoparticle surface in order to disrupt the ECM and showed improved penetration. Nanoparticle size has also been shown to mediate intratumor dispersion of nanoparticles [182]. In vivo time studies have shown that PEGylated gold nanoparticles of varying sizes were initially localized in the perivascular regions of tumors. Over a 24-hour time period, it was shown via microscopy that 20-nm particles diffused rather far from the initial blood vessels. In contrast, 60-nm particles migrated from the vessel to a lesser extent, and 100-nm particles were greatly hindered in their diffusion through tumors. Thus, a means of locally disrupting the tumor ECM while also targeting tumors for drug delivery may improve chemotherapy efficacy. Charged particles may create electrostatic interactions with charged elements in the interstitial matrix, such as positively charged collagen or negatively charged glycosaminoglycans [161]. These interactions may cause NP aggregates to form and also affect the stability of the NPs. Though these models provide an estimation of factors affecting nanoparticle diffusion through the ECM, in vivo tests will need to fully develop an understanding of nanoparticle diffusion into the ECM that takes into account all cellular/matrix interactions, mechanical forces, and fluid dynamics [164].

Employing nanoparticles to target tumor tissue is a current paradigm that aims to improve intratumor accumulation and localization of therapeutic agents. Thus, investigating antibody–tumor interaction is critical to understanding the most effective methods for not only targeting but improving the distribution of therapeutics within tumors. Graff et al. [176] have described, using mathematical models, the relative rates of antibody intratumor diffusion and parameters affecting tumor penetration. Through analysis and simulations, the key factors to increase tumor accumulation were defined as minimum molecular weight, high antigen expression, gradual antigen metabolism, high multi-bond interaction to antigen, bolus dosage, and optimized release kinetic. Interestingly, the correlation between antigen-turnover kinetics and antibody–antigen dissociation was defined to be one of the most important factors to balance. Details for protein engineering to improve tumor accumulation are very well discussed in a review [172]. Tumor targeting systems are separated in multiple compartments such as the targeting ligand molecular properties (size, charge, blood circulation half-life, and toxicity), the antigen properties (expression level, expression distribution, and recycling) and the interactions between the molecule and its antigen (specificity and affinity). It is suggested that a small molecule should be engineered with high blood-circulation half-life, low toxicity, and high antigen-binding specificity to bind to a homogeneously distributed antigen with a high expression level and affinity in the 1–10 nanomolar range. Lower-affinity characteristics have been shown to help diffusion farther into tissues owing to the relative ease with which they dissociate from antigens compared to higher-affinity binding. Also, lower-affinity antibodies, because of this dissociation, do not irreversibly bind to antigens and undergo less endocytic degradation. Conversely, retention time in the tumor decreases, leading to lower therapeutic efficacy [179, 180]. Low binding affinity entails a higher percent of unbound antibodies that are free to diffuse out of the tumor via convective clearance. Moreover, the vascular architecture heterogeneity of tissues and the reduced contact of cancer cells with blood flow [181] will affect the therapy and imaging efficacy.

Cellular Interactions and Uptake: Size, Charge, and Kinetics

Nanoparticles must first cross the plasma membrane to deliver drugs or agents to the cytosol. The cellular mechanisms that mediate nanoparticle internalization will be controlled primarily by the adsorption of proteins on the nanoparticle surface. Thus, the physical and chemical properties of the nanoparticle—size, shape, surface charge, hydrophobicity, surface functional groups, and targeting ligands—will determine nanoparticle–protein interaction and ultimately cellular response [183, 184]. Moreover, the mode by which nanoparticles enter cells is relevant because it will dictate the initial cellular microenvironment to which the nanoparticle will be exposed [185]. NPs will enter by a variety of methods, including clathrin-mediated endocytosis, caveolae-mediated endocytosis, clathrin- and caveolae-independent endocytosis, and macropinocytosis [185–187]. In clathrin-mediated endocytosis, receptor binding initiates the formation of a vesicle ~120 nm via the invagination of the cellular membrane [186, 187]. The cytoplasmic face of the membrane is coated with clathrin molecules, which aid in forming the budding vesicle. The clathrin coat is shed intracellularly, and these vesicles are further directed towards early endosomes, and then lysosomes or the trans-Golgi network. Caveolae are flask-shaped cavities in the plasma membrane that are ~60 nm in diameter and formed by membrane proteins identified as calveolins [186, 188]. When substrates bind to the surface of the calveolae, vesicle budding occurs and the substrates are internalized. These vesicles then fuse with the caveo-somes or multivesicular bodies, thereby possibly bypassing lysosomal incorporation and degradation [186]. Clathrin- and calveolae-independent endocytosis may occur where ~90 nm vesicles are internalized without the presence of these proteins. A special form of clathrin- and calveolae-independent endocytosis is macropinocytosis [186]. In macropinocytosis, the binding of receptor tyrosine kinases initiates the formation of ruffles in the plasma membrane, which then engulf fluids [189]. These ruffles may close off and form vesicles termed macropinosomes, which are directed towards lysosomes.

Size can be a factor in the endocytosis of nanocarriers. While particles >500 nm are phagocytosed by macrophages and those <5 nm are rapidly cleared by renal filtration and urinary excretion [190, 191], NPs <100 nm are able to exit circulation via extravasation and are internalized by endocytosis [185, 186, 191]. Chithrani et al. [26] have shown size- and shape-dependent uptake in HeLa cells, where 50-nm spherical gold nanoparticles were favorably endocytosed. DeSimone's group have developed nanofabricated nanoparticles and microparticles to study the effect of the size and shape on the cellular internalization pathway [28] and NPs' in vivo biodistribution [192]. Nanoparticles were more rapidly internalized through a caveolae-mediated endocytosis mechanism by Hela cells than were microparticles, and the fraction of nanoparticles taken up was substantially higher. The fastest uptake of nanoparticles (~150 nm) was found to be due to simultaneous internalization through multiple pathways. Interestingly, rod-like nanoparticles were internalized much more efficiently than their spherical counterparts. The influence of nanoparticle shape on cellular uptake is hypothesized to be dependent on the region of contact with cellular membranes. Various polystyrene ellipsoid and spherical microparticles were investigated for uptake in alveolar macrophages, and it was shown that for ellipsoid particles, internalization was favored when contacting the cell in a perpendicular manner [193]. Furthermore, cationic nanocarriers with a high aspect ratio will be internalized more easily [161]. Discher et al. [194] have shown that filomicelle (20–60 nm diameter and 2–18 μm length) circulation time was substantially longer than their spherical counterparts, with a strong dependence on their length, with a plateau at length >8 μm. In vivo results were correlated with in vitro fluid-flow experiments on macrophages, showing a slower cell-uptake kinetic caused by shear-inducing flow alignment and reduced cell–particle contact. In addition, filomicelles were found to accumulate in the lung and be rapidly taken up by epithelial cells. Paclitaxel-loaded filomicelles also increased apoptosis in human lung tumor xenograft mice models and reduced tumor size 7 days post injection. However, the data shows tumor reduction only at one time point, and comparative results with spherical particles were not published.

Because nanoparticles will first interact with the endothelial lining, Peetla et al. [195] used Langmuir films to provide a biomimetic endothelial cell model membrane (EMM) to understand the varied interactions between nanoparticles and lipid membranes resulting from changes in size and surface charge. Changes in surface pressure were used to identify interaction between particles and the membrane. It was found that positively charged, 60-nm polystyrene nanoparticles increased the surface pressure, indicating that the phospholipid layer had condensed. The positive charges may have created electrostatic interactions with the phosphate head groups. Negative charges had no effect, and neutral charged particles reduced the surface pressure. A reduction in surface pressure suggests a dispersion of the lipid layer. Likewise, size was shown to influence surface pressure: 20-nm particles always increased the surface pressure when compared to larger, 60-nm particles, regardless of surface charge. The authors stressed that this model is applicable only to endothelial cells, and interactions may be different based on the different chemical composition of the membranes of interest, especially the influence of membrane proteins, lipids, and carbohydrates. Moreover, they emphasize that this model is merely a lipid monolayer and might not be representative of the interaction of nanoparticles with a full plasma membrane lipid bilayer, complete with transmembrane proteins and extracellular matrix components.

Intracellular vs. Extracellular Drug Delivery

Finally, the nanocarriers will have to release the payload extracellularly or intracellularly. In general, intracellular delivery is a more efficient strategy for increasing cytotoxicity and in some cases reducing drug resistance [196, 197]. Impaired drug delivery, mutations in cellular genetics, and non-genetic environmental factors can result in multidrug resistance (MDR), a phenomenon of cancerous cells being resistant to structurally unrelated drugs that have discrete and separate modes of action [198–200]. There are several methods by which tumor cells can be resistant to drugs, namely activation of ATP-driven efflux pumps, inhibition of the influx of drugs to the cytoplasm, activation of DNA repair mechanisms, and activation of detoxifying agents [199]. Much of the research has focused on the efflux of hydrophobic drugs by ATP-binding cassette (ABC) transporters. These ATP-dependent transmembrane proteins are known to confer MDR to cancer cells, and cancer cell line cultures have been shown to overexpress certain ABC transporters [199, 201]. P-glycoprotein (Pgp), a twelve-pass transmembrane ABC transporter, is known to export drugs such as docetaxel, paclitaxel, doxorubicin, daunorubicin, etoposide, actinomycin D, methotrexate, mitoxantrone, and others [201–204]. It is hypothesized that loading nanoparticles with drugs will decrease drug recognition by efflux pumps, leading to higher intracellular concentrations and thus more efficient treatment [199, 205]. Moreover, the flexibility of nanocarrier engineering may be employed to circumvent MDR through coadministration of chemotherapeutic agents with drug efflux protein inhibitors [196, 199, 200,

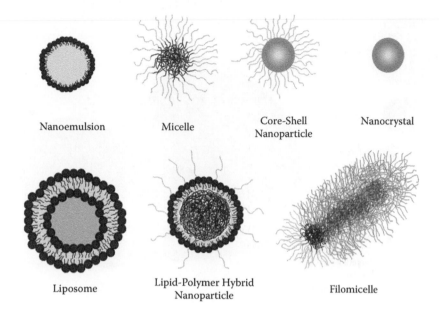

FIGURE 4.1

Examples of nanocarriers for drug delivery or medical imaging. By tuning the physiochemical properties and conjugating the particles with targeting ligands, site-specific delivery can be achieved. Particles can also be loaded with drugs or imaging agents to deliver a payload at the target site.

206, 207]. Also, combinatorial methods employing nanoparticles loaded with anti-cancer therapeutics and pro-apoptotic inhibitors have been investigated. Devalapally [197] loaded poly(caprolactone)-poly(ethylene glycol) (PCL-PEG) micelles with paclitaxel and C6-ceramide, a pro-apoptotic signaling messenger. In another study, nanoemulsions approximately 140 nm in diameter have also been developed which, when loaded simultaneously with paclitaxel and the pro-apoptotic MDR modifier curcumin, effectively inhibited multidrug-resistant SK-OV-3 ovarian adenocarcinoma cells [208].

The intracellular delivery of therapeutic agents may be most beneficial when directed towards specific cytoplasmic components or organelles, such as endosomes, lysosomes, mitochondria, or the nucleus (Figure 4.2) [209]. Thus, nanocarriers will require engineering in order to cross the plasma membrane and release their payload within organelles or in the cytoplasm. Partitioning across the cellular membrane is a described mechanism used for vesicular and positively charged dendrimeric vesicles to enter cells, thus passage into cells is governed mostly by receptor-mediated endocytosis [210]. Likewise, receptor ligands, such as folic acid and transferrin, can be employed to initiate receptor-mediated endocytosis [209], but endosomal/lysosomal escape is often necessary in order to avoid enzymatic degradation of the therapeutic agent. Intelligent carriers are engineered to escape

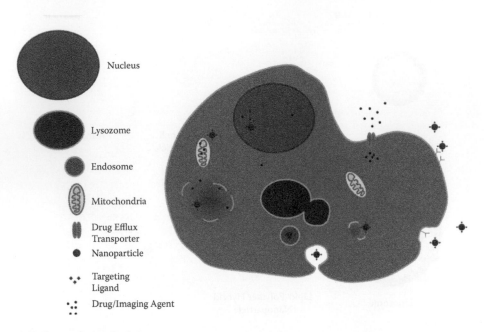

Nucleus

Lysozome

Endosome

Mitochondria

Drug Efflux
Transporter

Nanoparticle

Targeting
Ligand

Drug/Imaging Agent

FIGURE 4.2 (See color insert.)
Specific cellular or intracellular targets may increase nanoparticle treatment efficacy. To avoid the degradation of the payload, nanoparticles may need to actively escape endosomal or lysosomal compartments. Delivery of the payload to specific organelles may be mediated by the nanocarrier.

endosomes and/or release payloads intracellularly via pH-, temperature-, environment-, or light-responsive systems [211]. Endosomes have a characteristic pH between 5.0 and 6.5 [9], and nanoparticles have previously been designed to have pH-sensitive release. Shenoy et al. [212] synthesized poly(β-amino ester)-based 113-nm nanoparticles that escaped endosomes through polymer dissolution in acidic pH conditions. Polymeric cationic coatings have also allowed nanoparticles to escape endosomes [213]. Core shell fluorescent silica nanoparticles of 60 nm were coated with poly(ethylenimine) and shown by confocal microscopy to disperse in the cytosol. Controlled-release nanoparticles have been designed that utilize β-cyclodextrins as pH-sensitive caps covering 2-nm pores on ~100-nm silica nanoparticles. These delivered their payload in vitro by virtue of endosomal drop in pH [214].

Biomolecules can direct nanoparticles across cellular barriers and further target specific organelles such as the nucleus and mitochondria. Cell-penetrating peptides can be conjugated to nanoparticles to facilitate transcellular movement. HIV-1 trans-activating transcriptional activator peptide (TAT) is one such biomolecule that can successfully maintain cell-penetrating activity after conjugation to nanoparticles. In addition to cellular internalization, TAT-conjugated gold nanoparticles have been shown to localize in the nucleus [215]. This organelle-specific internalization could be customized to inhibit or promote organelle-specific activities. The mitochondria may be another opportune target for intracellular drug delivery because

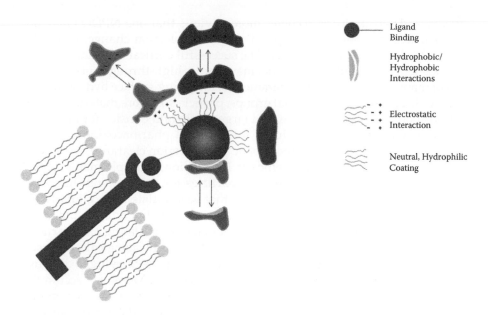

FIGURE 4.3 (See color insert.)
Once within the body, nanoparticle fate will be determined by interaction with proteins. The physiochemical properties and surface engineering of a nanoparticle will determine the particle/protein interaction. Electrostatic interactions between the nanoparticle surface charge and the positive and negative domains of proteins, along with non-polar interactions between hydrophobic regions, will determine protein adsorption. Hydrophilic coatings, such as PEG, can confer "stealth" properties. Various ligands attached to the nanoparticle surface can target cell-specific moeities and mediate cellular processes.

of its critical role in cell survival and unique characteristics. Properties of the mitochondria of primary interest for drug delivery are its internal pH of 8 and its negative membrane potential (–130 to –150 mV) [209]. Cationic, lipophilic moieties are favorable for mitochondrial targeting due in part to the negative membrane potential of the mitochondria. The chemical environment of the mitochondria may also be utilized for pH-sensitive controlled release as mentioned previously. Organelle-targeting nanoparticles have been investigated and described for improved therapeutic treatments [216–218]. The interaction of nanoparticles with intracellular proteins is another avenue of research that may improve the efficacy of nanomedicine technologies.

Conclusion

With the understanding of the many factors affecting the pharmacokinetic and biodistribution of nanoparticles, it is clear that optimization of formulations for specific applications are required to decrease cytoxicity and improve efficacy and sensitivity. It becomes more obvious that one NPDDS may not be efficient for all therapeutic applications, and molecular changes could significantly affect their performance. The successful clinical translation of any diagnostic or therapeutic nanocarrier may require high-throughput optimization of many physico-chemical parameters, including surface hydrophilicity, surface charge, surface functional groups, particle size, core materials, linker composition, nanoparticle shape, and targeting ligand density, for optimization of therapeutic efficacy, reduced toxicity, and pharmacokinetic parameters. These factors will affect the hepatic and excretion clearance, circulation time, tissue accumulation, and efficacy. In addition, multiple mathematical models are described to engineer nanoparticles for therapy and diagnosis. Certainly, non-degradable nanoparticles that accumulate intracellularly are likely to have a number of toxic effects [190, 219, 220]. In addition, the endothelium is a key tissue owing to its direct contact to circulating nanoparticles and is an important mediator of a large number of physiological responses, so in many cases it is important to prevent nanoparticle-carrying toxic drugs to be taken up by non-diseased endothelial cells [221]. Moreover, accumulation in tissues such as bone marrow and brain could be lethal depending on the toxicity of the nanomaterial, Therefore, non-degradable nanoparticles need to be targeted and designed for rapid excretion. More importantly, the progress of biology and genomics to discover and isolate new markers, as well as a better understanding of the disease microenvironments, is of critical importance to the development of more effective nanoscale systems for therapy and diagnosis.

In summary, nanomedicine has already resulted in numerous successful formulations in the clinic, and an increasing number of novel therapeutic and diagnostic modalities are in the early or late phases of development for many human diseases. It is expected that medicine will be practiced differently for years to come because of the benefits of nanotechnology.

Acknowledgments

The authors are grateful to the National Institutes of Health for funding the IDeA Networks of Biomedical Research Excellence (2008224).

References

1. M.C. Garnett, P. Kallinteri, Nanomedicines and nanotoxicology: Some physiological principles, *Occupational Medicine* 56 (August 2006) 307–311.

2. F. Alexis, J. Rhee, J.P. Richie, A.F. Radovic-Moreno, R. Langer, O.C. Farokhzad, New frontiers in nanotechnology for cancer treatment, *Urologic Oncology: Seminars and Original Investigations* 26 (2008) 74–85.

3. M. Sprintz, Editorial: Nanotechnology for advanced therapy and diagnosis, *Biomed. Microdevices* 6 (2004) 101–103.

4. X. Wang, L. Yang, Z., Chen, D.M. Shin, Application of nanotechnology in cancer therapy and imaging, *CA: A Cancer Journal for Clinicians* 58 (2008) 97–110.

5. S.S. Davis, Biomedical applications of nanotechnology: Implications for drug targeting and gene therapy, *Trends Biotechnol.* 15 (1997) 217–224.

6. O.C. Farokhzad, Nanotechnology for drug delivery: The perfect partnership, *Expert Opinion on Drug Delivery* 5 (2008) 927–929.

7. F. Alexis, E. Pridgen, L.K. Molnar, O.C. Farokhzad, Factors affecting the clearance and biodistribution of polymeric nanoparticles, *Molecular Pharmaceutics* 5 (2008) 505–515.

8. D. Peer, J.M. Karp, S. Hong, O.C. Farokhzad, R. Margalit, R. Langer, Nanocarriers as an emerging platform for cancer therapy, *Nat Nano* 2 (2007) 751–760.

9. M.E. Davis, Z. Chen, D.M. Shin, Nanoparticle therapeutics: An emerging treatment modality for cancer, *Nat Rev Drug Discov* 7 (2008) 771–782.

10. R. Paull, J. Wolfe, P. Hebert, M. Sinkula, Investing in nanotechnology, *Nat Biotech* 21 (2003) 1144–1147.

11. H. Devalapally, A. Chakilam, M.M. Amiji, Role of nanotechnology in pharmaceutical product development, *Journal of Pharmaceutical Sciences* 96 (2007) 2547–2565.

12. A. Sparreboom, C.D. Scripture, V. Trieu, P.J. Williams, T. De, A. Yang, B. Beals, W.D. Figg, M. Hawkins, N. Desai, Comparative preclinical and clinical pharmacokinetics of a cremophor-free, nanoparticle albumin-bound paclitaxel (ABI-007) and paclitaxel formulated in cremophor (Taxol), *Clinical Cancer Research* 11 (2005) 4136–4143.

13. W. Beecken, A. Fernandez, D. Panigrahy, E. Achilles, O. Kisker, E. Flynn, M.J. Antonia, J. Folkman, Y. Shing, Efficacy of antiangiogenic therapy with TNP-470 in superficial and invasive bladder cancer models in mice, *Urology* 56 (2000) 521–526.

14. T. Tanaka, H. Konno, I. Matsuda, S. Nakamura, S. Baba, Prevention of hepatic metastasis of human colon cancer by angiogenesis inhibitor TNP-470, *Cancer Research* 55 (1995) 836–839.

15. A.P. Kudelka, C.F. Verschraegen, E. Loyer, Complete remission of metastatic cervical cancer with the angiogenesis inhibitor TNP-470, *N. Engl. J. Med.* 338 (1998) 991–992.

16. R. Satchi-Fainaro, M. Puder, J.W. Davies, H.T. Tran, D.A. Sampson, A.K. Greene, G. Corfas, J. Folkman, Targeting angiogenesis with a conjugate of HPMA copolymer and TNP–470, *Nat. Med.* 10 (2004) 255–261.

17. R. Satchi-Fainaro, R. Mamluk, L. Wang, S.M. Short, J.A. Nagy, D. Feng, A.M. Dvorak, H.F. Dvorak, M. Puder, D. Mukhopadhyay, J. Folkman, Inhibition of vessel permeability by TNP-470 and its polymer conjugate, caplostatin, *Cancer Cell* 7 (2005) 251–261.

18. M.C. Garnett, Targeted drug conjugates: Principles and progress, *Adv. Drug Deliv. Rev.* 53 (2001) 171–216.

19. C. Li, Poly(glutamic acid)-anticancer drug conjugates, *Adv. Drug Deliv. Rev.* 54 (2002) 695–713.

20. R.B. Greenwald, Y.H. Choe, J. McGuire, C.D. Conover, Effective drug delivery by PEGylated drug conjugates, *Adv. Drug Deliv. Rev.* 55 (2003) 217–250.

21. C. Li, S. Wallace, Polymer-drug conjugates: Recent development in clinical oncology, *Adv. Drug Deliv. Rev.* 60 (2008) 886–898.

22. D.A. LaVan, D.M. Lynn, R. Langer, Moving smaller in drug discovery and delivery, *Nat Rev Drug Discov* 1 (2002) 77–84.

23. L.E. Euliss, J.A. DuPont, S. Gratton, J. DeSimone, Imparting size, shape, and composition control of materials for nanomedicine, *ChemInform* 38 (2007) 1095–1104.

24. R. Weissleder, K. Kelly, E.Y. Sun, T. Shtatland, L. Josephson, Cell-specific targeting of nanoparticles by multivalent attachment of small molecules, *Nat Biotech* 23 (2005) 1418–1423.

25. M. Grzelczak, J. Perez-Juste, P. Mulvaney, L. Liz-Marzan, Shape control in gold nanoparticle synthesis, *Chem. Soc. Rev.* 37 (2008) 1783–1791.

26. B.D. Chithrani, A.A. Ghazani, W.C.W. Chan, Determining the size and shape dependence of gold nanoparticle uptake into mammalian cells, *Nano Letters* 6 (2006) 662–668.

27. N.R. Jana, Shape effect in nanoparticle self-assembly, *Angewandte Chemie* 116 (2004) 1562–1566.

28. S.E.A. Gratton, P.A. Ropp, P.D. Pohlhaus, J.C. Luft, V.J. Madden, M.E. Napier, J.M. DeSimone, The effect of particle design on cellular internalization pathways, *Proceedings of the National Academy of Sciences* 105 (2008) 11613–11618.

29. J.P. Rolland, B.W. Maynor, L.E. Euliss, A.E. Exner, G.M. Denison, J.M. DeSimone, Direct fabrication and harvesting of monodisperse, shape-specific nanobiomaterials, *J. Am. Chem. Soc.* 127 (2005) 10096–10100.

30. V. Labhasetwar, C. Song, W. Humphrey, R. Shebuski, R.J. Levy, Arterial uptake of biodegradable nanoparticles: Effect of surface modifications, *Journal of Pharmaceutical Sciences* 87 (1998) 1229–1234.

31. Y. Yamamoto, Y. Nagasaki, Y. Kato, Y. Sugiyama, K. Kataoka, Long-circulating poly(ethylene glycol)-poly(-lactide) block copolymer micelles with modulated surface charge, *J. Controlled Release* 77 (2001) 27–38.

32. S.J. Clarke, C.A. Hollmann, F.A. Aldaye, J.L. Nadeau, effect of ligand density on the spectral, physical, and biological characteristics of CdSe/ZnS quantum dots, *Bioconjug. Chem.* 19 (2008) 562–568.

33. S. Nagayama, K. Ogawara, Y. Fukuoka, K. Higaki, T. Kimura, Time-dependent changes in opsonin amount associated on nanoparticles alter their hepatic uptake characteristics, *Int. J. Pharm.* 342 (2007) 215–221.

34. C. Fang, B. Shi, Y. Pei, M. Hong, J. Wu, H. Chen, In vivo tumor targeting of tumor necrosis factor-α-loaded stealth nanoparticles: Effect of MePEG molecular weight and particle size, *European Journal of Pharmaceutical Sciences* 27 (2006) 27–36.

35. J.B. Haun, D.A. Hammer, Quantifying nanoparticle adhesion mediated by specific molecular interactions, *Langmuir* 24 (2008) 8821–8832.

36. C. Chen, E.E. Dormidontova, Architectural and structural optimization of the protective polymer layer for enhanced targeting, *Langmuir* 21 (2005) 5605–5615.

37. F. Gu, L. Zhang, B.A. Teply, N. Mann, A. Wang, A.F. Radovic-Moreno, R. Langer, O.C. Farokhzad, Precise engineering of targeted nanoparticles by using self-assembled biointegrated block copolymers, *Proceedings of the National Academy of Sciences* 105 (2008) 2586–2591.

38. N. Desai, V. Trieu, Z. Yao, L. Louie, S. Ci, A. Yang, C. Tao, T. De, B. Beals, D. Dykes, P. Noker, R. Yao, E. Labao, M. Hawkins, P. Soon-Shiong, Increased antitumor activity, intratumor paclitaxel concentrations, and endothelial cell transport of cremophor-free, albumin-bound paclitaxel, ABI-007, compared with cremophor-based paclitaxel. *Clinical Cancer Research* 12 (2006) 1317–1324.

39. T. Kim, D. Kim, J. Chung, S.G. Shin, S. Kim, D.S. Heo, N.K. Kim, Y. Bang, Phase I and pharmacokinetic study of genexol-pm, a cremophor-free, polymeric micelle-formulated paclitaxel, in patients with advanced malignancies, *Clinical Cancer Research* 10 (2004) 3708–3716.

40. S.C. Kim, D.W. Kim, Y.H. Shim, J.S. Bang, H.S. Oh, S.W. Kim, M.H. Seo, In vivo evaluation of polymeric micellar paclitaxel formulation: Toxicity and efficacy, *J. Controlled Release* 72 (2001) 191–202.

41. A. Akinc, A. Zumbuehl, M. Goldberg, E.S. Leshchiner, V. Busini, N. Hossain, S.A. Bacallado, et al. A combinatorial library of lipid-like materials for delivery of RNAi therapeutics, *Nat Biotech* 26 (2008) 561–569.

42. M. John, R. Constien, A. Akinc, M. Goldberg, Y. Moon, M. Spranger, P. Hadwiger, J. Soutschek, H. Vornlocher, M. Manoharan, M. Stoffel, R. Langer, D.G. Anderson, J.D. Horton, V. Koteliansky, D. Bumcrot, Effective RNAi-mediated gene silencing without interruption of the endogenous microRNA pathway, *Nature* 449 (2007) 745–747.

43. D.G. Anderson, W. Peng, A. Akinc, N. Hossain, A. Kohn, R. Padera, R. Langer, J.A. Sawicki, A polymer library approach to suicide gene therapy for cancer, *Proceedings of the National Academy of Sciences of the United States of America* 101 (2004) 16028–16033.

44. E.A. Schellenberger, F. Reynolds, R. Weissleder, L. Josephson, Surface-functionalized nanoparticle library yields probes for apoptotic cells, *ChemBioChem* 5 (2004) 275–279.

45. J. Kim, J.E. Lee, J. Lee, Y. Jang, S. Kim, K. An, J.H. Yu, T. Hyeon, Generalized fabrication of multifunctional nanoparticle assemblies on silica spheres, *Angewandte Chemie* 118 (2006) 4907–4911.

46. A.G. Tkachenko, H. Xie, D. Coleman, W. Glomm, J. Ryan, M.F. Anderson, S. Franzen, D.L. Feldheim, multifunctional gold nanoparticle peptide complexes for nuclear targeting, *J. Am. Chem. Soc.* 125 (2003) 4700–4701.

47. L. Zhang, A.F. Radovic-Moreno, F. Alexis, F.X. Gu, P.A. Basto, V. Bagalkot, S. Jon, R.S. Langer, O.C. Farokhzad, Co-delivery of hydrophobic and hydrophilic drugs from nanoparticle-aptamer bioconjugates, *ChemMedChem* 2 (2007) 1268–1271.

48. A.Z. Wang, V. Bagalkot, C.C. Vasilliou, F. Gu, F. Alexis, L. Zhang, M. Shaikh, K. Yuet, M.J. Cima, R. Langer, P.W. Kantoff, N.H. Bander, S. Jon, O.C. Farokhzad, Superparamagnetic iron oxide nanoparticle-aptamer bioconjugates for combined prostate cancer imaging and therapy, *ChemMedChem* 3 (2008) 1311–1315.

49. R. Weissleder, M.J. Pittet, Imaging in the era of molecular oncology, *Nature* 452 (2008) 580–589.

50. S. Kumar, J. Aaron, K. Sokolov, Directional conjugation of antibodies to nanoparticles for synthesis of multiplexed optical contrast agents with both delivery and targeting moieties, *Nat. Protocols* 3 (2008) 314–320.

51. R.K. Jain, Normalization of tumor vasculature: An emerging concept in antiangiogenic therapy, *Science* 307 (2005) 58–62.

52. A. Hormigo, P.H. Gutin, S. Rafii, Tracking normalization of brain tumor vasculature by magnetic imaging and proangiogenic biomarkers, *Cancer Cell* 11 (2007) 6–8.

53. N. Ferrara, R.S. Kerbel, Angiogenesis as a therapeutic target, *Nature* 438 (2005) 967–974.
54. D.M. McDonald, P.L. Choyke, Imaging of angiogenesis: From microscope to clinic, *Nat. Med.* 9 (2003) 713–725.
55. V. Ntziachristos, C. Tung, C. Bremer, R. Weissleder, Fluorescence molecular tomography resolves protease activity in vivo, *Nat. Med.* 8 (2002) 757–761.
56. E. Mathiowitz, J.S. Jacob, Y.S. Jong, G.P. Carino, D.E. Chickering, P. Chaturvedi, C.A. Santos, K. Vijayaraghavan, S. Montgomery, M. Bassett, C. Morrell, Biologically erodible microspheres as potential oral drug delivery systems, *Nature* 386 (1997) 410–414.
57. L.X. Yu, E. Lipka, J.R. Crison, G.L. Amidon, Transport approaches to the biopharmaceutical design of oral drug delivery systems: Prediction of intestinal absorption, *Adv. Drug Deliv. Rev.* 19 (1996) 359–376.
58. I. Lambkin, C. Pinilla, Targeting approaches to oral drug delivery, *Expert Opinion on Biological Therapy* 2 (2002) 67–73.
59. M. Morishita, N.A. Peppas, Is the oral route possible for peptide and protein drug delivery? *Drug Discov. Today* 11 (2006) 905–910.
60. D.A. Edwards, J. Hanes, G. Caponetti, J. Hrkach, A. Ben-Jebria, M.L. Eskew, J. Mintzes, D. Deaver, N. Lotan, R. Langer, Large porous particles for pulmonary drug delivery, *Science* 276 (1997) 1868–1872.
61. J. Yu, Y.W. Chien, Pulmonary drug delivery: Physiologic and mechanistic aspects, *CRT* 14 (1997) 59.
62. I. Gonda, The ascent of pulmonary drug delivery, *Journal of Pharmaceutical Sciences* 89 (2000) 940–945.
63. H.M. Courrier, N. Butz, T.F. Vandamme, Pulmonary drug delivery systems: Recent developments and prospects, *CRT* 19 (2002) 64.
64. M. Kinoshita, K. Baba, A. Nagayasu, K. Yamabe, T. Shimooka, Y. Takeichi, M. Azuma, H. Houchi, K. Minakuchi, Improvement of solubility and oral bioavailability of a poorly water-soluble drug, TAS-301, by its melt-adsorption on a porous calcium silicate, *Journal of Pharmaceutical Sciences* 91 (2002) 362–370.
65. E. Khafagy, M. Morishita, Y. Onuki, K. Takayama, Current challenges in noninvasive insulin delivery systems: A comparative review, Adv. *Drug Deliv. Rev.* 59 (2007) 1521–1546.
66. S. Bisht, G. Feldmann, J.M. Koorstra, M. Mullendore, H. Alvarez, C. Karikari, M.A. Rudek, C.K. Lee, A. Maitra, A. Maitra, In vivo characterization of a polymeric nanoparticle platform with potential oral drug delivery capabilities, *Molecular Cancer Therapeutics* 7 (2008) 3878–3888.
67. Y.J. Yamanaka, K.W. Leong, Engineering strategies to enhance nanoparticle-mediated oral delivery, *Journal of Biomaterials Science, Polymer Edition* 19 (December 2008) 1549–1570(22).
68. S.P. Vyas, P.N. Gupta, Implication of nanoparticles/microparticles in mucosal vaccine delivery, *Expert Review of Vaccines* 6 (June 2007) 401–418(18).
69. W. Lu, Y. Tan, K. Hu, X. Jiang, Cationic albumin conjugated pegylated nanoparticle with its transcytosis ability and little toxicity against blood–brain barrier, *Int. J. Pharm.* 295 (2005) 247–260.
70. N.V.S. Madhav, A.K. Shakya, P. Shakya, K. Singh, Orotransmucosal drug delivery systems: A review, *J. Controlled Release* 140 (2009) 2–11.
71. B. Sarmento, S. Martins, D. Ferreira, E. Souto, Oral insulin delivery by means of solid lipid nanoparticles, *Int. J. Nanomedicine* 2 (2007) 743–749.

72. B.M. Koeppen, B.A. Stanton, *Berne & Levy Physiology*, Mosby Series, Philadelphia, PA, 2009, pp. 30–31.
73. J.C. Sung, B.L. Pulliam, D.A. Edwards, Nanoparticles for drug delivery to the lungs, *Trends Biotechnol.* 25 (2007) 563–570.
74. J.S. Patton, P.R. Byron, Inhaling medicines: Delivering drugs to the body through the lungs, *Nat Rev Drug Discov* 6 (2007) 67–74.
75. M. Smola, T. Vandamme, A. Sokolowski, Nanocarriers as pulmonary drug delivery systems to treat and to diagnose respiratory and non respiratory diseases, *Int. J. Nanomedicine* 3 (2008) 1–19.
76. M. McKinley, V. O'Loughlin, Respiratory System, in: *Anonymous Human Anatomy*, McGraw-Hill, New York, 2009, pp. 757–763.
77. M.M. Bailey, C.J. Berkland, Nanoparticle formulations in pulmonary drug delivery, *Medicinal Research Reviews* 29 (2009) 196–212.
78. B.N. Lambrecht, Alveolar macrophage in the driver's seat, *Immunity* 24 (2006) 366–368.
79. J.S. Patton, C.S. Fishburn, J.G. Weers, The lungs as a portal of entry for systemic drug delivery, *Proc Am Thorac Soc* 1 (2004) 338–344.
80. Y.Y. Zuo, R.A.W. Veldhuizen, A.W. Neumann, N.O. Petersen, F. Possmayer, Current perspectives in pulmonary surfactant: Inhibition, enhancement and evaluation, *Biochimica et Biophysica Acta (BBA) - Biomembranes* 1778 (2008) 1947–1977.
81. S.K. Lai, Y. Wang, J. Hanes, Mucus-penetrating nanoparticles for drug and gene delivery to mucosal tissues, *Adv. Drug Deliv. Rev.* 61 (2009) 158–171.
82. W. Yang, J.I. Peters, R.O. Williams III, Inhaled nanoparticles: A current review, *Int. J. Pharm.* 356 (2008) 239–247.
83. G. Oberdörster, Lung clearance of inhaled insoluble and soluble particles, *Journal of Aerosol Medicine* 1 (1988) 289–330.
84. H. Swai, B. Semete, L. Kalombo, P. Chelule, K. Kisich, B. Sievers, Nanomedicine for respiratory diseases, *Wiley Interdisciplinary Reviews: Nanomedicine and Nanobiotechnology* 1 (2009) 255–263.
85. J. Sung, D. Padilla, L. Garcia-Contreras, J. VerBerkmoes, D. Durbin, C. Peloquin, K. Elbert, A. Hickey, D. Edwards, Formulation and pharmacokinetics of self-assembled rifampicin nanoparticle systems for pulmonary delivery, *Pharm. Res.* (2009) 1847–1855.
86. N.R. Yacobi, L. DeMaio, J. Xie, S.F. Hamm-Alvarez, Z. Borok, K. Kim, E.D. Crandall, Polystyrene nanoparticle trafficking across alveolar epithelium, *Nanomedicine: Nanotechnology, Biology and Medicine* 4 (2008) 139–145.
87. C. Muhlfeld, B. Rothen-Rutishauser, F. Blank, D. Vanhecke, M. Ochs, P. Gehr, Interactions of nanoparticles with pulmonary structures and cellular responses, *Am. J. Physiol. Lung Cell. Mol. Physiol.* 294 (2008) L817–829.
88. J.S. Patton, J.G. Bukar, M.A. Eldon, Clinical pharmacokinetics and pharmacodynamics of inhaled insulin, *Clin. Pharmacokinet.* 43 (2004) 781.
89. J. Kwon, S. Hwang, H. Jin, D. Kim, A. Minai-Tehrani, H. Yoon, M. Choi, T. Yoon, D. Han, Y. Kang, B. Yoon, J. Lee, M. Cho, Body distribution of inhaled fluorescent magnetic nanoparticles in the mice, *Journal of Occupational Health* 50 (2008) 1–6.
90. P. Dames, B. Gleich, A. Flemmer, K. Hajek, N. Seidl, F. Wiekhorst, D. Eberbeck, I. Bittmann, C. Bergemann, T. Weyh, L. Trahms, J. Rosenecker, C. Rudolph, Targeted delivery of magnetic aerosol droplets to the lung, *Nat Nano* 2 (2007) 495–499.

91. S. White, D.B. Bennett, S. Cheu, P.W. Conley, D.B. Guzek, S. Gray, J. Howard, R. Malcolmson, J.M. Parker, P. Roberts, N. Sadrzadeh, J.D. Schumacher, S. Seshadri, G.W. Sluggett, C.L. Stevenson, N.J. Harper, EXUBERA®: Pharmaceutical development of a novel product for pulmonary delivery of insulin, *Diabetes Technology & Therapeutics* 7 (2005) 896–906.

92. J. Kling, Inhaled insulin's last gasp? *Nat Biotech* 26 (2008) 479–480.

93. N. Csaba, M. Garcia-Fuentes, M.J. Alonso, Nanoparticles for nasal vaccination, *Adv. Drug Deliv. Rev.* 61 (2009) 140–157.

94. D. Lee, S. Shirley, R. Lockey, S. Mohapatra, Thiolated chitosan nanoparticles enhance anti-inflammatory effects of intranasally delivered theophylline, *Respiratory Research* 7 (2006) 112.

95. T. Jung, W. Kamm, A. Breitenbach, E. Kaiserling, J.X. Xiao, T. Kissel, Biodegradable nanoparticles for oral delivery of peptides: Is there a role for polymers to affect mucosal uptake? *European Journal of Pharmaceutics and Biopharmaceutics* 50 (2000) 147–160.

96. P. Arora, S. Sharma, S. Garg, Permeability issues in nasal drug delivery, *Drug Discov. Today* 7 (2002) 967–975.

97. A. Vila, H. Gill, O. McCallion, M.J. Alonso, Transport of PLA-PEG particles across the nasal mucosa: Effect of particle size and PEG coating density, *J. Controlled Release* 98 (2004) 231–244.

98. X. Wang, N. Chi, X. Tang, Preparation of estradiol chitosan nanoparticles for improving nasal absorption and brain targeting, *European Journal of Pharmaceutics and Biopharmaceutics* 70 (2008) 735–740.

99. X. Gao, W. Tao, W. Lu, Q. Zhang, Y. Zhang, X. Jiang, S. Fu, Lectin-conjugated PEG-PLA nanoparticles: Preparation and brain delivery after intranasal administration, *Biomaterials* 27 (2006) 3482–3490.

100. A. Mistry, S.Z. Glud, J. Kjems, J. Randel, K.A. Howard, S. Stolnik, L. Illum, Effect of physicochemical properties on intranasal nanoparticle transit into murine olfactory epithelium, *J. Drug Target.* 17 (2009) 543–552.

101. I. Bravo-Osuna, C. Vauthier, H. Chacun, G. Ponchel, Specific permeability modulation of intestinal paracellular pathway by chitosan-poly(isobutylcyanoacrylate) core-shell nanoparticles, *European Journal of Pharmaceutics and Biopharmaceutics* 69 (2008) 436–444.

102. S.C. Corr, C.C.G.M. Gahan, C. Hill, M-cells: Origin, morphology and role in mucosal immunity and microbial pathogenesis, *FEMS Immunology & Medical Microbiology* 52 (2008) 2–12.

103. P.N. Gupta, K. Khatri, A.K. Goyal, N. Mishra, S.P. Vyas, M-cell targeted biodegradable PLGA nanoparticles for oral immunization against hepatitis B, *J. Drug Target.* 15 (2007) 701–713.

104. M. Garinot, V. Fiévez, V. Pourcelle, F. Stoffelbach, A. des Rieux, L. Plapied, I. Theate, H. Freichels, C. Jérôme, J. Marchand-Brynaert, Y. Schneider, V. Préat, PEGylated PLGA-based nanoparticles targeting M cells for oral vaccination, *J. Controlled Release* 120 (2007) 195–204.

105. I. Kadiyala, Y. Loo, K. Roy, J. Rice, K.W. Leong, Transport of chitosan–DNA nanoparticles in human intestinal M-cell model versus normal intestinal enterocytes, *European Journal of Pharmaceutical Sciences* 39 (2010) 103–109.

106. Y. Sato, Y. Kawashima, H. Takeuchi, H. Yamamoto, In vitro and in vivo evaluation of riboflavin-containing microballoons for a floating controlled drug delivery system in healthy humans, *Int. J. Pharm.* 275 (2004) 97–107.

107. M.R. Prausnitz, R. Langer, Transdermal drug delivery, *Nat Biotech* 26 (2008) 1261–1268.
108. R.W. Lee, D.B. Shenoy, R. Sheel, Micellar nanoparticles: Applications for topical and passive transdermal drug delivery, in Vitthal S. Kulkarni (Ed.), *Handbook of Non-Invasive Drug Delivery Systems*, William Andrew Publishing, Boston, 2010, pp. 37–58.
109. G. Cevc, U. Vierl, Nanotechnology and the transdermal route: A state of the art review and critical appraisal, *J. Controlled Release* 141 (2010) 277–299.
110. J.A. Bouwstra, P.L. Honeywell-Nguyen, G.S. Gooris, M. Ponec, Structure of the skin barrier and its modulation by vesicular formulations, *Prog. Lipid Res.* 42 (2003) 1–36.
111. E. Candi, R. Schmidt, G. Melino, The cornified envelope: A model of cell death in the skin, *Nat. Rev. Mol. Cell Biol.* 6 (2005) 328–340.
112. M.R. Prausnitz, S. Mitragotri, R. Langer, Current status and future potential of transdermal drug delivery, *Nat Rev Drug Discov* 3 (2004) 115–124.
113. A. Denet, R. Vanbever, V. Préat, Skin electroporation for transdermal and topical delivery, *Adv. Drug Deliv. Rev.* 56 (2004) 659–674.
114. Y.N. Kalia, A. Naik, J. Garrison, R.H. Guy, Iontophoretic drug delivery, *Adv. Drug Deliv. Rev.* 56 (2004) 619–658.
115. S. Mitragotri, Transdermal drug delivery using low-frequency sonophoresis, (2007) 223–236; 236.
116. T. Wong, C. Chen, C. Huang, C. Lin, S. Hui, Painless electroporation with a new needle-free microelectrode array to enhance transdermal drug delivery, *J. Controlled Release* 110 (2006) 557–565.
117. B. Chen, J. Wei, F. Tay, Y. Wong, C. Iliescu, Silicon microneedle array with biodegradable tips for transdermal drug delivery, *Microsystem Technologies* 14 (2008) 1015–1019.
118. D.V. McAllister, P.M. Wang, S.P. Davis, J. Park, P.J. Canatella, M.G. Allen, M.R. Prausnitz, Microfabricated needles for transdermal delivery of macromolecules and nanoparticles: Fabrication methods and transport studies, *Proceedings of the National Academy of Sciences of the United States of America* 100 (2003) 13755–13760.
119. J. Park, M.G. Allen, M.R. Prausnitz, Biodegradable polymer microneedles: Fabrication, mechanics and transdermal drug delivery, *J. Controlled Release* 104 (2005) 51–66.
120. P. Karande, A. Jain, S. Mitragotri, Discovery of transdermal penetration enhancers by high-throughput screening, *Nat Biotech* 22 (2004) 192–197.
121. T. Kuo, C. Wu, C. Hsu, W. Lo, S. Chiang, S. Lin, C. Dong, C. Chen, Chemical enhancer induced changes in the mechanisms of transdermal delivery of zinc oxide nanoparticles, *Biomaterials* 30 (2009) 3002–3008.
122. A. Arora, E. Kisak, P. Karande, J. Newsam, S. Mitragotri, Multicomponent chemical enhancer formulations for transdermal drug delivery: More is not always better, *J. Controlled Release* 144 (2010) 175–180.
123. Y. Chen, Y. Shen, X. Guo, C. Zhang, W. Yang, M. Ma, S. Liu, M. Zhang, L. Wen, Transdermal protein delivery by a coadministered peptide identified via phage display, *Nat Biotech* 24 (2006) 455–460.
124. Y. Kim, P.J. Ludovice, M.R. Prausnitz, Transdermal delivery enhanced by magainin pore-forming peptide, *J. Controlled Release* 122 (2007) 375–383.
125. H.W. Huang, F. Chen, M. Lee, Molecular mechanism of peptide-induced pores in membranes, *Phys. Rev. Lett.* 92 (2004) 198304.

126. V. Dubey, D. Mishra, M. Nahar, N.K. Jain, Elastic liposomes mediated transdermal delivery of an anti-jet lag agent: Preparation, characterization and in vitro human skin transport study, *Current Drug Delivery* 5 (July 2008) 199–206(8).

127. J. Fang, P. Liu, C. Huang, Decreasing systemic toxicity via transdermal delivery of anticancer drugs, *Curr. Drug Metab.* 9 (September 2008) 592–597(6).

128. V. Dubey, D. Mishra, M. Nahar, V. Jain, K.J. Narendra, Enhanced transdermal delivery of an anti-HIV agent via ethanolic liposomes, *Nanomedicine* 6 (2010) 590–596.

129. M. Whitefield, C.J.A. O'Kane, S. Anderson, Comparative efficacy of a proprietary topical ibuprofen gel and oral ibuprofen in acute soft tissue injuries: A randomized, double-blind study, *Journal of Clinical Pharmacy & Therapeutics* 27 (December 2002) 409–417.

130. E.B. Souto, R.H. Müller, Cosmetic features and applications of lipid nanoparticles (SLN®, NLC®), *International Journal of Cosmetic Science* 30 157–165.

131. J. Liu, W. Hu, H. Chen, Q. Ni, H. Xu, X. Yang, Isotretinoin-loaded solid lipid nanoparticles with skin targeting for topical delivery, *Int. J. Pharm.* 328 (2007) 191–195.

132. N.B. Vutla, G.V. Betageri, A.K. Banga, Transdermal iontophoretic delivery of enkephalin formulated in liposomes, *J. Pharm. Sci.* 85 (1996) 5–8.

133. O.V. Chumakova, A.V. Liopo, V.G. Andreev, I. Cicenaite, B.M. Evers, S. Chakrabarty, T.C. Pappas, R.O. Esenaliev, Composition of PLGA and PEI/DNA nanoparticles improves ultrasound-mediated gene delivery in solid tumors in vivo, *Cancer Lett.* 261 (2008) 215–225.

134. A.V. Badkar, G.V. Betageri, G.A. Hofmann, A.K. Banga, Enhancement of transdermal iontophoretic delivery of a liposomal formulation of colchicine by electroporation. *Drug Deliv.* 6 (1999) 111.

135. Y. Qiu, Y. Gao, K. Hu, F. Li, Enhancement of skin permeation of docetaxel: A novel approach combining microneedle and elastic liposomes, *J. Controlled Release* 129 (2008) 144–150.

136. J. Shim, H. Seok Kang, W. Park, S. Han, J. Kim, I. Chang, Transdermal delivery of minoxidil with block copolymer nanoparticles, *J. Controlled Release* 97 (2004) 477–484.

137. P. Lee, S. Hsu, J. Tsai, F. Chen, P. Huang, C. Ke, Z. Liao, C. Hsiao, H. Lin, H. Sung, Multifunctional core-shell polymeric nanoparticles for transdermal DNA delivery and epidermal Langerhans cells tracking, *Biomaterials* 31 (2010) 2425–2434.

138. S.A. Coulman, A. Anstey, C. Gateley, A. Morrissey, P. McLoughlin, C. Allender, J.C. Birchall, Microneedle mediated delivery of nanoparticles into human skin, *Int. J. Pharm.* 366 (2009) 190–200.

139. J.G. Rouse, J. Yang, J. Ryman-Rasmussen, A.R. Barron, N. Monteiro-Riviere, Effects of mechanical flexion on the penetration of fullerene amino acid-derivatized peptide nanoparticles through skin, *Nano Letters* 7 (2007) 155–160.

140. A.S. Zahr, C.A. Davis, M.V. Pishko, Macrophage uptake of core-shell nanoparticles surface modified with poly(ethylene glycol), *Langmuir* 22 (2006) 8178–8185.

141. B. Romberg, W. Hennink, G. Storm, Sheddable coatings for long-circulating nanoparticles, *Pharmaceutical Research* 25 (2008) 55–71.

142. S.M. Moghimi, A.C. Hunter, J.C. Murray, Long-circulating and target-specific nanoparticles: Theory to practice, *Pharmacological Reviews* 53 (2001) 283–318.

143. D. Bazile, C. Prud'homme, M. Bassoullet, M. Marlard, G. Spenlehauer, M. Veillard, Stealth Me.PEG-PLA nanoparticles avoid uptake by the mononuclear phagocytes system, *Journal of Pharmaceutical Sciences* 84 493–498.

144. R. Gref, M. Lück, P. Quellec, M. Marchand, E. Dellacherie, S. Harnisch, T. Blunk, R.H. Müller, 'Stealth' corona-core nanoparticles surface modified by polyethylene glycol (PEG): Influences of the corona (PEG chain length and surface density) and of the core composition on phagocytic uptake and plasma protein adsorption, *Colloids and Surfaces B: Biointerfaces* 18 (2000) 301–313.

145. K. Furumoto, K. Ogawara, S. Nagayama, Y. Takakura, M. Hashida, K. Higaki, T. Kimura, Important role of serum proteins associated on the surface of particles in their hepatic disposition, *J. Controlled Release* 83 (2002) 89–96.

146. W. Lu, X. Qi, Q. Zhang, R. Li, G. Wang, R. Zhang, S. Wei, A pegylated liposomal platform: Pharmacokinetics, pharmacodynamics, and toxicity in mice using doxorubicin as a model drug, *Journal of Pharmacological Sciences* 95 (2004) 381–389.

147. D. Liu, A. Mori, L. Huang, Role of liposome size and RES blockade in controlling biodistribution and tumor uptake of GM1-containing liposomes, *Biochimica et Biophysica Acta (BBA) - Biomembranes* 1104 (1992) 95–101.

148. T. Cedervall, I. Lynch, S. Lindman, T. Berggård, E. Thulin, H. Nilsson, K.A. Dawson, S. Linse, Understanding the nanoparticle–protein corona using methods to quantify exchange rates and affinities of proteins for nanoparticles, *Proceedings of the National Academy of Sciences* 104 (2007) 2050–2055.

149. M. Lundqvist, J. Stigler, G. Elia, I. Lynch, T. Cedervall, K.A. Dawson, Nanoparticle size and surface properties determine the protein corona with possible implications for biological impacts, *Proceedings of the National Academy of Sciences* 105 (2008) 14265–14270.

150. S. Li, L. Huang, Pharmacokinetics and biodistribution of nanoparticles, *Molecular Pharmaceutics* 5 (2008) 496–504.

151. F.M. Muggia, Doxorubicin-polymer conjugates: Further demonstration of the concept of enhanced permeability and retention, *Clinical Cancer Research* 5 (1999) 7–8.

152. Y.J. Jun, J.I. Kim, M.J. Jun, Y.S. Sohn, Selective tumor targeting by enhanced permeability and retention effect. Synthesis and antitumor activity of polyphosphazene–platinum (II) conjugates, *J. Inorg. Biochem.* 99 (2005) 1593–1601.

153. K. Greish, Enhanced permeability and retention of macromolecular drugs in solid tumors: A royal gate for targeted anticancer nanomedicines, *J. Drug Target.* 15 (2007) 457–464.

154. P. Carmeliet, R.K. Jain, Angiogenesis in cancer and other diseases, *Nature* 407 (2000) 249–257.

155. W. He, M.S. Ladinsky, K. Huey-Tubman, G.J. Jensen, J.R. McIntosh, P.J. Bjorkman, FcRn-mediated antibody transport across epithelial cells revealed by electron tomography, *Nature* 455 (2008) 542–546.

156. A.T. Florence, N. Hussain, Transcytosis of nanoparticle and dendrimer delivery systems: Evolving vistas, *Adv. Drug Deliv. Rev.* 50 (2001) S69–S89.

157. P. Oh, P. Borgstrom, H. Witkiewicz, Y. Li, B.J. Borgstrom, A. Chrastina, K. Iwata, K.R. Zinn, R. Baldwin, J.E. Testa, J.E. Schnitzer, Live dynamic imaging of caveolae pumping targeted antibody rapidly and specifically across endothelium in the lung, *Nat Biotech* 25 (2007) 327–337.

158. A.I. Minchinton, I.F. Tannock, Drug penetration in solid tumours, *Nat Rev Cancer* 6 (2006) 583–592.

159. D.P. McIntosh, X. Tan, P. Oh, J.E. Schnitzer, Targeting endothelium and its dynamic caveolae for tissue-specific transcytosis in vivo: A pathway to overcome cell barriers to drug and gene delivery, *Proceedings of the National Academy of Sciences of the United States of America* 99 (2002) 1996–2001.

160. J.E. Schnitzer, Caveolae: From basic trafficking mechanisms to targeting transcytosis for tissue-specific drug and gene delivery in vivo, *Adv. Drug Deliv. Rev.* 49 (2001) 265–280.

161. R.K. Jain, T. Stylianopoulos, Delivering nanomedicine to solid tumors, *Nat Rev Clin Oncol* 7 (2010) 653–664.

162. R. Cairns, I. Papandreou, N. Denko, Overcoming physiologic barriers to cancer treatment by molecularly targeting the tumor microenvironment, *Molecular Cancer Research* 4 (2006) 61–70.

163. R.K. Jain, Transport of molecules in the tumor interstitium: A review, *Cancer Research* 47 (1987) 3039–3051.

164. T.T. Goodman, C.P. Ng, S.H. Pun, 3-D tissue culture systems for the evaluation and optimization of nanoparticle-based drug carriers, *Bioconjug. Chem.* 19 (2008) 1951–1959.

165. S. Ramanujan, A. Pluen, T.D. McKee, E.B. Brown, Y. Boucher, R.K. Jain, Diffusion and convection in collagen gels: Implications for transport in the tumor interstitium, *Biophys. J.* 83 (2002) 1650–1660.

166. P. Calvo, B. Gouritin, H. Chacun, D. Desmaële, J. D'Angelo, J. Noel, D. Georgin, E. Fattal, J.P. Andreux, P. Couvreur, Long-circulating PEGylated polycyanoacrylate nanoparticles as new drug carrier for brain delivery, *Pharm. Res.* 18 (2001) 1157–1166.

167. V. Rousseau, B. Denizot, D. Pouliquen, P. Jallet, J. Le Jeune, Investigation of blood-brain barrier permeability to magnetite-dextran nanoparticles (MD3) after osmotic disruption in rats, *Magnetic Resonance Materials in Physics, Biology and Medicine* 5 (1997) 213–222.

168. R.K. Jain, Transport of molecules, particles, and cells in solid tumors, *Annu. Rev. Biomed. Eng.* 1 (1999) 241–263.

169. A. Pluen, Y. Boucher, S. Ramanujan, T.D. McKee, T. Gohongi, E. di Tomaso, E.B. Brown, Y. Izumi, R.B. Campbell, D.A. Berk, R.K. Jain, Role of tumor–host interactions in interstitial diffusion of macromolecules: Cranial vs. subcutaneous tumors, *Proceedings of the National Academy of Sciences of the United States of America* 98 (2001) 4628–4633.

170. S.K. Hobbs, W.L. Monsky, F. Yuan, W.G. Roberts, L. Griffith, V.P. Torchilin, R.K. Jain, Regulation of transport pathways in tumor vessels: Role of tumor type and microenvironment, *Proceedings of the National Academy of Sciences of the United States of America* 95 (1998) 4607–4612.

171. A. Krol, J. Maresca, M.W. Dewhirst, F. Yuan, Available volume fraction of macromolecules in the extravascular space of a fibrosarcoma: Implications for drug delivery, *Cancer Research* 59 (1999) 4136–4141.

172. G.M. Thurber, M.M. Schmidt, K.D. Wittrup, Antibody tumor penetration: Transport opposed by systemic and antigen-mediated clearance, *Adv. Drug Deliv. Rev.* 60 (2008) 1421–1434.

173. G.M. Thurber, K.D. Wittrup, Quantitative spatiotemporal analysis of antibody fragment diffusion and endocytic consumption in tumor spheroids, *Cancer Research* 68 (2008) 3334–3341.

174. A. Pluen, P.A. Netti, R.K. Jain, D.A. Berk, Diffusion of macromolecules in aga-rose gels: Comparison of linear and globular configurations, *Biophys. J.* 77 (1999) 542–552.
175. J.L. Horning, S.K. Sahoo, S. Vijayaraghavalu, S. Dimitrijevic, J.K. Vasir, T.K. Jain, A.K. Panda, V. Labhasetwar, 3-D tumor model for in vitro evaluation of antican-cer drugs, *Molecular Pharmaceutics* 5 (2008) 849–862.
176. C.P. Graff, K.D. Wittrup, Theoretical analysis of antibody targeting of tumor spheroids, *Cancer Research* 63 (2003) 1288–1296.
177. M.E. Ackerman, D. Pawlowski, K.D. Wittrup, Effect of antigen turnover rate and expression level on antibody penetration into tumor spheroids, *Molecular Cancer Therapeutics* 7 (2008) 2233–2240.
178. T.T. Goodman, P.L. Olive, S.H. Pun, Increased nanoparticle penetration in collagenase-treated multicellular spheroids, *Int. J. Nanomedicine* 2 (2007) 265–274.
179. L.S. Zuckier, E.Z. Berkowitz, R.J. Sattenberg, Q.H. Zhao, H.F. Deng, M.D. Scharff, Influence of affinity and antigen density on antibody localization in a modifiable tumor targeting model, *Cancer Research* 60 (2000) 7008–7013.
180. G.P. Adams, R. Schier, K. Marshall, E.J. Wolf, A.M. McCall, J.D. Marks, L.M. Weiner, Increased affinity leads to improved selective tumor delivery of single-chain Fv antibodies, *Cancer Research* 58 (1998) 485–490.
181. Y.S. Chang, E. di Tomaso, D.M. McDonald, R. Jones, R.K. Jain, L.L. Munn, Mosaic blood vessels in tumors: Frequency of cancer cells in contact with flow-ing blood, *Proceedings of the National Academy of Sciences of the United States of America* 97 (2000) 14608–14613.
182. S.D. Perrault, C. Walkey, T. Jennings, H.C. Fischer, W.C.W. Chan, Mediating tumor targeting efficiency of nanoparticles through design, *Nano Letters* 9 (2009) 1909–1915.
183. I. Lynch, K.A. Dawson, Protein-nanoparticle interactions, *Nano Today* 3 (2008) 40–47.
184. A.E. Nel, L. Madler, D. Velegol, T. Xia, E.M.V. Hoek, P. Somasundaran, F. Klaessig, V. Castranova, M. Thompson, Understanding biophysicochemical interactions at the nano-bio interface, *Nat Mater* 8 (2009) 543–557.
185. R.A. Petros, J.M. DeSimone, Strategies in the design of nanoparticles for thera-peutic applications, *Nat Rev Drug Discov* 9 (2010) 615–627.
186. G. Sahay, D.Y. Alakhova, A.V. Kabanov, Endocytosis of nanomedicines, *J. Controlled Release* 145 (2010) 182–195.
187. L.M. Bareford, P.W. Swaan, Endocytic mechanisms for targeted drug delivery, *Adv. Drug Deliv. Rev.* 59 (2007) 748–758.
188. R.G. Parton, K. Simons, The multiple faces of caveolae, *Nat. Rev. Mol. Cell Biol.* 8 (2007) 185–194.
189. J.A. Swanson, C. Watts, Macropinocytosis, *Trends Cell Biol.* 5 (1995) 424–428.
190. H. Soo Choi, W. Liu, P. Misra, E. Tanaka, J.P. Zimmer, B. Itty Ipe, M.G. Bawendi, J.V. Frangioni, Renal clearance of quantum dots, *Nat Biotech* 25 (2007) 1165–1170.
191. S. Mitragotri, J. Lahann, Physical approaches to biomaterial design, *Nat Mater* 8 (2009) 15–23.
192. S.E.A. Gratton, P.D. Pohlhaus, J. Lee, J. Guo, M.J. Cho, J.M. DeSimone, Nanofabricated particles for engineered drug therapies: A preliminary biodis-tribution study of PRINT™ nanoparticles, *J. Controlled Release* 121 (2007) 10–18.

193. J.A. Champion, S. Mitragotri, Role of target geometry in phagocytosis, *Proceedings of the National Academy of Sciences of the United States of America* 103 (2006) 4930–4934.

194. Y. Geng, P. Dalhaimer, S. Cai, R. Tsai, M. Tewari, T. Minko, D.E. Discher, Shape effects of filaments versus spherical particles in flow and drug delivery, *Nat Nano* 2 (2007) 249–255.

195. C. Peetla, V. Labhasetwar, Biophysical Characterization of Nanoparticle–Endothelial Model Cell Membrane Interactions, *Molecular Pharmaceutics* 5 (2008) 418–429.

196. L.E. van Vlerken, Z. Duan, M.V. Seiden, M.M. Amiji, Modulation of intracellular ceramide using polymeric nanoparticles to overcome multidrug resistance in cancer, *Cancer Research* 67 (2007) 4843–4850.

197. H. Devalapally, Z. Duan, M.V. Seiden, M.M. Amiji, Modulation of drug resistance in ovarian adenocarcinoma by enhancing intracellular ceramide using tamoxifen-loaded biodegradable polymeric nanoparticles, *Clinical Cancer Research* 14 (2008) 3193–3203.

198. M.M. Gottesman, T. Fojo, S.E. Bates, Multidrug resistance in cancer: Role of ATP-dependent transporters, *Nat Rev Cancer* 2 (2002) 48–58.

199. G. Szakacs, J.K. Paterson, J.A. Ludwig, C. Booth-Genthe, M.M. Gottesman, Targeting multidrug resistance in cancer, *Nat Rev Drug Discov* 5 (2006) 219–234.

200. L.S. Jabr-Milane, L.E. van Vlerken, S. Yadav, M.M. Amiji, Multi-functional nanocarriers to overcome tumor drug resistance, *Cancer Treat. Rev.* 34 (2008) 592–602.

201. J. Gillet, T. Efferth, J. Remacle, Chemotherapy-induced resistance by ATP-binding cassette transporter genes, *Biochimica et Biophysica Acta (BBA) – Reviews on Cancer* 1775 (2007) 237–262.

202. B.P. Sorrentino, Gene therapy to protect haematopoietic cells from cytotoxic cancer drugs, *Nat Rev Cancer* 2 (2002) 431–441.

203. E.M. Leslie, R.G. Deeley, S.P.C. Cole, Multidrug resistance proteins: Role of P-glycoprotein, MRP1, MRP2, and BCRP (ABCG2) in tissue defense, *Toxicol. Appl. Pharmacol.* 204 (2005) 216–237.

204. S. Nobili, I. Landini, B. Giglioni, E. Mini, Pharmacological strategies for overcoming multidrug resistance, *Curr. Drug Targets* 7 (July 2006) 861–879(19).

205. X. Dong, C.A. Mattingly, M.T. Tseng, M.J. Cho, Y. Liu, V.R. Adams, R.J. Mumper, Doxorubicin and paclitaxel-loaded lipid-based nanoparticles overcome multidrug resistance by inhibiting p-glycoprotein and depleting ATP, *Cancer Research* 69 (2009) 3918–3926.

206. M.D. Chavanpatil, Y. Patil, J. Panyam, Susceptibility of nanoparticle-encapsulated paclitaxel to P-glycoprotein-mediated drug efflux, *Int. J. Pharm.* 320 (2006) 150–156.

207. Y. Patil, T. Sadhukha, L. Ma, J. Panyam, Nanoparticle-mediated simultaneous and targeted delivery of paclitaxel and tariquidar overcomes tumor drug resistance, *J. Controlled Release* 136 (2009) 21–29.

208. S. Ganta, M. Amiji, Coadministration of paclitaxel and curcumin in nanoemulsion formulations to overcome multidrug resistance in tumor cells, *Molecular Pharmaceutics* 6 (2009) 928–939.

209. M. Breunig, S. Bauer, A. Goepferich, Polymers and nanoparticles: Intelligent tools for intracellular targeting? *European Journal of Pharmaceutics and Biopharmaceutics* 68 (2008) 112–128.

210. T. Tanaka, S. Shiramoto, M. Miyashita, Y. Fujishima, Y. Kaneo, Tumor targeting based on the effect of enhanced permeability and retention (EPR) and the mechanism of receptor-mediated endocytosis (RME), *Int. J. Pharm.* 277 (2004) 39–61.

211. D. Schaffert, E. Wagner, Gene therapy progress and prospects: Synthetic polymer-based systems, *Gene Ther.* 15 (2008) 1131–1138.

212. D. Shenoy, S. Little, R. Langer, M. Amiji, Poly(ethylene oxide)-modified poly(β-amino ester) nanoparticles as a pH-sensitive system for tumor-targeted delivery of hydrophobic drugs. 1. In Vitro Evaluations, *Molecular Pharmaceutics* 2 (2005) 357–366.

213. J.E. Fuller, G.T. Zugates, L.S. Ferreira, H.S. Ow, N.N. Nguyen, U.B. Wiesner, R.S. Langer, Intracellular delivery of core–shell fluorescent silica nanoparticles, *Biomaterials* 29 (2008) 1526–1532.

214. H. Meng, M. Xue, T. Xia, Y. Zhao, F. Tamanoi, J.F. Stoddart, J.I. Zink, A.E. Nel, Autonomous in vitro anticancer drug release from mesoporous silica nanoparticles by pH-sensitive nanovalves, *J. Am. Chem. Soc.* 132 (2010) 12690–12697.

215. L.M. de la.Fuente, C.C. Berry, Tat peptide as an efficient molecule to translocate gold nanoparticles into the cell nucleus, *Bioconjug. Chem.* 16 (2005) 1176–1180.

216. V. Weissig, G.G.M. D'Souza, S. Cheng, S. Boddapati, *Mitochondrial Nanotechnology for Cancer Therapy*, (2009) 265–279.

217. S.V. Boddapati, G.G.M. D'Souza, S. Erdogan, V.P. Torchilin, V. Weissig, Organelle-targeted nanocarriers: Specific delivery of liposomal ceramide to mitochondria enhances its cytotoxicity in vitro and in vivo, *Nano Letters* 8 (2008) 2559–2563.

218. Y. Yamada, H. Akita, K. Kogure, H. Kamiya, H. Harashima, Mitochondrial drug delivery and mitochondrial disease therapy: An approach to liposome-based delivery targeted to mitochondria, *Mitochondrion* 7 (2007) 63–71.

219. C. Medina, M.J. Santos-Martinez, A. Radomski, O.I. Corrigan, M.W. Radomski, Nanoparticles: Pharmacological and toxicological significance, *British Journal of Pharmacology* 150 (2007) 552–558.

220. A.M. Smith, H. Duan, A.M. Mohs, S. Nie, Bioconjugated quantum dots for in vivo molecular and cellular imaging, *Adv. Drug Deliv. Rev.* 60 (2008) 1226–1240.

221. H. Yamawaki, N. Iwai, Mechanisms underlying nano-sized air-pollution-mediated progression of atherosclerosis, *Circulation Journal* 70 (2006) 129–140.

222. S.H. Yalkowsky, J.F. Krzyzaniak, G.H. Ward, Formulation-related problems associated with intravenous drug delivery, *Journal of Pharmaceutical Sciences* 87 (1998) 787–796.

5

Biomaterials, Dental Materials, and Device Retrieval and Analysis

Jack E. Lemons

CONTENTS

Introduction

The disciplines associated with surgical implants in dentistry have evolved significantly during the period of the 1960s through the 2000s. In the 1960s, many different treatments were provided to restore intraoral function based on various implant designs (shape and size), biomaterials (implant), and dental (intraoral) materials. Within the profession, this situation was reviewed at many levels, and many opinions were provided about the factors influencing the clinical survival of implant versus other treatment modalities. Within each decade, consensus conferences were held where professionals provided analyses from the basic, applied, and clinical sciences [1–6]. From the various analyses, the device designs, the bio- and dental materials, and associated treatment options evolved from many to very few choices. This outcome was based on collaborations and a willingness to change the details of device selection and patient care. Importantly, clinical survival outcomes over the longer term (>5 years), when judged by objective criteria, evolved from about 50% to greater than 90%. Data from combinations of in vitro and laboratory in vivo investigations were correlated with information from clinical in vivo studies, and were then utilized to alter the discipline. One aspect of these studies was based on device retrieval and analysis (DRA), which will be the central theme of this chapter.

Methods and Procedures

Experience gained over the period since the late 1960s within the authors' program has focused on studies of surgical implant biomaterials and intra-oral dental materials, and were based on observations from retrieved and analyzed devices and prostheses from human clinical applications. This approach focused on functional conditions, primarily on two central themes: (1) interfacial transfers of elements, and (2) interfacial transfers of forces. The ever-evolving worldwide science from basic and applied viewpoints has strongly supported analyses utilizing these themes. The inclusion of individuals with materials, mechanical, and biomedical engineering backgrounds has permitted interactions within a multidisciplinary university program based in the school of dentistry departments and sections of biomaterials, biomechanics, and prosthodontics.

The key components needed for conducting this type of program include specimens and records; facilities for studies; expertise that evolves with experience; faculty, staff, and students; and funding. Specimens have been provided by dental professionals who were willing to share specimens and the details of treatments. During removal, and in strict compliance with the Institution Review Board (IRB) and Human Information Protection and Portability Act (HIPPA) programs, devices, tissues, and intraoral prostheses are treated the same as specimens for pathology. Most are placed in neutral buffered formalin and transferred using approved packaging and labeling. Upon receipt, a code is assigned, and all confidential information is maintained within limited access areas and information storage programs. All specimens are treated as infected, and all involved individuals must comply with barrier protection and confidentiality protocols.

After initial observation and photography for records, specimens are reviewed during a DRA meeting that includes individuals from physical, biological and clinical disciplines. As specimens are imaged for viewing by all participants, the clinical investigators review and present the history and clinical outcome. This is followed by other clinical discipline interactions, including oral surgery, restorative, periodontology, and pathology, along with engineering and other physical science–based investigators. Students participate, and the discussion addresses a central question: "What factors have contributed to this result and what might have been done to improve the outcome?" This evolves into a separation of key factors concerning (1) the patient, (2) the technology of treatment, and (3) the device/prosthesis. All follow-ups are then based on hypotheses and key factors, which then evolves into protocol-driven in vitro or in vivo–oriented research. Students become involved in reports and master of science (MS) or doctor of philosophy (PhD) project and degree activities. In most situations, abstracts, presentations, and publications are developed for peer review by all stakeholders. From this perspective, peer-review exchanges have provided the insights for progression within the various areas, contributing to longer-term stabilities and functionalities of dental implant–based restorative treatments.

Another aspect of peer review that involves most stakeholders is based within professional organizations that develop consensus standards, such as the American Dental Association (ADA) Standard Committee on Dental Products (SCDP) [7]; the American Society for Testing and Materials (ASTM) Committee F04 on Medical Devices [8]; and the International Standards Organization (ISO) Technical Committee (TC) 106 [9] on dental products, which includes standard subcommittees on restorative materials (SCIs) and the SC8 on dental implants. Examples of specialized workshops and symposia at ASTM that includes biomaterial and biomechanical information are listed in Table 5.1. In all situations, ASTM and ADA standards have provided rapidly evolving documents since the 1960s that permit consistent use of biomaterial and dental material products. Within the author's university program this integrates dentistry, medicine (orthopedic surgery), and engineering (materials and bioengineering). The program has received and analyzed more than 8,000 devices obtained from revision surgeries. Since 2005 [10] the program has evolved to now include postmortem in-situ specimens, including the implant and the anatomical region supporting the implant. Therefore, in most situations, this type of specimen—a device and prosthesis in place and functioning up to the time of death, followed by enbloc retrieval—and its clinical history provide opportunities to evaluate clinical success.

Results and Discussion

The bio- and dental materials utilized over the early decades (1950s–1980s) of dental implant treatments included metallics, ceramics, polymerics, and combinations/composites of these substances [11–17]. Examples of devices received for examinations are shown in Figures 5.1–5.4. Central within the metallics were the iron, cobalt, and titanium alloys. The ceramics were constituted primarily from aluminum and zirconium oxides plus calcium phosphate–based compounds. Ceramic biomaterials also included polycrystalline (vitreous) carbon and carbon-silicon. The polymerics primarily used for dental implant body components were polyethylene, polymethyl methacrylate, and polysulfone, with some use of porous combinations, including modified polytetrafluoroethylene (Proplast®).

Dental materials included everything that was available for crowns, bridges, bars, dentures, removable dentures, etc. This list is extensive, and readers are referred to several dental materials books [18–20] for the technical details. The metallics for crowns and bridges were often precious-grade (Au, Pd, Pt) alloys plus base-metal (Ni, Co, Cr, Mo, Be) alloys with

TABLE 5.1

List of Workshops and Symposia Held in Student Organizations that Included Information on Biomaterials and Biomechanics Based on Explant Device Retrieval and Analysis

Symposium on Bone Graft Substitutes, ASTM Book, 2003/2004
Symposium on Cross Linked and Thermally Treated Ultra-High Molecular Weight Polyethylene for Total Joint Replacements, ASTM STP 1445, 2004
Symposium on Spinal Implants: Are We Evaluating Them Appropriately? November 2004, ASTM STP 2004
Symposium on Titanium, Niobium, Zirconium, and Tantalum for Medical and Surgical Applications, November, 2004, ASTM STP, 2006
Symposium on Wear of Articulating Surfaces: Simulation and Clinical Measurements, November 2005
Symposium on Fatigue and Fracture of Medical Metallic Materials and Devices, Joint E-8 and SMST, November 2005
Workshop on Nanotechnology and Medical and Surgical Devices at Materials, November 2006
Workshop on Medical Devices Metrology and Standards Needs, November 2006
Workshop on Regenerative Medicine and Growing Cartilage, November 2007
Second Symposium on Fatigue and Fracture of Medical Metallic Materials and Devices, May 2008
Workshop on Future of Arthroplasty Standards: Planning for the Next 5 Years, November 2008
Workshop on Explant Shipping: The Black Hole between Explantation and Analysis, November 2008
Workshop on What's on What's Available in Fatigue Lifetime Prediction Software, E08-F04, November, 2008
Workshop on Developing a Proposed Regulatory Strategy for Neurotoxicity Testing Battery, May 2009
Workshop on Fretting Fatigue of Metallic Medical Devices and Materials, November 2009
Workshop on Biological and Synthetic Bone Grafts: Current Status and Future Directions, May 2010
Symposium on Mobile Bearing Total Knee (MBK) Replacement Devices, May 2010
Symposium on Static and Dynamic Spinal Implants: Are We Evaluating Them Appropriately? November 2010
Workshop on Toxicological Assessment of Residues on Implants: Effects – Toxicology – Limit Values, November 2010

some extensions to copper and other types of alloys. Alloys were utilized with porcelain and/or polymers such as acrylic and BisGMA composites for occlusal surfaces. Several issues evolved about biodegradation phenomena, especially when some base-metal alloys were combined with biomaterial-grade alloys and similar material abutments and bars.

During this period, the designs were as varied as the bio- and dental materials themselves, which in broad categories included subperiosteal, transosteal, endosteal (plate and root-form) and ramus frames [21, 22]. Each design utilized a different combination of dental materials and intraoral prosthesis constructions that were supported by clinical and company advocates. This multiplicity of applications, when considering DRA, provided opportunities

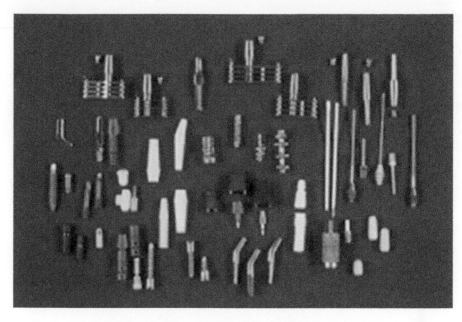

FIGURE 5.1
Examples of root-form dental implants received for investigation. (From Lemons, in Anusavice et al., eds, *Phillips' Science of Dental Materials*, 12 ed., Elsevier, Philadelphia, PA)

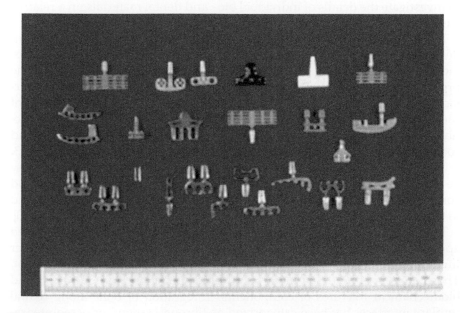

FIGURE 5.2
Examples of plate-form dental implants received for investigation. (From Lemons, in Anusavice et al. (eds), *Phillips' Science of Dental Materials*, 12 ed., Elsevier, Philadelphia, PA)

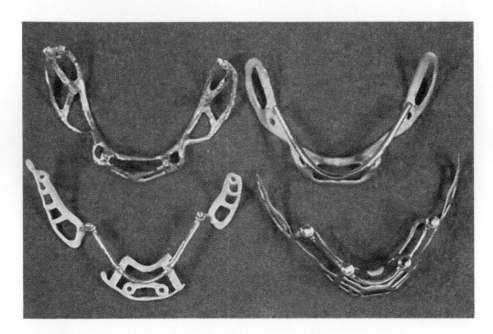

FIGURE 5.3
Examples of subperiosteal dental implants received for investigation. (From Lemons, in Anusavice et al. (eds), *Phillips' Science of Dental Materials*, 12 ed., Elsevier, Philadelphia, PA)

to investigate the details of individual bio- and dental materials on a relative basis specific to biocompatibility and clinical outcome.

The biomaterials have now evolved to a situation where most root-form designs are constructed from titanium and alloys with or without surface modifications, including calcium phosphate ($CaPO_4$) coatings [23]. Some properties of these biomaterials are summarized in Table 5.2. Multiple surface modifications have been introduced in recent years for dental implant root form designs that can be categorized broadly as as-processed; subtraction (e.g., acid etching and abrasive blasting); and addition (compounds and coatings). Some surface treatment descriptions are summarized in Table 5.3.

The dental implant types, in terms of design, have also evolved to now be primarily root-form systems. These shapes include mostly cylinders and cones, with and without threads and plateaus. Once again, readers are referred to the various dental implant textbooks for more detailed information [18–20].

Three areas will be used as examples of how the engineering-based approaches to device-interface conditions found during DRA and further investigated have contributed to available information for the profession. These areas will be tissue integration, biodegradation, and corrosion and biomechanical fractures of components. Integration of bone along dental implant interfaces has been categorized as fibrous, osseous, or osseointegration and mixed (called fibro-osseous integration) [22]. A question is, does the biomaterial surface chemistry and/or micro-topography influence the type

TABLE 5.2

Summary of Material–Biomaterial Properties Applicable for the Selection of Dental Implants

Properties	Unalloyed Titanium (I-IV) Wrought Condition	Titanium Alloy (Ti6A14V) Wrought Condition	Cobalt Alloy (Co-Cr-Mo Cast Condition)	Iron Alloy (316LSS Fe-Cr-Ni Wrought Condition)	
				Annealed	Cold Worked
Density (g/cc)	4.5	4.6	8.3	7.9	7.9
Yield strength (ys)					
MPa	170–485	795–827	490	240–300	700–800
(ksi)	(25–70)	(115–120)	(71)	(35–44)	(102–116)
Ultimate tensile strength (UTS)					
MPa	250–550	860–896	690	600–700	1000
(ksi)	(35–80)	(125–130)	(100)	(87–102)	(145)
Elastic modulus (e)					
GPa	96	105–117	200	200	200
(ksi x 10^3)	(14)	(15–17)	(29)	(29)	(29)
Endurance limit (fatigue)					
MPa		170–240	300	300	230–280
(ksi x 10^3)		(24.6–35)	(43)	(43)	(33.3–40.6)
Elongation (%)	15–24	10–15	8	8	7–22

TABLE 5.3

General Classification and Description of Surface Modifications for Surgical Implant Biomaterial

Classification	Condition	Comments
As Processed	Machined, cast, polished, molded, compacted, sintered, anodized, hot isostatically pressed (hipped) coined	Processed specific to metallic, ceramic and polymeric biomaterials (Al_2O_3, titanium, $CaPO_4$)
Addition	Sprayed (plasma or flame), deposit from solution, deposit from plasma vapor, irregular molding, sintering of particles, anodized and electron beam assisted deposition (ebad)	Very technique-dependent, standards about properties
Subtraction	Acid and dual acid etching, laser ablation, ion milling, blasting, sputtering	Variable micro- and nano-topographics and chemistries Al_2O_3, $CaPO_4$, SiO_2 and TiO_2

and amount of tissue integration along dental implant interfaces with bone? The answer of course can be yes or no, depending on conditions during healing and longer-term function. Studies have shown that all surfaces will show fibrous connective tissue under conditions of micromotions (>100 micrometers) during initial healing. Comparisons of surface oxides for alloys have shown that bone integration does not exist for chromium oxide on stainless steel and is limited for chromium oxides on cobalt alloys. Oxides on titanium and alloys with and without ion implantation as well as aluminum and zirconium oxide ceramics, carbons, and calcium phosphates have shown bone integration for a wide range of dental implant designs. Several of those biomaterials have also demonstrated bone integration along very smooth surfaces such as aluminum oxide (sapphire), although current systems now are finished with various shapes and sizes of nano-micro-topographies. Claims exist about enhanced rates of healing with various surface conditions. Our overall evaluations indicate most metallic and ceramic biomaterials exhibit bone integration after several weeks with maturity within a year. Overall, the magnitudes of bone-to-implant contact (BIC) range from about 20 to above 80% of the implant body section surface areas [24–26]. Our studies support high or magnitudes of BIC for calcium phosphate–coated (treated) surfaces; however, our results and those of other controlled studies are inadequate for correlation with clinical survival evaluations. One recent finding is that calcium phosphate coatings and post coating loss surfaces of cast cobalt alloy show bone integration after decades of in vivo function. This is called the custom osseous integrated implant (COII) system.

Biodegradation phenomena, primarily corrosion of metallics, have been an issue within some dental implant constructs, with several analyses initiating in the 1970s [27–37]. Subsequent studies provided guidelines on "acceptable combinations" from an electrochemical corrosion evaluation viewpoint. These data were expanded considerably, and recent studies now include alloy combinations tested under electrochemical (galvanic coupling) conditions. As a general guide (detailed information now available from product manufacturers), groupings of "acceptable and possible unacceptable" electrochemical combinations of metallics are summarized in Figures 5.4 and 5.5. Prior overviews have shown that high noble dental alloys (Au, Pt, Pd, Ir) when combined (coupled) with titanium alloys and cobalt alloys (Co-Cr-Mo) do not result in significant corrosion magnitudes of either part. However, some adverse conditions have been associated with combined titanium alloy and cobalt alloy compared with stainless steels, nonprecious (nickel-chromium) alloys (especially with Be), copper alloys, and amalgams.

A number of patient, technology, and biomaterial issues have been noted related to biomechanical fractures of intraoral prosthesis (primarily connectors and cantilevers), abutment components, and body sections of dental implants [38, 39]. Examples of such mechanisms usually present structural fatigue often caused by surface irregularities, inadequate structural dimensions, and/or fretting and corrosion processes. The extensive use of unalloyed titanium at grade I and II property magnitudes were shown to be at increased risk because of lower strength. The alloys and higher grades of titanium (III, IV) have shown limited numbers of biomechanical fractures. In some situations, such as the early root-form designs made from

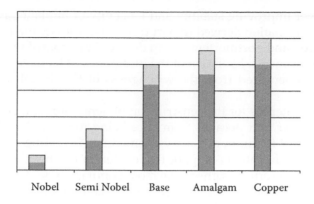

FIGURE 5.4
Corrosion characteristics of dental material alloys coupled electrically with surgical implant-grade titanium (T-IV), titanium alloys, or cobalt alloys (non-fretting conditions).

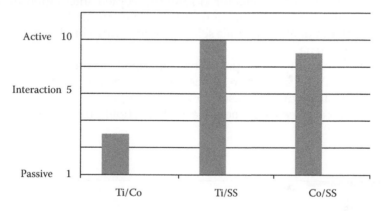

FIGURE 5.5
Corrosion of surgical-grade alloys versus one another when coupled alloys are electrically connected (non-fretting conditions).

polycrystalline aluminum oxide, biomechanical fractures at the transgingival region were identified as a significant issue.

The introduction of computer-assisted design and manufacturing (CAD/CAM) has reduced the incidence of misfit of components and the associated leakage fretting, corrosion, and fatigue fractures.

Summary and Conclusion

Advances over decades, based on multidisciplinary interactions between the disciplines associated with surgical implants in dentistry, have strongly influenced ever improving stability and longevity of implant-based clinical treatments. Information derived in part from observations made while analyzing retrieved and postmortem in-situ devices have contributed to aspects of these improvements. Evolution of the disciplines of biomaterials and biomechanics has benefited from the willingness of the dental profession to accept changes.

Overall, biomaterials for the construction of dental implant body sections have evolved over four decades from once being many and varied to now being metallics of titanium and alloys, sometimes including calcium phosphate surface modifications. In part, this evolution has been a result of the acceptance of root-form designs as the most popular implant design. Dental materials are now constituted and finished for combination with root-form constructs, with this area being enhanced by CAD/CAM technologies. Analyses of human device retrieval and analysis have often emphasized the biomaterial and biomechanical aspects of tissue-interface conditions, thereby supporting conditions for currently accepted integration by bone, where the construct and the force transfers are controlled by biomechanical principles and dental materials selected to minimize biodegradation and fracture phenomena.

Looking forward, it is anticipated that combined synthetic and biological biomaterials will be utilized to further enhance dental implant treatment opportunities. The affiliated research and development of imaging science and technology applied to regenerative medicine hold promise for the future.

References

1. Cranin N, ed. 1970. *Oral Implantology*. Springfield, IL: C. Thomas Publishing Co.
2. Natiella J, Armitage JJ, Meenaghan M, et al. 1972. Current evaluation of dental implants. *J Am Dent Assoc* 84: 1358.
3. Schnittman, P, and Shulman, L, eds. 1980. *Dental Implants: Benefit and Risk.* NIDCR/PHS Pub No. 81-531.
4. Lemons, J. 1988. Dental implant retrieval analyses. In Rizzo, AA, ed. 748–756, Proceedings of the 1988 Consensus Development Conference on Dental Implants. *J Dent Edu* 52: 678–827.
5. Weinstein, A, Gibbon, D, Brown, S, and Ruff, W, eds. 1981. *Implant Retrieval: Material and Biological Analysis.* NBS Spec. Pub 601, U.S. Government Printing Office, Washington, DC.
6. Cochran, D, and Fritz, M. 1996. Implant therapy I and II. *Annals of Periodontology* 1(1): 707–821.
7. American Dental Association, Standards Committee on Dental Products, Chicago, IL 2011.
8. American Society for Testing and Materials. 2011. Committee F04, Vol 13.01 *Medical Devices*, W. Conshohocken, PA.

9. International Standards Organization (ISO). 2011. Technical Committees TC150 and TC106 or Surgical Implants, Geneva, Switzerland.

10. Analyses of In Situ and Explanted Surgical Implant Devices; IRB X050823001. Laboratory research contract for student support and implant studies, UAB contract number 00320443, 2000-2011, NIBIB-BRP, ROIB001715, 2005-2010.

11. Roberts, H. 1971. *Surgical and Laboratory Procedures for Placement of Ramus, Single Tooth and Ramus Frame Implants*. College Place, WA: College Press.

12. McKinney, RV Jr (ed.). 1991. *Endosteal Dental Implants*. St. Louis: Mosby.

13. Fagan, M Jr., Ismail, J, Meffert, R, Fagan, M. 1990. *Implant Prosthodontics: Surgical and Prosthetic Techniques for Dental Implants*. Year Book Med. Pub, Chicago, IL.

14. Block, M, Kent, J, Guerra, L. 1997. *Implants in Dentistry*. Philadelphia: W. Saunders Co.

15. Branemark, P-I, Hansson, B, Ardell, R, Breine, U, Lindstrom, L, Hallen, O, Ohman, A. 1977. *Osseointegrated Implants in the Treatment of the Edentulous Jaw: Experience from a 10-Year Period*. Stockholm, Sweden: Almquist and Wilksel, Int.

16. Lee, J, Albrektsson, T, Branemark, P-I. 1982. Clinical applications of biomaterials. *Adv in Biomat*, Vol. 4. New York: J. Wiley.

17. Misch, CE, ed. 2008. *Contemporary Implant Dentistry*, Vol 3. St. Louis, MO: Mosby Inc.

18. Anusavice, K, ed., 2011. *Phillips' Science of Dental Materials*, 12th ed., Saunders Mosby and Churchill Livingston. Philadelphia: Elsevier Health.

19. Craig, R (ed). 1989. *Restorative Dental Materials*, Vol. 8, C.V. St. Louis: Mosby Co.

20. Leinfelder, K, and Lemons, J. 1988. *Clinical Restorative Materials and Techniques*. Washington, DC: Lee and Febiger.

21. Cranin, AN, Klein, M, and Simons, A. 1993. *Atlas of Oral Implantology*. New York: G. Thieme Verlag.

22. Weiss, C, and Weiss, A, eds. 2001. *Principles and Practice of Implant Dentistry*. St. Louis: C.V. Mosby.

23. Horowitz, E, and Parr, J. 1994. *Characterization and Performance of Calcium Phosphate Coatings for Implants*. ASTM STP 1196, ASTM, W. Conshohocken, PA.

24. Coelho, P, Granjeiro, J, Romanos, G, Suzuki, M, Silva, N, Cardaropoli, G, Thompson, V, and Lemons, J. 2008. Basic research methods and current trends in dental implant surfaces. *J Biomed Mater Res B*, 88B(2): 579–597.

25. Anabtawi, M, Breck, P, Bartolucci, A, Lemons, J. 2007. Histological analysis for forty-one retrieved dental implants, No. 1040 IADR, New Orleans, LA, March.

26. Morris, HF, Winkler, S, Ochi, S. 2001. A 48-month multicentric clinical investigation: Implant design and survival. *J of Oral Implantology* 27(4): 180–186.

27. Lemons, J, Sarver, D, Beck, P, Petersen, D and Eberhardt, A. 2011. Approach, rationale and examples of device retrieval and analysis as an educational program. *Clin Ortho and Rel Res* (in press).

28. Thompson, NG, Buchanan, RA, and Lemons, JE. 1979. In vitro corrosion of Ti-6Al-4V and Type 316L stainless steel when galvanically coupled with carbon. *Journal of Biomedical Materials Research* 13: 35–44.

29. Buchanan, RA, Lemons, JE, Griffin, CD, Thompson, NG, and Lucas, LC. 1981. Effects of carbon coupling on the in vivo corrosion of cast surgical cobalt-base alloy. *Journal of Biomedical Materials Research* 15: 611–614.

30. Lucas, LC, Buchanan, RA, and Lemons, JE. 1981. Investigations on the galvanic corrosion of multi-alloy total hip prostheses. *Journal of Biomedical Materials Research* 15: 731–747.

31. Lucas, LC, Buchanan, RA, Lemons, JE, and Griffin, CD. 1982. Susceptibility of surgical cobalt-base alloy to pitting corrosion. *Journal of Biomedical Materials Research* 16: 799–810.
32. Griffin, CD, Buchanan, RA, and Lemons, JE. 1983. In vivo electrochemical corrosion study of coupled surgical implants materials. *Journal of Biomedical Materials Research* 17: 489–500.
33. Johnson, BI, Lucas, LC, and Lemons, JE, 1989. Corrosion of copper and nickel dental casting alloys: An in vitro and in vivo study. *Journal of Biomedical Materials Research* 23: 349–362.
34. Johansson, BI, Lemons, JE, and Hao, SQ. 1989. Corrosion of dental copper, nickel and gold alloys in artificial saliva and saline solutions. *Journal of Dental Materials* 5: 324–328.
35. Lucas, LC, Dale, Buchanan, R, Gill, Y, Griddin, D, Lemons, JE. 1991. In vitro vs in vivo corrosion analysis of two alloys, *Journal of Investigative Surgery* 4: 13–21.
36. Lemons, JE, Lucas, LC, and Johansson, BI. 1992. Intraoral corrosion resulting from coupling dental implants and restorative metallic systems. *Implant Dentistry* 1: 107–112.
37. Venugopalan, R, Weimer JJ, George AM, and Lucas LC. 1999. Surface topography, microhardness and corrosion properties of nitrogen diffusion hardened Ti-6Al-4V alloy. *Biomaterials* 20(18): 1709–1716.
38. Goodacre, C, Bernam, G, Rungcharassaeng, K, et al. 2003. Complication with implants in implant prostheses. *J Prosthet Dent* 90: 121–132.
39. Lemons, JE. 1975. Biomaterial considerations for dental implants, Part I: Metals and alloys. Alabama Academy of General Dentistry Sponsored Symposium on Dental Implants, *Oral Implantology* 4, 503–515.

6

Biomaterials and the Central Nervous System: Neurosurgical Applications of Materials Science

Urvashi M. Upadhyay

CONTENTS

Introduction

A variety of biomaterials have been used in the treatment of various disorders of the central nervous system (CNS). Materials that have historically been used include both biodegradable and non-biodegradable materials, such as silicone, natural and synthetic polymers, and lipids. The desired biodegradability of these materials is in large part determined by their intended application. That is, those materials that should remain functional for long periods of time, such as deep brain stimulating electrodes, do not employ biodegradable materials, whereas devices implanted for the short-term purposes of tissue scaffolding or drug-delivery are made with degradable materials [1].

Regardless of a material's stability over time, it must be deemed biocompatible before it may be implanted intracranially. Biocompatibility with tissue outside the central nervous system does not always predict biocompatibility with brain and spinal cord tissue, nor does short-term biocompatibility predict long-term tolerance to implantation. For example, degradable poly(methylidene) malonate (PMM)–based microspheres have been implanted into rat striatum and acutely show no evidence of inflammation; however, several months post-implantation, inflammation, necrosis, and animal death occurred, presumably when PMM began to degrade [2]. Interestingly, many other biodegradable polymers do demonstrate tolerable long-term intracranial biocompatibility, highlighting that each new material's stability must be assessed individually before considering CNS implantation.

Understanding a material's biocompatibility requires an intimate understanding of the brain's immune system and inflammatory cascade. Within 24 hours of the tissue damage that is expected with the implantation of any device, microglia enter the damaged tissue site and begin the process of phagocytosis of dead cells [2–4]. Several groups have demonstrated that these microglia then secrete neurotrophic factors, such as brain-derived neurotrophic factor (BDNF) or glial cell line–derived neurotrophic factor (GDNF) [3, 4] that may exert some neuroprotective role. In the days following, astrocytes arrive at the injured site; they have also been shown to release neurotrophic factors [5, 6]. With activated macrophages/microglia, a release of certain chemical factors such as the chemokine monocyte chemotactic protein (MCP-1) and the pro-inflammatory cytokine tumor necrosis factor-alpha (TNFα) is seen [7]. TNFα, in turn, has demonstrable neurotoxic activity, suggesting that released inflammatory factors may reduce neuronal viability surrounding the site of an implanted device [7–11]. Certainly the limitation of many implanted intracranial devices may be the failure of the device to evade the host's immune system over the long term. Some have even stated that the major clinical limitation of brain-machine-interface (BMI) technology is the inability to consistently record from a single neuron over time as a result of the brain's robust foreign body response to implanted electrodes [11, 12].

Though identifying materials suitable for intracranial implantation may appear daunting, many materials have been used historically with great success for varied clinical CNS applications. Applications of biomaterials currently in use include shunt systems used to treat hydrocephalus, intracranial drug-delivery vehicles, hydrogel scaffolds for CNS repair, microelectrodes for deep brain stimulation, and vehicles for delivery of neural stem cells [1]. In the following text, we will explore the various biomaterials in use in the CNS at the experimental and clinical levels. We will explore their relative safety, limitations, and strengths and, in turn, will identify the ideal characteristics of materials that may be used safely in the neurosurgical theater.

Cerebrospinal Fluid Shunt Systems

One of the most common clinical entities encountered in neurosurgery is that of hydrocephalus, a condition characterized by excessive accumulation of cerebrospinal fluid (CSF) due to either excessive production or inadequate clearance. The most common treatment for hydrocephalus is placement of a shunt system that diverts excess CSF from the cerebral ventricles or subarachnoid space to another potential space in the body where the fluid will be readily absorbed. This method of diverting CSF has been used for over fifty years [1, 13, 14]. The typical shunt system comprises a proximal catheter (which sits in the ventricular or subarachnoid space) that connects to a valve that controls drainage of CSF and a distal catheter that carries the CSF to a distal location (Figure 6.1). The most common targets for CSF reabsorption are the peritoneal cavity, the right atrium of the heart, the pleural space, and more rarely the gall bladder [15, 16]. CSF shunt systems are made from medical-grade silicone, which would appear to be an ideal material given its low toxicity, stability, and minimal biological reactivity [1]. While shunt systems are highly effective in treating hydrocephalus, the hardware failure rates from shunt malfunctions or infections are extremely high. Eighty percent of newly placed shunts will succumb to obstruction within

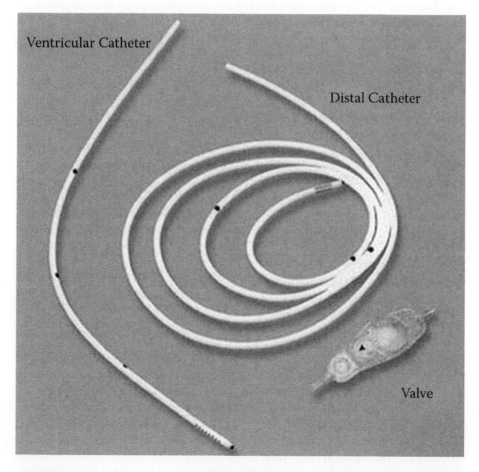

FIGURE 6.1
Ventriculoperitoneal shunt system. (Photograph courtesy of Codman © 2007 Codman & Shurtleff, Inc.)

ten years of implantation; half of these will fail from obstruction within one year of placement [1, 17]. Inflammation appears to play a key role in causing many shunt obstructions, be it at the ventricular catheter, valve, or distal catheter. Certainly, even routine placement of a shunt system requires passing a rigid ventricular catheter through brain tissue, resulting in cell death, tissue injury, and disruption of the blood–brain barrier (BBB) [1]. The resultant tissue damage then begets a stereotyped inflammatory cascade around the implanted shunt. Microscopic analysis of the outer and inner surfaces of retrieved shunts does reveal biofilm formation. Further, gross examination of many shunt obstructions reveals choroid plexus or glial tissue ingrowth into the ventricular catheter [18, 19]. Other groups have reported collections of clotted blood, lymphocytes, macrophages, necrotic brain tissue, and ependymal cells among the material clogging the proximal ventricular catheter [18]. Examination of obstructed valves reveals particulate matter made of fibroblasts, blood cells, and pathogenic bacteria [1, 18]. The other major cause of shunt malfunction is infection of the hardware, which occurs in as many as 10% of shunts within four months [15, 18]. A majority of infections are caused by normal skin flora, such as *Staphylococcus* species, which are introduced into the shunt system at the time of original insertion and often result in biofilm formation [18]. The inner lumen of catheters represents a slightly sequestered environment where skin flora bacteria cannot be readily cleared by the brain immune system, resulting in a potential nidus for infection [19–21]. Approaches to reduction of shunt failure take either of two tacks: change in operative insertional technique, or modification of shunt material surfaces.

While rates of infection may decrease through changes in operative technique, shunt manufacturers have made efforts to modify the materials used in these systems to reduce rates of infection. One strategy involves antibiotic coating of shunt materials [20, 22]. On follow-up, patients who have had placement of antibiotic-coated shunt systems have not shown significant reduction in infection rates, leading some to postulate that the antibiotic may be inactivated with time or simply diffuse away from the site [14, 23]. Further attempts at surface modification include coating the surface with a hydrophilic polymer or impregnating the material with antimicrobial agents [23]. The rationale for modifying the surface to be more hydrophilic is that it will reduce bacterial adhesion to the materials [22]. However, surface analysis of these materials revealed that they did have diminished bacterial adhesion but not bacterial colonization [22]. Efforts at manufacturing antimicrobial-impregnated silicone shunt systems have been promising and are thought to be more effective than mere coating as the antimicrobial agents may achieve a sustained release and will have access to both the inner and out luminal surfaces of the shunt system [23]. Data on the long-term efficacy of these antibiotic-impregnated shunts is forthcoming. What is clear is that there is a comfort level with implanting silicone-based devices intracranially, though there remains much room for improvement in biocompatibility.

Drug Delivery Systems

One of the significant challenges in the pharmacologic treatment of CNS conditions is the necessity of crossing the BBB. Because of the selective permeability of the endothelial cells in brain capillaries, many drugs do not cross the BBB with high efficiency [24, 25]. As a result, CNS drugs may require administration at high systemic doses to achieve appropriate intracranial concentrations. These high doses may result in a worsening of systemic side effects. Efforts in neuropharmacology have focused largely on improving a drug's ability to hone in on the brain and cross the BBB more efficiently, or on strategies to circumvent the BBB, such as through intracranial delivery.

Systemic administration of drugs intended for the CNS may be ameliorated by the use of certain drug carriers that target the CNS tissue and/or facilitate crossing of the BBB [1]. There are many examples of drug carriers, including nanoparticles, polymeric micelles, liposomes, and dendrimers [26]. Liposomes are small vesicles whose cores are enclosed by phospholipid bilayer membranes, which are fully biocompatible [1]. Liposomes have shown the most clinical promise, though traditional liposomes do little to improve crossing of the BBB without vector-mediated delivery [1, 25]. Without surface modification, these liposomes are rapidly cleared by the reticuloendothelial system (RES); coating their surface with hydrophilic polymers greatly reduces this problem and prolongs their circulation time [27]. Further surface modification with targeting vectors improves the tissue-specificity of liposomal drug formulations [27]. Doxil® is a pegylated (polyethylene glycol) liposomal formulation of doxorubicin, an anthracycline antibiotic cum chemotherapy used to treat cutaneous tumors. The pegylated surface modification allows better skin targeting and reduces the cardiotoxicity usually seen with its administration [28]. Intracranial drug targeting is largely in experimental stages but hinges on the understanding that intracranial tumors (both intrinsic and metastatic) demonstrate a breakdown in the BBB, thereby facilitating the crossing of liposomal drugs across the BBB [1, 29].

Other drug carriers that have shown promise include polymeric nanoparticles, which are particles less than 100 nm in size attached to natural or synthetic polymers that encapsulate or are attached to pharmacologic agents [30]. The polymers most commonly used are polyalkylcyanoacrylates (PACAs), polyacetates, polysaccharides, and copolymers [1, 31]. These polymeric NPs are thought to cross the BBB through receptor-mediated endocytosis or by the process of diffusion [25]. Release of drug from the polymer is carried out through a process of hydrolytic degradation of the polymer [32]; this method of degradation allows for sustained and controlled release of pharmacologic agents. Two opiod agonists, loperamide and dalargin, have profound CNS pharmacologic effects, but do not readily cross the BBB. Using polybutylcyanoacrylate (PBCA) nanoparticles overcoated with surfactant and loaded with each of these agents, investigators were able to demonstrate good pharmacologic effect in a rodent model [33, 34]. Animals had a more rapid onset of anesthesia and likely achieved less systemic exposure of these opiod agents [33, 34].

While strategies to improve the systemic administration of CNS drugs have been promising, patients are still exposed to potential side effects from non-specific drug delivery. To address this limitation, several groups have explored strategies for delivering pharmacologic agents locally, within the brain. Most efforts have focused on the treatment of brain tumors, since chemotherapeutic agents often cause significant side effects, such as bone marrow suppression, nausea, and alopecia. Polymers that have been used intracranially include polylactide (PL), polyglycolide (PG), copolymers of lactide/glycolide (PLGA), and polyanhydride poly-bis-propane-sebacic acid (PCPP:SA). All of these polymers have been safely implanted intracranially and have demonstrated good biocompatibility and long-term drug release [35]. Gliadel® is a polymer-chemotherapy composite wafer approved for intracranial implantation for recurrent glioma tumors and has been demonstrated to be safe and non-immunogenic when administered intracranially [35] (Figure 6.2). Unfortunately, increases in survival in patients who have been treated with Gliadel® have been modest at best, with survival improved on the order of months [36]. Some have postulated that these modest increases in survival are due to inferior tumoricidal activity of carmustine (BCNU), the chemotherapy in Gliadel®, or perhaps to the low drug payload achieved in these wafers. Others have suggested that the tumor target is not appropriate for intracranial implantation, as high-grade glioma tumors are diffuse and infiltrative by nature. However, in spite of these limitations, the addition of Gliadel® to the neurooncology armamentarium has been significant as it has demonstrated that high concentrations of chemotherapy may be delivered intracranially and that such concentrations may be well tolerated and efficacious.

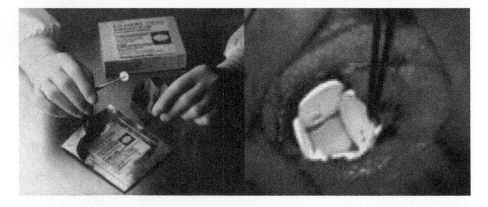

FIGURE 6.2
(a) GLIADEL® Wafer (polifeprosan 20 with carmustine implant) is a registered trademark of Eisai Inc. GL188R1 © 2010 Eisai Inc. All rights reserved. September 2010. (b) Gliadel Wafers being implanted in the brain. (Courtesy of NCI 1997)

Tissue Scaffolds

Site-specific therapeutic delivery has become more widely accepted as the means of delivering pharmacologic agents or placing extracellular matrices for neural growth, migration, and axonal regeneration in the central nervous system (CNS) [12]. Hydrogels are a network of cross-linked insoluble polymers that serve as excellent tissue scaffolds for new nerve growth or support matrices for grafted cells into the CNS [1, 12]. Hydrogels have the unique property of swelling with water, and it is this property that gives them the rheologic properties necessary for CNS implantation (Figure 6.3). That is, swollen hydrogels can readily incorporate into CNS tissue given their elasticity [12, 35, 37]. In vitro work has demonstrated the polyethylene glycol (PEG)–based hydrogels support the growth of human neural progenitor cells [37, 38]. Further work examining the biocompatibility of PEG-based hydrogels implanted into nonhuman primate striatum and cerebral cortex revealed that they generated no T-cell response, minimal gliotic scar at site of implantation, and low-density glial fibrillary acidic protein (GFAP) staining [35, 37]. Interestingly, investigators found that the magnitude of the inflammatory response was highly correlated with the degradation rate of the hydrogel. That is, the slower degrading and nondegrading hydrogels showed an attenuated inflammatory response, suggesting that it is the degradation

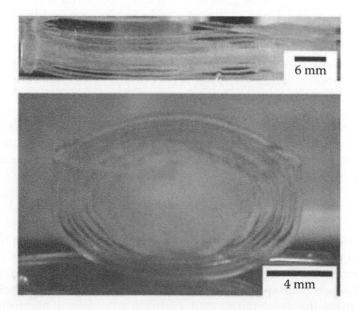

FIGURE 6.3
Swollen hydrogel following exposure to water may take several forms such as a tubular structure (above) or onion-ring structure (below). (From Elisseeff, J 2008. Hydrogels: Structure starts to gel. *Nature Materials* 7: 271–273)

of these materials, rather than intrinsic properties of the materials, that make them less biocompatible [37]. These results taken together suggest that implantation of slowly degrading or nondegradable PEG-based hydrogels may be well tolerated in humans.

The relative static nature of these types of hydrogels suggests that they may not be useful in all CNS clinical applications. There may be situations in which a specific hydrogel's mechanical properties change over time. For example, amphiphilic diblock copolypeptide hydrogels (DCHs) are synthetic materials whose mechanical properties may be altered by altering the polypeptide backbone to gain certain functionality [12]. Specifically, this polypeptide may be altered to impart enzymatic degradability, cell–cell adhesion, and molecular signaling. Furthermore, DCHs may be thinned and deformed to allow for injection through a small-bore cannula; following injection, these are rapidly reassembled into rigid gel networks [12] (Figure 6.4). This property allows for their minimally invasive deployment into the CNS. Investigators have also shown the DCHs induce little inflammatory response, glial scarring, or neurotoxicity in the CNS host tissue compared with saline injection controls, suggesting that this material may be both a safe and versatile tissue scaffold for nerve regeneration.

Recent work on determining the role of matrix stiffness in promoting neurite and axon outgrowth reveals that soft materials appear to be more successful in promoting tissue growth in the CNS compared with materials to higher stiffness [39]. Fibrin gels, composed of fibrinogen and thrombin (the final two factors in the coagulation cascade), have been shown to support the growth of cells *in vitro* and *in vivo*. Some have suggested that fibrin in particular is useful in supporting neuronal regrowth because it does not support glial proliferation [39]. The fibrin gel's preferential support of neuronal cell ingrowth is thought to result from the different ways neurons and glia respond to matrix stiffness and from the fibrin gel's low elastic modulus, which promotes neuronal differentiation [39, 40]. Several groups have shown that the neurons appear to prefer to grow on soft surfaces and that the *in vitro* work utilizing soft materials such as fibrin gels holds promise for future *in vivo* work [39–41].

Microelectrode Systems

Much of the current understanding of systems neuroscience, that is, the study of neural networks and circuits, has developed with the help of implantable microelectrode technology, which allows for recording and stimulation of a single neuron or a group of neurons. It is this understanding that has informed the development of deep brain stimulator (DBS) technology. DBSs are stimulating electrodes implanted traditionally into the subthalamic nucleus, where stimulating signals are thought to feed into basal ganglia circuitry and override some of the motor symptoms associated with conditions such as Parkinson's disease [42]. Other deep brain targets, such

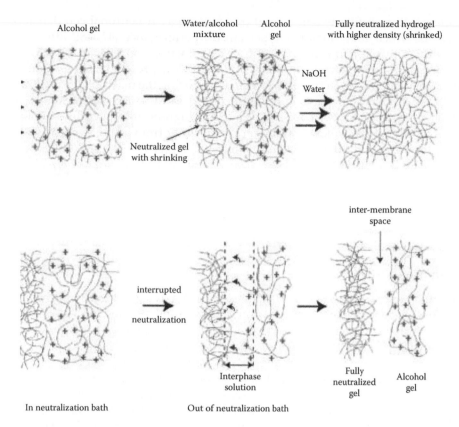

FIGURE 6.4
Neutralized hydrogel is formed from the repeated treatment of an alcohol gel with NaOH and water, which shrinks as a result of the physical crosslinks formed between the chains of the gel. With treatment, the hydrogel's physical properties may be controlled and changed. From Elisseeff, J 2008. Hydrogels: Structure starts to gel. *Nature Materials* 7: 271–273)

as the ventral intermediate thalamic nucleus, may be targeted to alleviate debilitating tremors [43]. Breakthrough technology in human motor cortex electrophysiology has also shown great promise. Researchers have been able to utilize neural recordings to control limb prostheses to perform meaningful movements in paralyzed patients; this brain–machine interface is the topic of much focus and study [44]. While BMI technology appears promising, one of the long-term challenges to its widespread clinical use is the inability to consistently record from a single neural unit over time [42]. Several groups have demonstrated a robust foreign body response that develops over time in the area around implanted electrodes [42, 45–47]. Macrophages have been observed at the electrode–brain interface; this inflammatory response is elaborated by macrophage-secreted factors, such as TNFα and MCP-1 [7, 8, 42]. Furthermore, some investigators have demonstrated a correlation between inflammatory infiltrate and astrogliosis (a decreased neuronal cell density), both of which may impair microelectrode performance [42, 48]. Efforts to minimize this inflammatory response have focused on modifying surface chemistry. Investigators have found that using low protein-binding coatings on implanted microelectrodes may help in reducing the macrophage infiltrate [42]. With further advances and modifications in the microelectrode insulation materials, one can envision designing high-performance implantable electrodes that may allow for long-term recording from a single neuron. The leap from this technologic advance to clinical application is a short one; with less biofouling of electrodes, the widespread use of BMI in paralyzed patients may be achieved.

Conclusions

Identifying properties of materials that are the least likely to incite the inflammatory cascade has been the goal of much work in biomaterials. The desired characteristics of an intracranial device require materials with rheologic properties different from those of other areas of the body. These include nontoxicity, facile integration with host tissue, and tunable mechanical properties (such as rigidity and porosity). Materials currently under investigation share many but not all of these properties and are therefore limited in application. Naturally occurring polymers tend to elicit less immune response and are often degradable, but lack optimal mechanical properties, such as rigidity. Other synthetic polymeric materials may have more tunable rigidity but may lack other properties such as causing minimal tissue inflammation or structural damage. What is clear is that recent advances in the field of materials science have informed work in neurosurgical material clinical applications. A better understanding of which materials are less likely to incite inflammatory reactions, or less likely to succumb to biofilm formation, will certainly improve varied areas of neurosurgery from hydrocephalus to drug delivery.

References

1. Zhong Y, Bellamkonda RV. 2008. Biomaterials for the central nervous system. *J. R. Soc. Interface* 5: 957–975, 13 May.
2. Fournier E, Passirani C, Colin N, Sagodira S, Menei P, Benoit JP, Montero-Menei CN. 2006. The brain tissue response to biodegradable poly(methlidene malonate 2.1.2)-based microspheres in the rat. *Biomaterials* 27: 4963–4974.

3. Batchelor PE, Liberatore GT, Wong JY, Porritt MJ, Fredrichs F, Droman GA, Howells DW. 1999. Activated macrophages and microglia induce dopaminergic sprouting in the injured striatum and express brain-derived neurotrophic factor and glial cell line-derived neurotrophic factor. *J. Neurosci.* 19: 1708–1716.

4. Imai F, Suzuki H, Ota J, Nimomia T, Ono K, Sano H, Sawada M. 2007. Neuroprotective effect of exogenous microglia in global brain ischemia. *J Cereb. Blood Flow Metab.* 27: 488–500.

5. Basu S, Yang ST. 2005. Astrocyte growth and glial cell line derived neurotrophic factor secretion in three-dimensional polyethylene terephthalate fibrous matrices. *Tissue Eng.* 11: 940–952.

6. Bresjanac M, Antauer G. 2000. Reactive astrocytes of the quinolinic acid-lesioned rat striatum express GFRalphaI as well as GDNF in vivo. *Exp Neurol* 164: 53–59.

7. Biran R, Martin DC, Tresco PA. 2005. Neuronal cell loss accompanies the brain tissue response to chronically implanted silicon microelectrode arrays. *Exp Neurol* 195: 115–126.

8. Biran R, Martin DC, Tresco PA. 2007. The brain tissue response to implanted silicon microelectrode arrays is increased when the devices is tethered to the skull. *J Biomed Mater Res A.* 82: 169–78.

9. Colias JC, Maneurlidis EE. 1957. Histopathological chances produced by implanted electrodes in cat brains; comparison with histopathologic changes in human and experimental puncture wounds. *J Neurosurg* 14: 302–28.

10. Dymond AM, Kaechele LE, Jurist JM, Crandall PH. 1970. Brain tissue reaction to some chronically implanted metals. *J Neurosurg* 33: 574–80.

11. Leung BK, Biran R, Underwood CJ, Tresco PA. 2008. Characterization of microglial attachment and cytokine release on biomaterials of differing surface chemistry. *J Biomaterials* 29: 3289–3297.

12. Yang CY, Song B, Ao Y, Nowak AP, Abelowitz RB, Korsak RA, Havton LA, Deming TJ, Sofroniew MV. 2009. Biocompatibility of amphiphilic diblock copolypeptide hydrogels in the central nervous system. *Biomaterials* 30: 2891–98.

13. Vandevord PJ, Gupta N, Wilson RB, Vinua RZ, Schaefer CJ, Canady AI, Wooley PH. 2003. Immune reactions associated with silicone-based ventriculo-peritoneal shunt malfunctions in children. *Biomaterials* 25: 3853–60.

14. Liang X, Wang A, Chao T, Tang H, McAllister II JP, Sulley SO, Ng KY. 2006. Effect of cast molded rimpicin/silicone on *Staphylococcus epidermidis* biofilm formation. *J Biomed Mater Res A* 76: 580–88.

15. Kalousdian S, Karlan MS, Williams MA. 1998. Silicone elastomer cerebrospinal fluid shunt systems. Council on Scientific Affairs, American Medical Association. *Neurosurgery* 42: 887–92.

16. Garton HJ, Piatt, JH Jr. Hydrocephalus. 2004. *Pediatri Clin North Am* 51: 305–25.

17. Sgouras S, Dipple SJ. An investigation of structural degradation of cerebrospinal fluid shunt valves performed using scanning electron microscopy and d energy-dispensive X-ray microanalysis. 2004. *J Neurosurg* 100: 534–540.

18. Del Bigio MR. Biological reactions to cerebrospinal fluid shunt devices: A review of the cellular pathology. 1998. *Neurosurgery* 42: 319–25.

19. Browd SR, Ragel RT, Gottfried ON, Kostle JR. Failure of cerebrospinal fluid shunts: Part I: Obstruction and mechanical failure. 2006. *Pediatr Neurol.* 34: 83–92.

20. Bayston R, Ashraf W, Bhundia C. 2004. Mode of action of an antimicrobial biomaterial for use in hydrocephalus shunts. *J Antimicrob Chemother* 52: 7780–82.

21. Bayston R, Bhundia C, Ashraf WW. 2005. Hydromer-coated catheters to prevent shunt infection? *J Neurosurg* 102 (suppl 2): 207–212.

22. Cagavi F, Akalan N, Celik H, Gur D, Guciz B. 2004. Effect of hydrophilic coating on microorganism colonization in silicone tubing. *Acta Neurochir* (Wien) 146: 603–10.

23. Furno F, et al. 2004. Silver nanoparticles and polymeric medical devices: A new approach to prevention of infection? *J Antimicrob. Chemother.* 54: 1019–24.

24. Wang PP, Frazier J, Brem H. 2002. Local delivery to the brain. *Adv Drug Deliv. Res.* 54: 987–1013.

25. Miisra A, Ganesh S, Shalhiwala A, Shah SP. 2003. Drug delivery to the central nervous system: A review. *J Pharm Pharmacol Sc* 6: 252–273.

26. Kaur LP, Bhander R, Bhander S, Kakkar V. 2008. Potential of solid lipid nanoparticles in brain targeting. *J Control Rel* 127: 97–109.

27. Siwak DR, Tari AM, Lopex-Brenestein G. 2002. The potential of drug-carrying immunoliposomes as anticancer agents. *Clin Cancer Res* 8: 955–56.

28. Lown JW. 1993. Anthracycline and anthraquinone anticancer agents: Current status and recent developments. *Pharmacol. Ther.* 60 (2): 185–214.

29. Tyler B, Grossman R, Patta Y, Upadhyay U, Huey L, Azadi J, Cima MJ, Brem H. Systemically delivered doxil increases survival in both an intracranial gliosarcoma model and an intracranial metastatic breast cancer model in rodents. Poster submission: American Association of Neurological Surgeons, Annual Meeting Spring 2011.

30. Lockman PR, Mumper RJ, Kgan MA, Allen DD. 2002. Nanoparticle technology for drug delivery across the blood-brain barrier. *Drug Dev Ind Pharm* 28: 11–13.

31. Garcia-Garcia F, Andreiux K, Gill S, Couvereur P. 2005. Colloidal carriers and blood-brain barrier translocation: A way to deliver drugs to the brain? *Int J Pharm* 298: 274–292.

32. Anderson JM, Shive MS. 1997. Biodegradation and biocompatibility of PLA and PLGA microspheres. *Advanced Drug Deliv Reviews* 28: 5–24.

33. Alyautdin RN, Petrov VE, Langer K, Berthold A, Kharkevich DA, Kreuter J. 1997. Delivery of loperamide across the blood brain barrier with polysorbate-80 coated polybutylcyanoacrylate nanoparticles. *Pharmaceutical Research* 14(3).

34. Kreuter J, Alyautdin RN, Kharkevich DA, Ivanov A. 1995. Passage of peptides through the blood-brain barrier with colloidal polymer particles (nanoparticles). *Brain Res* 674: 171–4.

35. Bjustad KB, Redmond DE, Lampe KJ, Kern DS, Sladek JR, Mahoney MJ. 2008. Biocompatibility of PEG-hydrogels in primate brain. *Cell Transplantation* 17: 409–15.

36. Perry J, Chambers A, Spithof KF, Laperriere N. 2007. Gliadel wafers in the treatment of malignant glioma: A systematic review. *Curr Oncol.* October 14(5): 189–194.

37. Bjustad KB, Lampe K, Kern DS, Mahoney M. 2010. Biocompatibility of poly(ethylene glycol)-based hydrogels in the brain: An analysis of the glial response across space and time. *J Biomed Mat Res A* Oct, Vol 96A (1).

38. Thornhoff JR, Lou DJ, Jordan PM, Zhao X, Wu P. 2008. Compatibility of human fetal neural stem cells with hydrogel biomaterials in vitro. *Brain Res.* 1187: 42–51.

39. Uibo R, Laidmae I, Sawyer ES, Flanagan LA, Penelope C, Winder JP, Janmey PA. 2009. Soft materials to treat central nervous system injuries: Evaluation of the suitability of non-mammalian fibrin gels. *Biochim Biophys Acta.* May 1793(5): 924–30.

40. Babensee JH. Interaction of dendritic cells with biomaterials. 2008. *Demin Immunol* 20: 101–108.

41. Balgude AP, Yu X, Szymanski A, Bellakonda RV. 2001. Agarose gel stiffness determines rate of DRG neurite extension in 3D cultures. *Biomaterials* 22: 1077–84.
42. Leung RK, Biran R, Underwood CJ, Tresco PA. 2008. Characterization of microglial attachment and cytokine release on biomaterials of differing surface chemistry. *Biomaterials* 29: 3289–97.
43. Benabid AL, Pollak P, Cervasson C, Hoffman D, Gao DM, Hommel M, et al. 1991. Long-term suppression of tremor by chronic stimulation of the ventral intermediate thalamic nucleus. *Lancet* 337: 403–406.
44. Hochberg LR, Serroya MD, Priehs CM, Mukand JA, Saleh M, Caplan AH, et al. 2006. Neuronal ensemble control of prosthetic devices by a human with tetraplegia. *Nature* 442: 164–71.
45. Bickford RG, Fischer H, Sayre GP. 1957. Histologic changes in the cat's brain after introduction of metallic and plastic coated wire used in electro-encephalography. *Mayo Clin Prox* 32: 14–21.
46. Colias JC, Manuelidis EE. 1957. Histopathological changes produced by implanted electrodes in cat brains: Comparison with histopathological changes in human and experimental puncture wounds. *K Neurosurg* 14: 302–28.
47. Dymond AM, Kaechele LE, Jurist JM, Crandall PH. 1970. Brain tissue reaction to some chronically implanted metals. *J Neurosurg* 33: 574–80.
48. Balasingham C, Dickson L, Brade A, Young VW. 1996. Astrocyte reactivity in neonatal mice: Apparent dependence on the presence of reactive microglia/macrophages. *Glia* 18: 11–26.
49. Elisseeff J. 2008. Hydrogels: Structure starts to gel. *Nature Materials* 7: 271–273.

7

Biomaterials in Obstetrics and Gynecology

David Shveiky, Yael Hants

CONTENTS

Introduction

The field of obstetrics and gynecology is dedicated to women's health. We treat medical and surgical conditions throughout a woman's lifetime. In addition to assisting in the creation of new lives through infertility treatment and obstetrics, we treat serious conditions such as infections and tumors of the female reproductive tract. Also, attention is being turned to improving women's quality of life. As populations age, we understand that our role as physicians continues through life in treating these conditions.

In this chapter we will discuss the role of various biomaterials in obstetrics and gynecology. We chose to present this on the time axis of a modern woman's lifetime. In early reproductive years, many women try to postpone childbirth; biomaterials used for contraception are discussed. While most women conceive spontaneously, some will need assistance with infertility treatment. Here also, biomaterials are being used, especially in preventing pelvic adhesions that block the fallopian tubes. We briefly discuss the concerns regarding the presence of biomaterials, such as hair dye and breast implants, during pregnancy and describe natural and synthetic biomaterials used for the induction of labor when indicated. In later reproductive age, uterine fibroids are a major health problem. We present the use of small-particle injection to block blood supply to these tumors. Finally, a separate section is dedicated to the use of biomaterials in the growing field of female pelvic reconstructive surgery in order to improve women's quality of life.

Trying to Postpone Childbirth

Biomaterials Used in Contraception

Decreased fertility is a known characteristic of affluent Western societies. This is mainly a consequence of effective family planning measures and a desire to postpone childbearing. Despite that, it was estimated that about 54% of all pregnancies and up to 78% of teenage pregnancies in the United States are unintended [1]. The most common family planning measures include periodic abstinence, withdrawal, barrier methods such as condom or diaphragm use, spermicides, oral contraceptives ("the pill"), intrauterine devices (IUDs), and male or female sterilization.

In order to demonstrate the use of biomaterials for contraception, we chose to focus on the intrauterine device (IUD) as a prototype and to describe as well the spermicides and implantable devices.

Intrauterine Devices (IUDs)

The intrauterine device (IUD) is one of the most effective methods for preventing unwanted pregnancy. It lacks the systemic metabolic effects associated with oral contraceptives, and it provides a long-term protection. In Western European countries, 15–30% of married couples use the IUD, while less than 1% in the United States do [2]. According to a frequently told urban legend, the first IUD users were desert caravan drivers who used intrauterine stones to protect their female camels from pregnancy [3]. This story, despite being not well documented, is a demonstration of an early use of biomaterials as a contraceptive method. In 1909, Richter reported his experience with a silkworm catgut ring with coils of nickel and bronze [4]. This preliminary IUD was the base of the 1930s Grafenberg ring, made of coiled silver and gold.

Modern IUDs include the copper IUD and progesterone (or levonorgestrel LNG)-eluding IUDs (Mirena®).

The copper IUD is made of a plastic stem and arms, and is covered with coiled copper. The TCu380A IUD (Paragard®), currently approved by the Food and Drug Administration, is widely used in the United States. This IUD creates a copper surface area of 300 mm^2 in the vertical arm and 40 mm^2 on each of the transverse arms. The lifespan of this IUD is at least 10 years, with reported efficacy even after 12 years of use.

Mechanism of action of the copper IUD is the release of copper ions into the endometrium. This creates a local inflammatory response. There is a dramatic increase in endometrial leukocyte count with local secretion of inflammatory cytokines. Pregnancy is prevented mainly by the spermicidal effect of this inflammatory process. In addition to phagocytosis by endometrial macrophages, some of the products of these leukocytes create a toxic environment to the spermatozoa. Additionally, copper has the ability to impede sperm transport through the cervical mucous and sperm viability. A third mechanism of action is debatable: it has been claimed that the environment around the IUD may be hostile also to the blastocyte, thus interfering with early embryo implantation.

The LNG-releasing IUD (Mirena®) was approved by the FDA in 2000. This T-shaped device contains a reservoir of LNG on its vertical arm. LNG is a progestin (a synthetic derivative of testosterone that activates progesterone receptors) that is released at a rate of 20 micrograms per 24 hours by this intrauterine system. It acts directly on the endometrium, interfering with the endometrial maturation required for implantation. In addition, it has an inhibitory effect on ovulation as well as an effect of thickening the cervical mucous. The inhibitory effect on endometrial maturation is responsible for a significant reduction in menstrual blood flow and dysmenorrhea (pain with menstruation). Indeed, Mirena® was approved by the FDA in 2009 for the indication of heavy menstrual bleeding. It reduces menstrual blood loss by 90% one year after insertion.

Both the copper IUD and the LNG-releasing IUD have excellent efficacy, with a 5-year pregnancy rate of only 1.4% for the TCu380A IUD and 0.7% for the Mirena®.

When comparing efficacy of contraceptive devices, one must take into consideration that there is a difference between "typical failure rate" and "perfect failure rates." For example, if a couple is relying on condoms for contraception, the typical failure rate will be higher than the perfect rate, because couples tend too often to forget to use the condom. In the case of IUD, the typical failure rate almost equals the perfect failure rate.

Adverse effects of IUDs include the risk for an infection and pelvic inflammatory disease, especially around insertion, heavy menstrual bleeding (with the copper IUD) and menstrual abnormalities, uterine perforation at insertion, and septic abortion in case of pregnancy in the presence of IUD. Despite that, the IUD remains a safe and effective contraception method utilizing biomaterials to increase women's liberty and choice.

Implantable Devices

Norplant is a levonorgestrel (LNG) subdermal implant. It is implanted under the skin in a simple office procedure under local anesthesia and provides contraceptive protection for 5 years. It releases LNG to the blood stream at a rate of 80 mg/day in the first year, then 30–35 mg/day for another 4 years. The blood levels of this progestin remain stable at 0.25–0.35 ng/dL, sufficient for its contraceptive activity. LNG blocks the LH surge required for ovulation. It also thickens the cervical mucous to prevent transport of sperm to the uterus.

Norplant is a very effective contraceptive agent, with only 1% pregnancy over 5 years. One product that has been used consists of 6 rods measuring 34 x 2.4 mm, each containing 36 mg of LNG. Its use involves insertion under the skin at the upper arm using a special trocar, with removal after 5 years of use [5].

Spermicides

A spermicidal agent is a material introduced into the vagina before intercourse that immobilizes or kills sperm cells. Spermicides typically contain a surfactant, nonoxynol 9, that is spermicidal and also provides a mechanical barrier that blocks sperm transfer into the cervical canal.

Spermicides are available in the form of foam, cream, and suppositories. A contraceptive sponge made of polyurethane that contains nonoxynol 9 was available in the United States until 1994 and is still available in some European countries. The advantage of this device is that it remains effective for 24 hours after insertion and does not need to be inserted right before intercourse.

In addition to its contraceptive activity, nonoxynol 9 is known to reduce the risk of vaginal infections such as bacterial vaginosis and other sexually transmitted diseases, including HIV. Being toxic to some of the normal flora of the vagina, spermicides may increase the risk for colonization of pathogenic bacteria, such as *E. coli*, thus increasing the risk of urinary tract infection.

Although older studies raised concerns about potential teratogenicity of nonoxynol 9, several large studies found no greater risk of congenital malformations in embryos conceived despite the use of this contraceptive agent.

Finally, spermicides are much more effective when used along with mechanical barriers, such as a condom or a diaphragm [5].

Permanent Tubal Sterilization

Essure®, Conceptus Inc.

After completion of the family, many women seek a permanent measure of contraception. Until recently, the most popular permanent contraception was tubal ligation. This operative procedure, usually done laparoscopically, requires general anesthesia and short hospital stay. New emerging technology, first approved by the FDA in 2002, has revolutionized permanent sterilization. Essure® is a trans-cervical contraceptive device that offers women seeking permanent contraception in a non-surgical non-hormonal solution. In this office procedure, flexible inserts are inserted into the fallopian tube through the cervix and the uterus via hysteroscopy. The inserts are made of inner polyethylene terephthalate (PET) fibers to induce tissue fibrotic reaction and are held in place by a flexible stainless steel inner coil and a dynamic outer nickel titanium alloy coil. PET fibers are being used because of their ability to induce tissue ingrowth into medical devices in other procedures, such as arterial grafts. The physician performing the hysteroscopy identifies the fallopian tube ostia and inserts these coils. During the next three months, there is an inflammatory response caused by the invasion of macrophages, fibroblasts, giant cells, and plasma cells, resulting in fibrosis in the fallopian tube around the insert. This natural "plug" prevents sperm from reaching the eggs. After three months, a hystero-salpingogram is performed by injecting dye through the cervix to ensure a complete blockade of both fallopian tubes. A follow-up study of 5 years reported excellent results with zero pregnancies and a 99.74% success rate. Failures in this study were to the result of the inability to insert or of the expulsion of the device. The risks include pain and cramping, uterine perforation, inability to insert, allergic reaction to the materials, and vasovagal response (fainting) during insertion. In the first 3 months there is risk of failure as well as ectopic pregnancy, and therefore a complementary contraceptive method is recommended [6].

Adiana® Hologic, Inc.

This permanent sterilization method is a combination of controlled thermal damage to the lining of the fallopian tube followed by insertion of a non-absorbable biocompatible silicone elastomer matrix into the tubal lumen. This procedure is done under hysteroscopic guidance as well. A catheter is inserted into the tubal ostium and delivers radiofrequency (RF) energy for 1 minute, causing a lesion in the fallopian tube lumen. Then, a 3.5-mm silicone matrix is placed in the injured area. During the next few weeks, tubal occlusion is achieved by fibroblast ingrowth into the matrix, which serves as permanent scaffolding and allows for space-filling. As with the Essure® device, tubal occlusion is assessed 3 months after device placement. Adiana® matrix is not visible by X-ray, but can be seen by ultrasound [7, 8].

Trying to Conceive

Biomaterials Used for Prevention and Treatment of Infertility

Infertility is generally defined as one year of unprotected intercourse without conception. This is a common condition that affects 10–15% of couples and has important psychological and medical, as well as economic and demographic, implications. During the last three decades there was no change in the prevalence of infertility, but the demand for infertility treatment has grown substantially. During this period, new emerging technologies, mainly ART (assisted reproductive technologies), inflicted a dramatic change in the field of infertility treatments and improved the prognosis for many infertile couples.

Pelvic Adhesions

About 60% of infertility cases are attributed to the female factor. One of the most common identifiable female causes of infertility is the presence of pelvic adhesions causing tubal blockage. This factor constitute up to 23% of female causes. Pelvic adhesions occur in 60–90% of women following major gynecologic surgery. They cause infertility by preventing the normal transport of the oocyte, sperm, or fertilized egg through the fallopian tube. Apart from preventing conception, adhesions can cause considerable acute or chronic pelvic and abdominal pain and small bowel obstruction. They may also complicate future surgeries by causing difficulties in access and dissection, prolongation of operative time, increase in blood loss, and predisposition to injury to the bowel or urinary system [9–11].

Adhesions usually result from the normal peritoneal inflammatory wound healing response and develop in the first five to seven days after surgery [12]. They are composed of fibrous tissue but also contain blood vessels, fat, and nerves. Several factors involved in mechanisms of adhesion formation include peritoneal injury, the attendant inflammation, imbalance in the plasmin system, and the proximity of injured surfaces [9].

Hence, preventive strategies have been designed in order to target these steps individually or in combination. Limiting the injured area, a vital aspect of prophylaxis, is largely a surgical endeavor achieved by meticulous attention to operative technique and tissue handling as well as by using minimally invasive techniques when possible.

However, since injury can only be minimized but not completely avoided, adjuncts to injury limitation techniques are necessary to reduce the risk of adhesions [9].

Barriers agents are one of the methods traditionally employed for this purpose. Barriers may be either liquid or solid, and the latter can be absorbable or non-absorbable. Several synthetic barriers with different characteristics are commercially available, but the evidence for the use of these products is not adequate for definite conclusions to be drawn, and further research in this field is warranted.

The following biomaterials are examples of products being used for adhesion prevention [10, 11, 13, 14].

Interceed®, Ethicon Women's Health & Urology, Somerville, NJ, USA

Interceed® is an oxidized regenerated cellulose. It was the first degradable barrier used in clinical practice to cover traumatized peritoneum in the pelvis. Interceed® can be cut as necessary, requires no suturing, and is completely absorbable. It is applied over raw tissue surfaces at the end of surgery after hemostasis has been achieved. It forms a gelatinous protective coat within eight hours of application, and is broken down into its monosaccharide constituents and absorbed within two weeks. In order to evaluate the efficacy of intercede in the prevention of the development of post-surgical adhesions, many studies have been carried out. A meta-analysis of 11 randomized controlled trials has shown that the barrier is safe and significantly reduces the incidence of de novo adhesions, as well as the reformation of adhesions that were previously lysed, compared with no treatment in laparoscopy. Rather than acting systemically, this product is site specific. Therefore, its efficacy is limited to surgical situations where raw surfaces can be completely covered with the product and its benefit is limited to the site of barrier placement. The main disadvantage of this product is that it is not effective unless the entire area is completely hemostatic. The presence of small amounts of blood in the peritoneal cavity or post-operative bleeding results in blood permeating the mesh, fibrin deposition and, finally, adhesion formation. Therefore, there is no substitute for meticulous hemostasis and good surgical techniques and tissue handling to maximize the benefit from this product.

Intercoat®, Ethicon Women's Health & Urology, Somerville, NJ, USA

Intercoat® is an absorbable adhesion barrier gel used to prevent adhesion formation in women. Its gel formulation makes it easy to apply on surgical areas with precision. It is an effective adjunct to peritoneal surgery and is intended to reduce the incidence, extent, and severity of postoperative adhesions at the surgical site. This barrier is composed of a combination of polyethylene oxide (PEO) and sodium carboxymethylcellulose (CMC), which is stabilized with calcium and made isotonic through the use of sodium chloride. As oppose to the solid alternative, Intercoat® is injected into the peritoneal cavity with a syringe that is provided with the product, instead of being placed directly over the exposed area. It allows for application in one single layer to the traumatized tissue, thus creating a temporary barrier during the healing process.

The effectiveness of Intercoat® was proved in two clinical trials. Using the American Fertility Society (AFS) adnexal score as a measure of the severity of pelvic adhesions, it was shown that application of Intercoat® improved or did not worsen in 91% of cases as opposed to the 63% in the control group [15, 16].

Seprafilm®, Genzyme, Cambridge, Massachusetts, USA

Seprafilm® is an adhesion barrier composed of hyaluronic acid and carboxymethylcellulose. It is a membrane applied to the traumatized tissue during surgery and is absorbed from the peritoneal cavity within seven days.

Seprafilm® is a site-specific agent and acts as a mechanical barrier, preventing opposite tissue surfaces from sticking to each other. Its function lasts for 7 days and is completely excreted from the body within 28 days. It is a brittle film that has a tendency to fracture when bent, thus making it hard to use in laparoscopic surgery. In a blind prospective, randomized, multicenter study, the treatment of patients after myomectomy (resection of uterine fibroids) with Seprafilm® significantly reduced the extent and area of post-operative adhesions. Potential side effects include induced foreign body reaction, higher incidence of pulmonary emboli, and intraperitoneal abscess formation, but these findings were not statistically significant in the relevant trials. High cost is another limitation because, for an effective protection from intestinal obstruction, a mean of 4.5 sheets per patient is required. Seprafilm® was approved by the FDA for use in open surgery in the United States in 2006 [10, 11, 13].

Fallopian Tube Blockage

There are instances where fallopian tube blockage is desired. Hydrosalpinx is a condition of the fallopian tube being dilated and filled with fluid. This condition may be secondary to pelvic inflammatory disease or surgery of the reproductive system and results in a dilated, non-functional fallopian tube. Hydrosalpinx appears as an irregular cystic mass in the pelvis and may cause no symptoms, but in some instances may get infected and cause a tubo-ovarian abscess. In cases of severe infertility requiring assisted reproductive technologies, in-vitro fertilization (IVF) is commonly performed. During this treatment, after ovarian hyper-stimulation with hormones (gonadotropins), eggs are retrieved from the ovaries. These eggs are fertilized in vitro with the male partner's sperm to create embryos. These embryos are then transferred into the uterus for implantation of pregnancy. In the past two decades, studies have shown lower implantation rates in women with hydrosalpinx. Further studies demonstrated that the fluid in these dilated fallopian tubes is toxic to the embryos. In addition, this fluid may mechanically wash out the recently transferred embryos from the uterus. Therefore, disconnection of the fallopian tubes from the uterus is desired in these cases. While traditionally performed surgically by tubal ligation or resection of the fallopian tubes, this can be now achieved by using the Essure® or Adiana® procedure, previously discussed in detail [6, 7]. Thus, using the biomaterials originally designed for contraception, we are able to assist reproduction.

The Use of Biomaterials during Pregnancy

During pregnancy, women are advised not to be exposed to chemicals or fumes. This is because certain chemicals tend to interfere with the biochemical reactions that occur during fetal development and thus may cause genetic defects or birth defects in the fetus. These are several examples of controversial materials in common use. In this chapter we will focus on two examples of biomaterials that are commonly used by women and may raise controversies about their safety during pregnancy.

Hair Dye

Although hair coloring is a common practice in women of childbearing age, only few studies have addressed the issue of hair dying safety before or during pregnancy. These substances can enter the body via direct skin absorption as well as inhalation of fumes, resulting in an elevated systemic concentration associated with a teratogenic effect. Of major concern are permanent hair dyes that contain ammonia, which has a strong chemical fume, and many health care providers recommend avoiding their use during the first trimester.

During the last decades, studies attempting to link the use of hair dying materials and various childhood cancers, including neuroblastoma, have yielded inconsistent results.

In teratologic studies on pregnant rats following topical application of 12 hair-dying formulations, no significant soft tissue or skeletal changes were noted. Other studies showed no evidence of teratogenic effects in rats or rabbits after administration of hair dye by gavage.

Future studies of hair-dye use during pregnancy should be designed to gather necessary information for more specific analysis of association with teratogenic effects or childhood tumors. It is important to emphasize that at this time, there is no conclusive evidence to claim that the use of hair dyes is unsafe during pregnancy [17, 18].

Breast Implants

A breast implant is a prosthesis mostly used in post-mastectomy breast reconstruction and for cosmetic breast augmentation. The breast implants that are approved by the Food and Drug Administration (FDA) are composed of an elastic silicone rubber shell filled with sterile saline solution or cohesive silicone gel.

Ever since the beginning of breast implant use, there have been concerns raised regarding potential damaging effects on children born of mothers with breast implants. Whether silicone crosses the placenta has not been evaluated in women, but there is little evidence of any elevation of blood or serum silicon concentrations in women with silicone breast implants. There are also no studies of teratologic effects of silicone in humans. A review of the published literature suggests that the information is insufficient to show definitive conclusions [19].

"A Time to Be Born"

The Use of Biomaterials during Labor and Delivery

Labor and delivery is a natural process that we strive to keep natural. Nevertheless, in the developed world it is usually done under medical supervision in order to be able to provide medical assistance and to respond promptly to complications during or after labor. As a natural process with minimal intervention, very few biomaterials are used during labor. In this chapter we will present one biomaterial used for induction of labor.

Induction of labor is a medical intervention meant to initiate the process of labor in order to deliver the baby earlier than expected. Indications for induction may include high-risk conditions in pregnancy, such as post-term, gestational hypertensive disorders and pre-eclampsia, and diabetes. Non-medical and sometimes controversial indications may include planned delivery because of the patient's or doctor's wishes.

Parturition, or the initiation of labor, is a complex process involving both maternal and fetal signals that is not fully understood. It does include a local secretion of prostaglandins, mainly prostaglandin E2 (PGE2), in the cervix in a process called *cervical ripening*. Cervical ripening includes softening, some shortening, and initial dilatation of the cervix preparing for labor. Prostaglandins also increase the concentration of receptors to the hormone oxytocin in the uterine muscle to enable effective contractions.

Prostaglandins can be given directly into the cervix or vagina in a form of gel, tablets, or a vaginal insert of controlled released PGE2 (Propess®). While drugs are beyond the scope of this chapter, we will focus on a mechanical method for cervical ripening: *Laminaria*.

Laminaria japonicum is a type of kelp or seaweed that is extensively cultivated in China, Japan, and Korea. It acts as a hygroscopic or osmotic cervical dilator by absorbing local fluids in the cervix and expanding, thus providing controlled mechanical pressure inside the cervical canal. This, in turn, increases local production of prostaglandins, causing cervical ripening. A synthetic product with similar hygroscopic ability, Lamicel®, is commercially available. Apart from its use for induction of labor, *Laminaria* is used for cervical dilation prior to dilatation and curettage (D&C) of the uterus for the treatment of missed abortion or in cases of medical abortion. In fact, it is widely used in order to reduce complications related to cervical dilation by its ability to promote controlled cervical ripening [20].

Improving Quality of Life

The Use of Biomaterials in the Treatment of Benign Gynecologic Conditions

Uterine Artery Embolization

Uterine fibroids (leiomyomas, myomas) are the most common tumors of the female genital tract. According to recent studies, the lifetime risk of fibroids in a woman over the age of 45 is more than 60%, with incidence higher in blacks than in whites. Although most fibroids cause no symptoms, fibroid uterus remains a leading cause of hysterectomy. Fibroid tumors are regulated by many factors, mainly ovarian steroids, estrogen and progesterone, growth factors, and angiogenic factors. Black race, nulliparity, obesity, polycystic ovary syndrome, hypertension, and diabetes are associated with increased risk of fibroids. The genetic basis of uterine fibroids has not been elucidated yet; however, recent studies demonstrated the role of key genes in the pathogenesis of these tumors, including genes related to alcohol metabolism and apoptosis. Additionally, a few familial syndromes of uterine fibroids were described, raising the possible role of the gene coding to the protein fumarate hydratase, a Krebs cycle enzyme.

Until recently, the mainstay of treatment of uterine fibroids was hysterectomy. In the past two decades, minimally invasive techniques have been utilized to improve symptoms of fibroids without hysterectomy. One of the most effective uterine-sparing methods is uterine fibroid embolization (UFE). Embolization is a minimally invasive means of blocking the arteries that supply blood to the fibroids. In this procedure, a catheter in introduced into the uterine arteries using angiographic techniques. Then, small particles of polyvinyl alcohol (PVA) are injected into the arteries, which results in their blockage. This procedure was first used to reduce blood loss during resection of fibroids (myomectomy). Surprisingly, patients who were treated with UFE prior to myomectomy showed reduction in their fibroid size and improvement of symptoms while awaiting surgery. Recent studies have shown that UFE is a safe and effective uterine-sparing treatment for uterine fibroids.

The same technique is being used to treat severe bleeding after childbirth. Despite modern medicine, bleeding remains a leading cause of mortality during labor. When possible to transfer a bleeding patient to the interventional radiology suite, this treatment may save the patient's life without the need for a hysterectomy. In conclusion, UFE, with the injection of PVA particles into the uterine arteries, is a minimally invasive uterine-sparing treatment for fibroids that may also be utilized as a life-saving measure [21–24].

Female Pelvic Reconstructive Surgery

Pelvic floor dysfunction consists of a variety of conditions affecting women in later reproductive years and after menopause. These include urinary incontinence and pelvic organ prolapse as well as bowel symptoms and sexual dysfunction. As the population ages, more and more patients present with these symptoms. Moreover, many more women are now willing to share their complaints with their gynecologists with an increased demand for quality of life. Epidemiological studies have shown that up to 50% of women may experience one or more of these conditions in their lifetime [25]. Many of these conditions are caused by damage to the vaginal supporting structures and the fibromuscular layer of the vaginal walls. Vaginal birth, aging, and conditions that increase abdominal pressure are the main risk factors. The field of female pelvic medicine and reconstructive surgery utilizes cutting edge technology in the treatment of these conditions. Biomaterials used in this field include biologic and synthetic grafts and slings, as well as injectable bulking agents to treat urinary incontinence. This growing field is probably the greatest consumer of biomaterials in modern gynecology. Owing to the limited scope of this chapter, we will discuss only a few examples of biomaterials used in this field.

Mid-urethral Slings for Stress Urinary Incontinence [25, 26]

Stress urinary incontinence (SUI) is a condition of urine leakage that occurs with increased intra-abdominal pressure such as coughing, sneezing, and laughing. It is a common condition that affects millions of women worldwide at all ages. It prevalence increases with age, and parity is a known risk factor, especially in the younger age group. The pathogenesis of this condition is related to the loss of the support provided by the connective tissue under the urethra, causing urethral hypermobility. In a smaller portion of the patients, there is an intrinsic defect in the sealing mechanism within the urethra. Until two decades ago, the surgical treatment of SUI included an abdominal surgery requiring a lower abdomen incision and recreation of a hammock-like support under the urethra. Mid-urethral slings have revolutionized the current treatment of SUI. In this procedure, a synthetic polypropylene sling is implanted between the vagina and mid-urethra via a vaginal incision of 2 cm. This procedure can be performed under local anesthesia or sedation, requires no hospitalization, and is very successful, with cure rates of over 80% (completely dry) and over 90% significantly improved. The sling can be passed behind the pubic bone (TVT) or through the obturator foramens (TVT-O or TOT).

Vaginal Mesh for Pelvic Organ Prolapse (POP) [27–31]

Failure of the vaginal fibromuscular tissue, also known as the endopelvic fascia, to maintain integrity of the vaginal walls causes pelvic organ prolapse. When the weakness is in the apex of the vagina and the utero-sacral and cardinal ligaments that normally support the uterus in place are stretched, uterine prolapse or vaginal vault prolapse occurs. When the anterior vaginal connective tissue is weakened, the bladder comes down, causing cystocele, and when the bowel is pushing through a weakened posterior vaginal wall, rectocele occurs. The main symptoms of POP are bulge, pressure, obstructed urination and defecation, sexual dysfunction, and loss of self-image. Surgical repair of PAP can be performed abdominally (sacral colpopexy and paravaginal repair) or by using the less invasive vaginal approach. Vaginal repair can be done using the patient's native tissue in a procedure called colporrhaphy, in which the vaginal fibromuscular tissue is plicated with the correction of the connective tissue defect and reduction of the prolapsed organ, or using a graft material to augment tissue strength. POP can be looked at as a hernia of the pelvic organs through a weakened vagina. The experiences from abdominal hernia show that repair can be improved by mesh interposition. In the last three decades, many clinical and scientific efforts have been targeted to find the ideal graft to use in vaginal repair. The ideal graft should be chemically and physically inert, non-carcinogenic and non-immunogenic, mechanically strong, generally available and inexpensive, resistant to infection and shrinkage, pliable, and easy to be applied surgically. Since this ideal mesh does not exist, many products are being widely used for vaginal prolapse repair. The sources of grafts include autografts (grafts that are harvested from various parts of the body, such as the fascia lata and gracilis fascia) and xenografts (cadaveric fascia lata from humans or collagen matrix laminates from porcine dermis or porcine small intestine submucosa (SIS)). The most obvious advantage of biologic grafts is avoidance of morbidity to the donor site. Several studies have shown better cure rate of cystocele with the use of biologic grafts compared to traditional repair. Synthetic grafts were developed to provide the advantages of availability, promotion of postoperative fibroblast activity in the tissue, less infection, and no risk of rejection. These grafts are built as a meshwork that serves as a scaffold for the patient's connective tissue to grow over. Many biomaterials have been used in vaginal mesh, including absorbable polyglycolic acid and polyglactin 910 and the non-absorbable polypropylene, polyester, polytetrafluoroethylene (PTFE), and its expansion (Gore-Tex). There are also composite mesh grafts that contain both biologic and synthetic materials. Mesh grafts are also classified by the size of the pores and by being multifilamentous or monofilamentous. The bigger the size of the pore (as in type 1 mesh, >75 μm) the better its quality, since macrophages can then pass through these pores to fight bacteria.

Several studies have shown the potential advantage of VMC with synthetic mesh over traditional anterior colporrhaphy for anterior vaginal wall prolapse repair. There is significant debate as to the relative safety of these devices. Complications of synthetic vaginal mesh include erosion to the vagina, rare erosions to pelvic organs such as bladder and rectum, infection, pain during sex, and constant pelvic pain. The Society of Gynecologic Surgeons workgroup had developed evidence-based guidelines for vaginal mesh repair. It was suggested that native tissue repair remains appropriate compared with biologic graft use. Nonabsorbable synthetic graft use may improve anatomic outcomes of anterior vaginal wall repair, but there are trade-offs in regard to additional risks. Recently it was concluded that there is limited evidence to guide decisions regarding whether to use graft materials in transvaginal prolapse surgery.

Other biomaterials used in urogynecology and pelvic reconstructive surgery include injection of botulinum toxin into the bladder wall to treat refractory cases of overactive bladder with urinary frequency, urgency, and urge incontinence; bulking agents, such as collagen or carbon particles, injected to the bladder neck to treat stress urinary incontinence in patients who are poor surgical candidates; porcine small intestine submucosa (SIS) to treat rectovagina fistulae; and many more that are beyond the scope of this chapter.

Conclusion

In this chapter we described the use of biomaterials in obstetrics and gynecology through the life course of a modern woman. We demonstrated their use earlier in life, when pregnancy is not desired, in fertility treatment, during pregnancy and delivery, and in the treatment of common gynecologic problems later in life. Biomaterials have been used for thousands of years in this field, and it is expected that their use will keep growing exponentially, utilizing top materials to make women's lives better.

References

1. Henshaw SK. 1998: Unintended pregnancy in the United States. *Fam Plan Perspect* 30: 24.
2. Mishell DR. 2001. Family planning. In *Comprehensive Gynecology* 4th edition, Stenchever MA. et al., eds. St. Louis, MO: Mosby.

3. Speroff L. 2005. Intrauterine device (IUD). In *Clinical Gynecologic Endocrinology and Infertility* 7th edition. Speroff L, et al., eds. Lippincott Williams and Wilkins.
4. Gamble CJ. 1957. Spermicidal times as aids to the clinician's choice of contraceptive materials. *Fertile Steril* 8: 174.
5. Stubblefield PG, Carr-Ellis S, Kapp N. Family planning. In *Berek and Novak's Gynecology* 14th edition, pp. 247–312. Lippincott, Williams and Wilkins.
6. www.essure.com
7. www.adiana.com
8. Palmer SN, Greenberg JA. 2009. Transcervical sterilization: A comparison of Essure® Permanent Birth Control System and Adiana® Permanent Contraception System. *Rev Obstet Gynecol* 2(2): 84–92.
9. Kumar S. 2009. Intra-peritoneal prophylactic agents for preventing adhesions and adhesive intestinal obstruction after non-gynaecological abdominal surgery. *Cochrane Database Syst Rev.*
10. Ahmad G. 2010. Barrier agents for adhesion prevention after gynecological surgery. *Cochrane Database Syst Rev.*
11. Pados G. 2010. Prevention of intra-peritoneal adhesions in gynecological surgery: Theory and evidence. *Reproductive BioMedicine* 21: 290–303.
12. DeChemey AH. 2010. Preventing postoperative peritoneal adhesions. *UpToDate.*
13. Monk BJ. 1994. Adhesions after extensive gynecologic surgery: Clinical significance, etiology, and prevention. *Am J Obstet Gynecol* 170(5 Pt 1): 1396–1403.
14. Holmdahl L. 1997. Adhesions. Pathogenesis and Prevention. Panel discussion and summary. Department of Surgery, Ostra Hospital, University of Göteborg, Sweden.
15. Lundorf P, J Donnez, M Korell, AJ Audeburt, K Block, GS diZegega. 2005. Clinical evaluation of a viscoelastic gel for reduction of adhesions following a gynaecological surgery by laparoscopy in Europe. *Human Reproduction* 20(2): 514–520.
16. Young P, A Johns, C Templeman, C Witz, B Webster, R Ferland, M Diamond, K Block, GS diZerega. 2005. Reduction of postoperative adhesions after laparoscopic gynaecological surgery with oxiplex/AP gel: A pilot study. *Fertility and Sterility* 845: 1450–1456.
17. Koren G. 2007. Medication safety. In *Pregnancy and Breastfeeding.*
18. Erin E. 2005. Maternal hair dye use and risk of neuroblastoma in offspring. *Cancer Causes and Control* 16: 743–748.
19. Bondurant S. 1999. *Safety of Silicone Breast Implants.* Chapter 11 (pp. 248–263).
20. ACOG Practice Bulletin Number 107. 2009. Induction of labor. *Obstet. Gynecol* 114 (2) Part 1: 386–397.
21. Shveiky D, Shushan A, Ben Bassat H, Klein BY, Ben Meir A, Levitzky R, Rojansky N. 2009. Acetaldehyde differentially affects the growth of uterine leiomyomata and myometrial cells in tissue cultures. *Fertil Steril* Feb 91(2): 575–79.
22. Shveiky D, Rojansky N, Ben Bassat H, Ben Meir A, Klein B, Shushan A. 2009. Family history of uterine fibroids associated with low level of fumarate hydratase in leiomyomata cells. *Eur J Obstet Gynecol Reprod Biol* 146(2): 234–35.
23. Goodwin SC, Spies JB. 2009. Uterine fibroid embolization. *N Engl J Med* Aug 361(7): 690–97.
24. Bloom AI, Verstandig A, Gielchinsky Y, Nadiari M, Elchalal U. 2004. Arterial embolisation for persistent primary postpartum haemorrhage: Before or after hysterectomy? *BJOG.* 111(8): 880–84.

25. Nygaard I, Barber MD, Burgio KL, Kenton K, Meikle S, Schaffer J, Spino C, Whitehead WE, Wu J, Brody DJ. 2008. Pelvic Floor Disorders Network. Prevalence of symptomatic pelvic floor disorders in US women. *JAMA.* 300(11): 1311–16.

26. Ward KL, Hilton P, 2007. UK and Ireland TVT Trial Group. Tension-free vaginal tape versus colposuspension for primary urodynamic stress incontinence: 5-year follow up. *BJOG.* Jan 2008, 115(2): 226–33. Epub Oct 25.

27. Fatton B, Amblard J, Debodinance P, et al. 2007. Transvaginal repair of genital prolapse: Preliminary results of a new tension-free vaginal mesh (Prolift™ technique): Case series multicentric study. *Int Urogynecol J Pelvic Floor Dysfunct* 18(7): 743–52.

28. Hiltunen R, Nieminen K, Takala T, et al. 2007. Low-weight polypropylene mesh for anterior vaginal wall prolapse: A randomized controlled trial. *Obstet Gynecol* 110(2 Pt 2): 455–62.

29. Meschia M, Pifarotti P, Bernasconi F, et al. 2007. Porcine skin collagen implants to prevent anterior vaginal wall prolapse recurrence: A multicenter, randomized study. *J Urol* 177(1): 192–95.

30. Nguyen JN, Burchette RJ. 2008. Outcome after anterior vaginal prolapse repair: A randomized controlled trial. *Obstet Gynecol* 111(4): 891–98.

31. Paraiso MFR. 2007. The use of biologic tissue and synthetic mesh in urogynecology and reconstructive pelvic surgery. In Walters MD and Karram MM, *Urogynecology and Reconstructive Pelvic Surgery,* 3rd edition, pp. 295–330. Mosby Elsevier.

8

Tissue Engineering: Focus on the Cardiovascular System

Ayelet Lesman, Shulamit Levenberg

CONTENTS

Introduction

Owing to limited regenerative capacity of the adult mammalian heart, any significant myocardial cell loss is mostly irreversible and can lead to progressive loss of ventricular function and eventual heart failure. Tissue-engineered cardiac constructs can be applied to repopulate scar tissue with a new pool of contractile cells, restoring function to the failing heart. Appropriate selection of cells and biomaterials is the key factor in the construction of viable and clinically relevant engineered tissue for myocardial regeneration. Various stem cell types have been proposed for use in treatment of injured heart tissue. However, the optimal cell type has yet to be established.

In this chapter, we will focus on derivation of human stem cells for use in clinical applications of this nature. Implementation of human embryonic stem cell and induced pluripotent stem cell (iPS)–based therapies will be discussed, along with a review of the potential of adult stem cells, including bone marrow cells (hematopoietic and mesenchymal stem cells), resident cardiac stem cells, and skeletal muscle stem cells (satellite stem cells), in such clinical protocols.

The biomaterials used for creation of three-dimensional (3D) engineered tissues dictate the scaffolding capacities necessary for organizing cells into appropriate tissue structures both in vitro and in vivo. The various applications of biomaterials for such purposes will be discussed. Recent breakthroughs in tissue engineering disciplines allowing for the design of biomaterial-based heart tissue constructs have transformed this avenue into a promising approach toward advancing myocardial repair. As the quality of engineered tissues continues to be optimized, vascularization of engineered tissues in efforts to augment construct viability, thickness, and architecture will also be introduced.

Human Cell Types for Cardiac Regeneration

Human Embryonic Stem Cells (hESCs)

The extensive proliferative and differentiative capacities of early embryonic blastocyte-derived hESCs render them one of the most promising sources of human cells for repair of injured tissue. Ever since successful isolation of hESCs [1], which occurred twenty years after first reports of mouse ESC (mESC) derivation [2, 3], extensive research efforts have been invested toward inducing cardiomyocyte (CM) population propagation and purification from hESC pools. The hESC in vitro–derived CMs (hES-CMs), first described by Gepstein and colleagues [4] and later by others [5], have been extensively characterized. Their unique early-stage cardiac potential features sarcomeric structural patterning, spontaneous contractility, capacity to both structurally and functionally integrate with preexisting cardiac tissue, responsiveness to pharmacological agents, and expression of cardiac-specific genes [4–8]. Moreover, hES-CMs exhibit significant proliferative capacity both in vitro (15–25% BrdU+ cells) and in vivo, when compared with that of mESC-CMs (<1% BrdU+ cells) [9, 10]. Our works have shown that the number of proliferating hES-CMs can be significantly augmented when cultured together with endothelial and fibroblast cells on 3D polymeric scaffolds (Figure 8.1), presumably via paracrine signaling [11]. Furthermore, upon transplantation

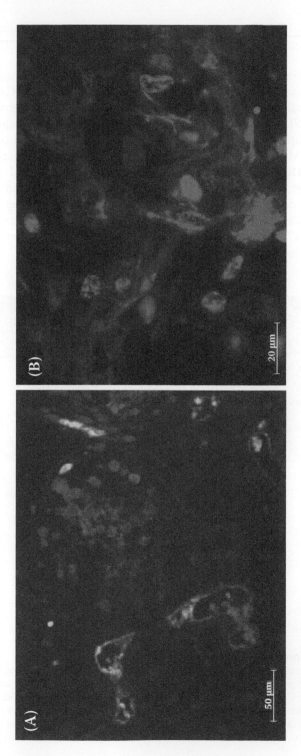

FIGURE 8.1 (See color insert.)
In-vitro construction of engineered vascularized cardiac muscle using a multi-cellular strategy of hES-CMs, endothelial cells, and embryonic fibroblasts seeded within a porous polymer scaffold of PLLA/PLGA. (A) The endothelial cells (vWF, green) within the scaffold self-organized to lumen vessel structures located in close proximity to the hES-CMs (troponin I red). (B) Higher magnification reveals that the hES-CMs matured to a certain degree, presenting developed cytoplasm and sarcomeric pattering (troponin I, red). Nuclei are stained with DAPI (blue).

into animal hearts, hES-CMs underwent a maturing process, as observed through their elongation and development of sarcomeric patterning [12] (Figure 8.2a, b).

While the potential of hES-CMs toward repair of the human heart has not been put to clinical tests, the contribution of differentiated hESCs to the reversal of myocardial infarction has been evaluated in a number of pre-clinical trials [13–17]. Upon injection of hES-CMs to the infracted rat heart together with a unique prosurvival protein cocktail, Laflamme and cowork-ers [14] reported CM survival after transplantation, cell proliferation, graft area expansion with time, attenuation of ventricular dilation, and preser-vation of contractile function. Similarly, formation of stable cardiomyocyte grafts providing functional benefit have been described by Caspi et al. [13] after injection of hES-CMs to infarcted rat hearts. In line with these reports, Van Laake et al. [17] demonstrated transient functional improvement of infarcted mouse hearts following hES-CMs transplantation.

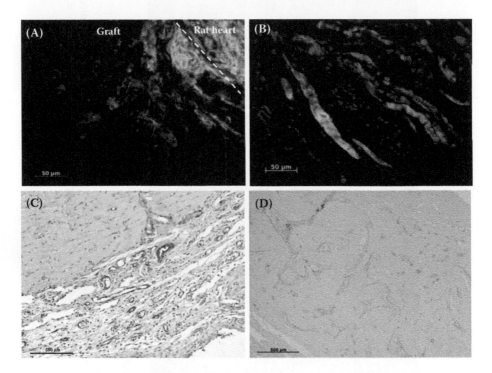

FIGURE 8.2 (See color insert.)
Transplantation of the engineered human vascularized cardiac tissue demonstrating localiza-tion of the hES-CMs (troponin I, red) in the graft area next to the myocardium (A), and struc-tural maturation of the CMs in the graft area (B). The graft area was occupied with intense vascularization, as detected by staining with aSMA antibody (host and human derived vessels, brown, C) and with human specific endothelial CD31 antibody (human implanted vessels, brown, D). Nuclei were stained with DAPI (A, B in blue) or with hematoxylin (C, D in blue).

Induced Pluripotent Stem Cells (iPS Cells)

The first successful reprogramming of mouse fibroblasts toward cells bearing characteristics similar to those of mESCs was reported at 2006, as described by Takahashi and Yamanaka [18]. Shortly after, similar protocols were applied toward human cells, to yield human iPS cells [19, 20]. Reprogramming was achieved by overexpression of the Oct3/4, Sox2, Klf4, and c-Myc transcriptional factors [19] or a combination of Oct4, Sox2, NANOG, and LIN28 [20], using retroviruses. Subsequent studies confirmed that mouse and human somatic cells can be reprogrammed to the pluripotent state by means of the same or similar sets of factors to yield cells similar to hESCs in morphology, gene expression, and differentiation potential, both in vitro and in vivo [21–25]. Later studies demonstrated that introduction of Oct4 and Sox2 alone suffice for achieving reprogramming [22]. Derivation of iPS cells from somatic cells enabling preparation of patient-specific cell samples for cell therapies is dependent on viral induction of the required gene expression and, therefore, limits its clinical practicality. Thus, current research efforts concentrate on stimulating pluripotency without the use of viruses [26–28]. To date, the use of transposons [28], transfection of expression plasmids [27], and delivery of reprogramming proteins [26] has already been introduced as alternative means of inducing cell fate while avoiding use of viruses. Generation of stable iPS cells via delivery of proteins has been described for Oct4, Sox2, Klf4, and c-Myc fused to a cell-penetrating peptide (CPP), but demonstrated significantly reduced efficiency (~0.001% of input cells) compared with virus-based protocols (~0.01% of input cells) [26].

In-vitro differentiation of iPS toward CMs has been the focus of much attention in recent years. Researchers have already demonstrated similar behavior between mouse iPS-derived CMs (iPS-CMs) and CMs derived from well-established mESCs lines [29–31]. More recently, the generation of human iPS-CMs has also been described [25, 32–36] and reviewed [37–39], with claimed responses to chronotropic agents and the ability to functional syncytium in vitro. Moreover, as in the case of hESC-CMs, their response to cardiovascular drugs in vitro [33, 34] render them powerful models for in-vitro cardiac electrophysiological and drug screening studies.

While their capacity to generate CMs in vitro has been proven, reports of iPS-CMs transplantation studies will be necessary to provide further information regarding their true potential and safety in treatment of the infarcted myocardium. Clinical application of iPS-CMs will provide for genetically appropriate tissue grafts, eliminate many political and ethical obstacles, and usher in a new era of stem cell research.

Bone Marrow Stem Cells

The heterogeneous bone marrow (BM) tissue comprises a number of stem cell subpopulations, including the hematopoietic stem cells (HSCs; 0.001–0.01% of total BM cell population) and the mesenchymal stem cells (MSCs; ~0.01% of the total BM cell population). Endothelial precursor cells have also been isolated from human BM, and effectively improved heart function upon transplantation into ischemic myocardium [40]. The possibility of directing BM cell differentiation toward generation of CM pools, either in vitro or in vivo, has been intriguing researchers for many years.

Successful in-vitro induction of CMs from BM cell sources [41–47] has encouraged researchers to assess their therapeutic potential toward repair of myocardial infarction. Orlic and collaborators were the first to demonstrate that direct injection of BM stem cells into ischemic cardiac regions can lead to a rise in CM levels and enhanced heart function [48, 49], a finding subsequently substantiated by later works of other groups [50, 51]. In contrast, several groups have reported limited BM cell plasticity in vivo, suggesting paracrine activity lacking cell therapeutic value [52–54]. In these studies, BM cells were reported to differentiate into mature hematopoietic-forming blood cells, but no CMs were detected in the treated region. Such varied results may be the result of BM population heterogeny, disparities in cell harvesting techniques and yields, or the timing of cell injection. However, while the CM-generating potential of BM cells injected into the infarcted myocardium remains controversial [48, 49, 52–54], the induced improvement in cardiac function remains unequivocal.

A review of the clinical status of injected BM stem cells uncovers significantly discrepant results (see review [55] for a summary table) with regards to effectiveness of clinically applied BM stem cells [55]. More specifically, the REPAIR-AMI [56] and BOOST [57] trials reported improved left ventricular ejection fraction upon BM cells transplantation, where any differences noted between control and experimental groups became insignificant by the 18-month follow-up examination [58]. Similarly, a meta-analysis summarizing the results of 18 studies, involving a total of 999 patients undergoing BM cell transplantation in the cardiac tissue [59], concluded that such treatment leads to improved left ventricular ejection fraction, decreased infarct size, and reduced left ventricular end systolic volume. In contrast, a doubled-blinded, randomized and controlled trial involving transplantations of similar nature [60] demonstrated a reduction in infarct size, but no change in left ventricular ejection fraction. These contrasting reports can be rooted in disparities in cell-harvesting techniques, the timing of cell injection, administration methods, patient demographics, and the heterogeny of the BM population used for injection. Despite the inconclusive degree of efficacy of BM-based transplantation toward renewing cardiac function, all works demonstrate feasibility, safety, and at least partial efficacy.

Mesenchymal stem cells (MSCs), typically isolated from BM stroma and capable of differentiating to form osteoblasts, adipocytes, chondrocytes, skeletal muscle cells, [61] and others, have recently also been purified from adipose tissue as well [62]. Their isolation is significantly facilitated by their extensive adhesiveness and lack of hematopoietic markers. Works demonstrating development of CMs from MSCs stimulated in vitro [45, 63], together with the less immunogenic nature of MSCs, have initiated transplantation studies exploring the potential of MSC-derived CMs in treatment of the infarcted heart. Preclinical studies yielded improved ventricular function upon MSC transplantation to the infarcted heart [64–70], with a seemingly paracrine-dependent mechanism [64, 66, 68]. Moreover, improved left ventricular function was achieved within six months of intracoronary delivery of autologous bone marrow MSCs in 69 patients with acute myocardial infarction [71]. Several ongoing trials are evaluating administrations of allogenic or autologous MSCs, and are certain to influence MSC-based heart repair applications.

Resident Cardiac Stem Cells

Recent reports have described cardiac progenitor cell pools embedded within the heart, which feature proliferative and differentiative properties. These cardiac progenitor cells have been classified into four groups and include cardiac side population (SP) cells, c-kit+ cells, Islet-1+ cells and stem cell antigen-1 (Sca-1+) cells. It is still unclear whether these cellular groups represent distinct cell types or progressive stages of the same cell source. Cardiosphere-forming stem cells, derived from primary cultures of human biopsy patients, include about 30% subpopulation of c-kit–expressing cells, and 20% exhibiting full cardiogenic potential have been also described [72, 73]. SP cells, constituting ~1% of the adult heart [74, 75] are characterized by exclusion of metabolic dyes such as Hoechst and Rhodamine. The c-kit+ cells, making up approximately 0.01% of the entire myocyte population, express a c-Kit-specific receptor [76]. Islet-1+ cells have been successfully isolated from both infantile rat hearts [77, 78] and human adult hearts [79], with enriched c-kit+ cells and islet-1+ cell pools in the right atrium [79]. To date, Sca-1+ cells [74] have been isolated from the murine but not from the human heart. The prospect of applying cardiac progenitor cell-derived CMs toward myocardial repair has been the focus of many recent investigations. While SP cells have been shown to differentiate into CMs both in vitro and in vivo [80–82], their therapeutic potential toward the infarcted heart has not been extensively evaluated. c-kit+ cells have been described to regenerate infarcted myocardium and improve cardiac function after transplantation [79, 83]. Islet-1+ cells were shown to differentiate in vitro into CMs upon co-culture with neonatal CMs. Sca-1+ cells isolated from murine hearts differentiated into CMs both in vitro and, after cell transplantation, in vivo [84, 85]. Although very unique, resident cardiac stem cells are rare and insufficient to naturally regenerate the infarcted myocardium. Future research will require evaluation of their reliable production in sufficient quantities and their autologous potential in cell therapy of the heart.

Adult Skeletal Muscle Stem Cells

Adult skeletal muscle stem cells, residing below the basal membrane of skeletal muscle tissue, proliferate and differentiate to give rise to myoblasts that first form myotubes and then mature skeletal muscle fibers. While these cells, also termed *satellite cells*, are further committed than ESCs, they have been reported to give rise to cardiogenic, neurogenic, osteogenic, and adipogenic lineages [86, 87]. Their notable and advantageous harvestability, proliferative capacities in vitro, lower tendency to form teratomas, and minimal sensitive to ischemic conditions render them ideal candidates for cell therapy protocols. Thus, it was no surprise that skeletal muscle stem cells were among the first stem cell sources to be considered for heart repair and were the first to be put to the test in clinical trials. However, their inability to form gap junctions and to couple with host myocardium raises deep concerns for development of arrhythmias post-transplantation. In addition, these cells fail to form cardiomyocytes in vivo and remain committed to the skeletal muscle lineage [88].

Preclinical transplantation studies of skeletal muscle stem cells to the rat hearts resulted in improved ventricular contractile functioning [89, 90]. Unfortunately, these cells did not mange to couple directly with the myocardium, and the risk of ventricular arrhythmias exists. Interestingly, the transplantation of skeletal muscle stem cells overexpressing Cox-43, by gene manipulation, improved electrical coupling to the myocardium [91]. Clinical studies applying cells of such source toward cardiac repair have been thoroughly reviewed by Murry and colleagues [92], who conclude that hundred of autologous skeletal muscle stem cells can be generated after expansion in vitro and can be effectively engrafted in the scar myocardium area. Conclusions regarding functional heart improvement as well as safety parameters remain to be fully determined. The recent MAGIC clinical trial reported disappointing results regarding the benefit of autologous skeletal muscle cells in reversal of impaired heart functioning [93]. While the study demonstrated technical feasibility, it failed to determine functional efficacy and avoidance of arrhythmia [93].

Biomaterial Strategies for Cell Delivery into the Heart

Combinations of cells with biocompatible scaffolds, forming engineered tissues, can reduce cell loss and preserve cells in the desired engraftment position in vivo, compared to their administration in an aqueous solution. Moreover, formation of engineered tissue ex vivo allows for tight control of tissue-construct properties, including differentiation, maturation, and organization, which may result in improved functionality following implantation.

This section will review the most commonly employed biomaterial-based tissue engineering techniques. Polymeric, biodegradable porous scaffolds into which cells are seeded, adhere, and proliferate will be described. These scaffolds are said to gradually degrade, leaving space for further cell growth and development of the new tissue. Non-injectable and injectable hydrogels will be introduced, and their cell-entrapping qualities will be detailed. The high water and natural proteins and/or synthetic polymer content of hydrogels provide an extracellular matrix (ECM)-like environment supportive of prolonged cell survival. Scaffold-"free" tissue engineering approaches will be introduced, including the cell sheets and cell aggregation techniques. In the cell sheet approach, confluent and intact cell layers are harvested and overlaid to create 3D tissue constructs. The cell aggregation technique, also void of biomaterials, encourages cellular self-assembly within rotating shakers, eventually leading to 3D tissue patch construction. Use of natural, acellular ECM components derived from native tissues will also be described as a common scaffolding method. Lastly, protein-engineered biomaterials, composed entirely of recombinant proteins, will be described, as well as their advantageous design flexibility.

Polymeric Porous Scaffolds

Porous matrices are commonly employed as scaffolds for tissue-engineering purposes. They allow for direct cell seeding where cells can fill the micropores, proliferate, adhere to scaffold walls, and assume their shape. Particulate leaching methods using microspheres or salt grains offer regulation over scaffold microstructure, porosity, and interpore connectivity [94, 95]. Synthetic polymers, such as poly-lactic acid (PLA), poly-glycolic acid (PGA), and the PLA-PGA co-polymer [96, 97], serve as typical scaffold materials in such protocols. Polyglycerol sebacate (PGS) [98] and polycaprolactone (PCL) [94] represent an additional two commonly used scaffold reagents. Synthetic polymers allow for simple tailoring of the scaffold's mechanical, morphological, and degradative properties. In many cases, scaffolds are supplemented with natural ECM matrix components (matrigel, collagen, or fibronectin) to foster cell adhesion. Porous scaffolds can also be designed using naturally occurring proteins, such as collagen, gelatin, fibrin, and alginate [99], all of which support cell adhesion and proliferation.

In a collaborative work with Gepstein et al. [11], we have demonstrated that hES-CMs, endothelial, and embryonic fibroblast cells seeded within a PLLA-PLGA (50/50) scaffold successfully form 3D human, vascularized cardiac muscle constructs (Figure 8.1). The patch was occupied with differentiated CMs arranged in a sarcomeric pattern and exhibited synchronic beating, as well as contractions responsive to both positive and negative chronotropic agents [11]. The endothelial and embryonic fibroblast cells were shown to be responsible for formation of the intense inter-CM vascular networks observed in vitro (Figure 8.1a). Transplantation of these vascularized human cardiac constructs into rat myocardium led to intense graft vascularization and formation of functional human blood vessels [12] (Figure 8.2c, d). Implanted hES-CMs continued to thrive and mature and underwent elongation and directed alignment [12] (Figure 8.2a, b).

Hydrogels

A hydrogel is a network of polymers that are water insoluble and contains significant water content (usually more then 90%). As such, they closely imitate natural ECM environment. Hydrogels can be prepared from synthetic polymers (e.g., poly-ethylene-glycol (PEG)), or natural materials (e.g., fibrin, collagen, alginate, and chitosan), or a combination of the two [100]. Hydrogels can be polymerized in vitro, cultured for the desired time ex vivo and then sutured to the infarcted area (non-injectable hydrogels), or injected with cells into the body as a liquidous materials (injectable hydrogels) and then be polymerized in situ (via UV light, temperature, or pH differences).

Non-Injectable Hydrogels

Zimmermann et al. [101] have reported use of the non-injectable hydrogels for preparation of engineered heart tissue (EHT). Collagen-matrigel matrix was used to culture neonatal CMs or embryonic chick CMs and subsequently subjected to mechanical stretch during in-vitro cultivation. The EHT displayed myocardium-like properties, and contained interconnected, longitudinally oriented cardiac cells. A large (thickness/diameter: 1–4 mm/15 mm), force-generating EHT prepared from neonatal rat heart cells, using the same scaffolding techniques, has been recently described as undergoing electrical coupling with native myocardium upon transplantation, with no evidence of arrhythmia [102]. Moreover, the grafted tissue prevented further ventricular dilation, and induced systolic wall thickening and improved fractional area shortening of infarcted hearts. Guo et al. [103] generated cardiac tissue using mESC-derived CMs. mESC-CMs were embedded in a ring shape within a Type I collagen and matrigel matrix, and after being stretched in vitro for 7 days displayed spontaneous beating movements and response to physical and pharmaceutical stimulation in vitro. No teratomas were detected upon its subcutaneous transplantation into nude mice after 4 weeks. Simpson and associates [104] used a Type I collagen matrix embedded with human MSC to produce a cardiac patch that was transplanted to the infarcted rat heart. Cardiac remodeling of the infracted heart was achieved, where 23% of the engrafted cells were still detectable one week after implantation, but no longer detectable after 4 weeks. Shapira-Schweitzer et al. [100] recently described the use of hybrid PEG and fibrinogen hydrogel to support rat neonatal cardiac cells and hESC-CMs. The PEG-fibrinogen hydrogel polymerized after exposure to UV light and supported maturation of embedded hES-CM cells within 10–14 days in culture. The matured cells expressed cardiac-specific markers and responded to pharmacological agents.

Injectable Hydrogels

The less invasive delivery of injectable hydrogels renders them more clinically appealing than preformed tissue-constructs. These hydrogels solidify at body temperature, assume the infarcted zone geometry, and provide a matrix for cell retention, migration, proliferation, and neovascularization in vivo. The Leor and Cohen groups [105, 106] have been pioneers in the use of injectable alginate hydrogels for repair of infarcted myocardium, and have demonstrated improved rat cardiac function upon its delivery to the heart [105]. More recently, these researchers reported reversed left ventricular enlargement and increased scar thickness [106] upon intracoronary injection of alginate biomaterial into the swine heart. Kofidis and colleagues [107] successfully delivered mESCs by injectable matrigel to the infarcted mouse heart and observed enhanced heart functioning when compared with controls animals. In parallel, Lu et al. [108] considered an injectable, temperature-sensitive chitosan hydrogel for delivery of mESCs to the infarcted rat heart and reported effective cell transfer that correlated with improved cardiac functioning.

Biomaterial-"Free" Tissue Engineering

Cell Aggregation

As biomaterials can introduce undesirable and/or toxic by-products, stimulate unfavorable host responses, or interrupt critical cell–cell interactions, biomaterial-free tissue engineering techniques, employing only cells and their naturally secreted ECM for development of 3D tissues, have been attracting much attention [109]. Murry et al. have practiced this method in creating human cardiac patches of fully controllable sizes, formed by forcing hES-CM cells to aggregate within a rotating shaker [109]. Cardiac populations became increasingly enriched with mature CMs over time, and the addition of endothelial and fibroblast cells led to enhanced in vitro performance (higher force generation), and to improved rat heart engraftment outcomes, as determined by graft sizes and viability after transplantation [110].

Cell Sheets

The cell sheet biomaterial-free tissue engineering approach, first described in 2002 by Shimizu et al. [111, 112], can be implemented toward preparation of 3D cardiac tissue by harvesting confluent CM layers from culture dishes and then laying them one over another to form 3D cardiac tissue. This process utilizes temperature-sensitive cell culture dishes made of poly(N-isopropylacrylamide) (PIPAAm), which become hydrophilic and non-adhesive at reduced temperatures. Alternatively, stimulation of protease activity can induce detachment of whole cell layers. This approach to construct generation preserves critical cell-to-cell contacts, maintains expression of adhesion proteins, and stimulates secretion of extracellular matrix components natural to the tissue, but lacks the mechanical stiffness required for maintenance of physiological conditions. Cardiomyocyte sheets have been reported to synchronously and spontaneously beat when partly overlaid one over another [111]. A keynote paper published by Miyahara and colleagues [67] described the construction of a cell sheet from adipose tissue–derived MSCs. After transplantation into the infarcted rat heart, the cell sheet gradually evolved to form a thick, vascularized graft, which contained CMs and undifferentiated MSCs and stimulated improved heart functioning.

Decellularized Matrix

Tissue-engineered constructs can also be generated from decellularized cadaveric tissue to then serve as the basis for engineering whole tissue organs or segmental patches. Following treatment with detergents, only ECM and vascular network components remain, which can then be re-seeded with functional cells and cultured under physiological conditions. This approach takes advantage of natural blood supply networks, intended to enhance graft viability. Proof of concept studies have been performed by Ott and colleagues [113], who demonstrated feasibility of rat heart decellularization while preserving its natural chambers, valves, and vasculature. The acellular heart construct, re-seeded with CMs or endothelial cells and sustained in a bioreactor that provided pulsatile flow and pacing, reached ~34% recellularization and exhibited pumping function equivalent to ~2% of adult rat heart potential within 8 days in culture.

Tissue constructs formed of decellularized matrices embedded with stem cells have been evaluated by Tan et al. [114] in the context of heart repair. Decellularized small intestinal submucosa (SIS) embedded with MSCs and cultivated in vitro for 5–7 days was then transplanted into an infarcted rabbit heart and monitored for one month. Heart function was more significantly upgraded following engraftment SIS-MSC grafts, compared to acellular SIS. Moreover, MSCs migrated toward the infarcted area, where they differentiated into CMs and smooth muscle cells.

Singelyn et al. have proposed combination of decellularized and injectable biomaterial matrices for formulation of a myocardial matrix more closely mimicking the natural tissue [115]. They decellularized porcine myocardial tissue, which was then processed to form a viscous myocardial matrix gel-able at 37°C. Intense neovascularization was detected within the graft area, following its injection into the rat myocardium, with a significant increase in the number of mature blood vessels at 11 days post-transplantation. Similarly, Wei et al. combined the decellularization and cell sheet engineering approaches [116] to generate thick cardiac patches. MSC cell sheets inserted between slices of acellular bovine pericardium scaffold layers were transplanted four weeks after infraction in rats, underwent integration, and became enriched with neo-vessels and neo-muscles, where blood vessel density increased four-fold, compared to the untreated infarct group. In addition, a small fraction of transplanted MSCs expressed mature CM markers.

Valve engineering via decellularized matrices is a rapidly growing discipline, yet extends beyond the scope of this chapter.

Protein-Engineered Biomaterials

Protein-engineered biomaterials, composed of genetically engineered protein domains, provide biopolymers with exact molecular-level sequence specification [117–120] and offer significant advantages over both traditional natural and synthetic polymers. The molecular design, which outlines the specification of a chain structure, can be altered by modifying the sequence of amino acids to form new classes of engineered proteins with adjustable mechanical properties, self-assembly features, degradation profiles, and biological interactions. The sequence is then encoded into an artificial gene, and then expressed in an appropriate microbial host. In recent years, Tirrell and his group [119–123] have applied protein-engineering techniques toward generation of ECM protein domain biomaterials. Sophisticated protein-engineered hydrogels composed of intracellular peptide domains, which gel upon the mixing of two separate components, have been recently described [124]. Mechanical and biological properties of protein-engineered biomaterials determine cellular activity and tissue regeneration potential. Mechanical properties of protein-based biomaterials can be controlled using several techniques, including incorporation of elastin-like peptides, cross-linking, and manipulation of biomaterial degradability. The latter can be regulated by incorporation of amino acid sequences susceptible to specific cellular proteases. Biological considerations of protein-based materials include the density and presentation of cell-adhesive peptide domains (RGD and CS5 peptide sequences). Careful selection of cell-binding domains and their spatial density are critical to the design of successful protein-based biomaterials.

In the context of the cardiovascular system, it was shown that protein-engineered biomaterials derived from the ECM domains of CS5 and elastin are suitable for generation of small diameter vascular grafts, as they encourage endothelial cell adhesion while providing the necessary physical strength and elasticity [125]. Response of endothelial cells to RGD domain density has also been described, and has been recognized as a means of modulating cellular function [121].

The scope for using protein-engineered biomaterials in tissue engineering can be further expanded to include unnatural (non-canonical) amino acids. Incorporation of amino acid analogues into the biomaterial design can introduce new chemical functionality, providing a great deal of design versatility and creativity and expanding the potential applications of these materials. Examples include incorporation of photoactive [126], fluorinated [127], and unsaturated amino acid analogues that enable photo-patterning, enhanced protein stability, and chemical tethering, respectively.

Vascularization Strategies in Tissue Engineering

Vascularization is the process whereby new blood vessels assemble within tissues, providing the nutritive support crucial for extended tissue survival. Incorporation of functional blood vessels is therefore essential to the viability of thick and metabolically demanding tissue constructs, such as the heart [128, 129], whether sustained in vitro or in vivo. Upon implantation, pre-existing blood vessels are expected to integrate with host vasculature, enhancing graft perfusion and accelerating host neovascularization via paracrine signaling pathways. Therefore, in addition to considering appropriate selection of biomaterials and cell types for tissue construction, vascularization techniques must also be addressed [130].

To date, several vascularization approaches are available, and one or a combination of any of the following three major techniques are typically utilized:

- In-vitro multi-cellular culturing strategy
- In-vivo prevascularization protocols
- Growth factor (GF)–induced vascularization

In-vitro Multicellular Culturing Strategy

The multicellular strategy entails co-seeding of endothelial cells and vascular mural cells with cells specific to the tissue of interest within 3D scaffolds. Cells are then allowed to self-assemble to form vascular networks embedded within the engineered tissue of choice [128, 131]. Vascular mural cells provide physical support to endothelial cells and release angiogenic growth factors (GFs), which stimulate vascularization. Embryonic fibroblasts and mesenchymal precursor cells are extensively used in these protocols because of their high differentiation capacity toward mural cells. In recent years, our group has employed and calibrated the multicellular strategy for induction of vascular network formation within engineered muscle tissues [11, 131]. Engineered vascularized human cardiac tissue was successfully constructed by co-seeding hESC-CMs and endothelial and embryonic fibroblast cells onto 3D biodegradable, highly porous PLLA/PLGA polymeric scaffolds [11] (Figure 8.1). The resulting tissue exhibited synchronic beating, as detected by both visual inspection and confocal laser calcium imaging studies. Ultrastructural analysis and immunostaining for cardiac-specific proteins confirmed the presence of differentiated cardiomyocytes arranged in a typical sarcomeric pattern, T-tubules, sarcoplasmic reticulum (SR), and gap junctions between neighboring cardiomyocytes. In addition, cardiac construct contractions were responsive to both positive and negative chronotropic agents. Transplantation of the vascularized cardiac patches into rat hearts demonstrated evolution of a stable, relatively mature grafts in vivo [12] (Figure 8.2a, b). We have also demonstrated that a multicellular preparation of hESC-CMs, endothelial cells, and embryonic fibroblasts results in increased graft vascularization (Figure 8.2c, d) and in anastomosis of the pre-existing human vessels with host rat vascular networks.

In-vivo Prevascularization Protocols

In-vivo prevascularization protocols recruit the body as a bioreactor for induction of construct vascularization in vivo. Constructs are implanted into environments rich in vascular supply (intra-abdominal, subcutaneous, or intramuscular regions), where they can be invaded with newly formed vascular networks. Within a few days of implantation, host-derived blood vessels penetrate the graft, forming a stable and functional vascular network within the construct. Subsequently, the vascularized constructs are transferred into infarcted hearts to determine functionality. In a recent work pioneered by Dvir et al. [132], a cardiac patch was first engineered in vitro and then vascularized upon implantation onto the omentum, a blood vessel–enriched region. After graft vascularization and subsequent implantation into infracted hearts, the cardiac patches underwent structural and electrical coupling with the host myocardium, leading to beneficial effects on both systolic and diastolic left ventricular function. The arteriovenous loop (AV loop), another attractive in-vivo configuration model, has been employed to prevascularize cardiac patches as well. In this model, intrinsic vascularization is induced in an isolated and protected space, created by a polycarbonate chamber in which a macrovascular arteriovenous shunt loop (AVL) is enclosed [133], and has been successfully applied toward production of contractile 3D cardiac tissue [134]. For this purpose, neonatal cardiac myocytes embedded within fibrin gel were cultured in vivo in silicone chambers in proximity to the femoral artery and vein of adult rats. At 3 weeks post-transplantation, cardiac cells were found to be organized, vascularized, and functional within the chambers. In a subsequent publication, Morritt et al. successfully designed thick, vascularized cardiac tissue constructs (maximum thickness of ~2 mm) by placing an AV loop inside a semi-sealed polycarbonate chamber later implanted into the groin of a rat [135]. The chamber was seeded with cardiomyocytes in matrigel and contained differentiated, spontaneously contracting cardiomyocytes and abundant vascularization within a few weeks of implantation.

GF-Induced Vascularization

Scaffolds designed to release one or a combination of angiogenic GFs have proven an exciting strategy for vascularization induction both in vitro and in vivo. GFs, such as VEGF, PDGF, angiopoietin, and FGF, all critical to angiogenesis [136], all present a unique platform for accelerating vascularization. GFs can be incorporated into scaffolds by their simple addition to scaffold polymer solutions, or can be encapsulated in microspheres to enable sustained and controlled release. Administration of multiple GFs has been demonstrated to stimulate mature and functional blood vessel formation in vivo [137]. Richardson et al. described a unique scaffold system delivering VEGF and PDGF-BB, each with distinct release kinetics. Only their co-administration resulted in a synergistic effect in vivo, yielding mature, stable vessels covered with smooth muscle cells [137]. Peters et al. demonstrated that PLG scaffolds incorporating VEGF and seeded with human microvascular endothelial cells significantly increased the density of host blood vessels penetrating to the graft site within 7 days of implantation [138].

Conclusions

Engineering cardiac tissue represents a multidisciplinary, emerging field that promises to regenerate diseased heart tissue. These techniques comprise novel and experimental means for repairing infarcted myocardium and enhancing cardiac function. Preclinical and clinical studies have determined the feasibility and efficacy of the proposed methods yearning to enhance myocardial functioning. However, significant levels of cell death or loss following cell injection constitute one of the major obstacles facing researchers attempting to design productive engraftment techniques. In this regard, the creation of preformed cardiac construct or injectable-based biomaterial systems in tissue engineering protocols is expected to have a notable impact on cell-delivery efficiency. Biomaterials will allow for optimization of cell delivery and retention while also providing for improved tissue formation in vitro, with enhanced graft survival and functionality following implantation. Vascularization has emerged as a prerequisite for designing large tissues in vitro and for enhancing graft survival following implantation by providing a robust source of oxygen and nutrient supply as well as intracellular signaling critical to further tissue development. While various vascularization strategies have been described, the optimal method has yet to be defined. In summary, combination of the leading stem cell type, scaffolding biomaterial, and vascularization techniques will allow for effective cardiac graft construction and cell therapy.

References

1. Thomson JA, Itskovitz-Eldor J, Shapiro SS, Waknitz MA, Swiergiel JJ, Marshall VS, Jones JM. Embryonic stem cell lines derived from human blastocysts. *Science.* Nov 6 1998; 282(5391): 1145–1147.
2. Evans MJ, Kaufman MH. Establishment in culture of pluripotential cells from mouse embryos. *Nature.* Jul 9 1981; 292(5819): 154–156.

3. Martin GR. Isolation of a pluripotent cell line from early mouse embryos cultured in medium conditioned by teratocarcinoma stem cells. *Proc Natl Acad Sci USA.* Dec 1981; 78(12): 7634–7638.

4. Kehat I, Kenyagin-Karsenti D, Snir M, Segev H, Amit M, Gepstein A, Livne E, Binah O, Itskovitz-Eldor J, Gepstein L. Human embryonic stem cells can differentiate into myocytes with structural and functional properties of cardiomyocytes. *J Clin Invest.* Aug 2001; 108(3): 407–414.

5. Xu C, Police S, Rao N, Carpenter MK. Characterization and enrichment of cardiomyocytes derived from human embryonic stem cells. *Circ Res.* Sep 20 2002; 91(6): 501–508.

6. Mummery C, Ward-van Oostwaard D, Doevendans P, Spijker R, van den Brink S, Hassink R, van der Heyden M, et al. Differentiation of human embryonic stem cells to cardiomyocytes: Role of coculture with visceral endoderm-like cells. *Circulation.* Jun 3 2003; 107(21): 2733–2740.

7. Caspi O, Itzhaki I, Arbel G, Kehat I, Gepstien A, Huber I, Satin J, Gepstein L. In vitro electrophysiological drug testing using human embryonic stem cell derived cardiomyocytes. *Stem Cells Dev.* May 29 2008.

8. Kehat I, Gepstein A, Spira A, Itskovitz-Eldor J, Gepstein L. High-resolution electrophysiological assessment of human embryonic stem cell-derived cardiomyocytes: A novel in vitro model for the study of conduction. *Circ Res.* Oct 18 2002; 91(8): 659–661.

9. Snir M, Kehat I, Gepstein A, Coleman R, Itskovitz-Eldor J, Livne E, Gepstein L. Assessment of the ultrastructural and proliferative properties of human embryonic stem cell-derived cardiomyocytes. *Am J Physiol Heart Circ Physiol.* Dec 2003; 285(6): H2355–2363.

10. McDevitt TC, Laflamme MA, Murry CE. Proliferation of cardiomyocytes derived from human embryonic stem cells is mediated via the IGF/PI 3-kinase/ Akt signaling pathway. *J Mol Cell Cardiol.* Dec 2005; 39(6): 865–873.

11. Caspi O, Lesman A, Basevitch Y, Gepstein A, Arbel G, Habib IH, Gepstein L, Levenberg S. Tissue engineering of vascularized cardiac muscle from human embryonic stem cells. *Circ Res.* Feb 2 2007; 100(2): 263–272.

12. Lesman A, Habib M, Caspi O, Gepstein A, Arbel G, Levenberg S, Gepstein L. Transplantation of a tissue-engineered human vascularized cardiac muscle. *Tissue Eng Part A.* Jan; 16(1): 115–125.

13. Caspi O, Huber I, Kehat I, Habib M, Arbel G, Gepstein A, Yankelson L, Aronson D, Beyar R, Gepstein L. Transplantation of human embryonic stem cell-derived cardiomyocytes improves myocardial performance in infarcted rat hearts. *J Am Coll Cardiol.* Nov 6 2007; 50(19): 1884–1893.

14. Laflamme MA, Chen KY, Naumova AV, Muskheli V, Fugate JA, Dupras SK, Reinecke H, et al. Cardiomyocytes derived from human embryonic stem cells in pro-survival factors enhance function of infarcted rat hearts. *Nat Biotechnol.* Sep 2007; 25(9): 1015–1024.

15. Leor J, Gerecht S, Cohen S, Miller L, Holbova R, Ziskind A, Shachar M, Feinberg MS, Guetta E, Itskovitz-Eldor J. Human embryonic stem cell transplantation to repair the infarcted myocardium. *Heart.* Oct 2007; 93(10): 1278–1284.

16. Tomescot A, Leschik J, Bellamy V, Dubois G, Messas E, Bruneval P, Desnos M et al. Differentiation in vivo of cardiac committed human embryonic stem cells in postmyocardial infarcted rats. *Stem Cells.* Sep 2007; 25(9): 2200–2205.

17. van Laake LW, Passier R, Monshouwer-Kloots J, Verkleij AJ, Lips DJ, Freund C, den Ouden K, et al. Human embryonic stem cell-derived cardiomyocytes survive and mature in the mouse heart and transiently improve function after myocardial infarction. *Stem Cell Res.* Oct 2007; 1(1): 9–24.

18. Takahashi K, Yamanaka S. Induction of pluripotent stem cells from mouse embryonic and adult fibroblast cultures by defined factors. *Cell.* Aug 25 2006; 126(4): 663–676.

19. Takahashi K, Tanabe K, Ohnuki M, Narita M, Ichisaka T, Tomoda K, Yamanaka S. Induction of pluripotent stem cells from adult human fibroblasts by defined factors. *Cell.* Nov 30 2007; 131(5): 861–872.

20. Yu J, Vodyanik MA, Smuga-Otto K, Antosiewicz-Bourget J, Frane JL, Tian S, Nie J, et al. Induced pluripotent stem cell lines derived from human somatic cells. *Science.* Dec 21 2007; 318(5858): 1917–1920.

21. Aasen T, Raya A, Barrero MJ, Garreta E, Consiglio A, Gonzalez F, Vassena R, et al. Efficient and rapid generation of induced pluripotent stem cells from human keratinocytes. *Nat Biotechnol.* Nov 2008; 26(11): 1276–1284.

22. Giorgetti A, Montserrat N, Rodriguez-Piza I, Azqueta C, Veiga A, Izpisua Belmonte JC. Generation of induced pluripotent stem cells from human cord blood cells with only two factors: Oct4 and Sox2. *Nat Protoc* 5(4): 811–820.

23. Park IH, Zhao R, West JA, Yabuuchi A, Huo H, Ince TA, Lerou PH, Lensch MW, Daley GQ. Reprogramming of human somatic cells to pluripotency with defined factors. *Nature.* Jan 10 2008; 451(7175): 141–146.

24. Yamanaka S. A fresh look at iPS cells. *Cell.* Apr 3 2009; 137(1): 13–17.

25. Haase A, Olmer R, Schwanke K, Wunderlich S, Merkert S, Hess C, Zweigerdt R, et al. Generation of induced pluripotent stem cells from human cord blood. *Cell Stem Cell.* Oct 2 2009; 5(4): 434–441.

26. Kim D, Kim CH, Moon JI, Chung YG, Chang MY, Han BS, Ko S, Yang E, Cha KY, Lanza R, Kim KS. Generation of human induced pluripotent stem cells by direct delivery of reprogramming proteins. *Cell Stem Cell.* Jun 5 2009; 4(6): 472–476.

27. Okita K, Nakagawa M, Hyenjong H, Ichisaka T, Yamanaka S. Generation of mouse induced pluripotent stem cells without viral vectors. *Science.* Nov 7 2008; 322(5903): 949–953.

28. Yusa K, Rad R, Takeda J, Bradley A. Generation of transgene-free induced pluripotent mouse stem cells by the piggyBac transposon. *Nat Methods.* May 2009; 6(5): 363–369.

29. Kuzmenkin A, Liang H, Xu G, Pfannkuche K, Eichhorn H, Fatima A, Luo H, Saric T, Wernig M, Jaenisch R, Hescheler J. Functional characterization of cardiomyocytes derived from murine induced pluripotent stem cells in vitro. *Faseb J.* Dec 2009; 23(12): 4168–4180.

30. Narazaki G, Uosaki H, Teranishi M, Okita K, Kim B, Matsuoka S, Yamanaka S, Yamashita JK. Directed and systematic differentiation of cardiovascular cells from mouse induced pluripotent stem cells. *Circulation.* Jul 29 2008; 118(5): 498–506.

31. Schenke-Layland K, Rhodes KE, Angelis E, Butylkova Y, Heydarkhan-Hagvall S, Gekas C, Zhang R, Goldhaber JI, Mikkola HK, Plath K, MacLellan WR. Reprogrammed mouse fibroblasts differentiate into cells of the cardiovascular and hematopoietic lineages. *Stem Cells.* Jun 2008; 26(6): 1537–1546.

32. Gai H, Leung EL, Costantino PD, Aguila JR, Nguyen DM, Fink LM, Ward DC, Ma Y. Generation and characterization of functional cardiomyocytes using induced pluripotent stem cells derived from human fibroblasts. *Cell Biol Int.* Nov 2009; 33(11): 1184–1193.

33. Tanaka T, Tohyama S, Murata M, Nomura F, Kaneko T, Chen H, Hattori F, et al. In vitro pharmacologic testing using human induced pluripotent stem cell-derived cardiomyocytes. *Biochem Biophys Res Commun.* Aug 7 2009; 385(4): 497–502.

34. Yokoo N, Baba S, Kaichi S, Niwa A, Mima T, Doi H, Yamanaka S, Nakahata T, Heike T. The effects of cardioactive drugs on cardiomyocytes derived from human induced pluripotent stem cells. *Biochem Biophys Res Commun.* Sep 25 2009; 387(3): 482–488.

35. Zhang J, Wilson GF, Soerens AG, Koonce CH, Yu J, Palecek SP, Thomson JA, Kamp TJ. Functional cardiomyocytes derived from human induced pluripotent stem cells. *Circ Res.* Feb 27 2009; 104(4): e30–41.

36. Zwi L, Caspi O, Arbel G, Huber I, Gepstein A, Park IH, Gepstein L. Cardiomyocyte differentiation of human induced pluripotent stem cells. *Circulation.* Oct 13 2009; 120(15): 1513–1523.

37. Freund C, Mummery CL. Prospects for pluripotent stem cell-derived cardio-myocytes in cardiac cell therapy and as disease models. *J Cell Biochem.* Jul 1 2009; 107(4): 592–599.

38. Shiba Y, Hauch KD, Laflamme MA. Cardiac applications for human pluripotent stem cells. *Curr Pharm Des.* 2009; 15(24): 2791–2806.

39. Yuasa S, Fukuda K. Recent advances in cardiovascular regenerative medicine: The induced pluripotent stem cell era. *Expert Rev Cardiovasc Ther.* Jul 2008; 6(6): 803–810.

40. Kocher AA, Schuster MD, Szabolcs MJ, Takuma S, Burkhoff D, Wang J, Homma S, Edwards NM, Itescu S. Neovascularization of ischemic myocardium by human bone-marrow-derived angioblasts prevents cardiomyocyte apoptosis, reduces remodeling and improves cardiac function. *Nat Med.* Apr 2001; 7(4): 430–436.

41. Belema Bedada F, Technau A, Ebelt H, Schulze M, Braun T. Activation of myogenic differentiation pathways in adult bone marrow-derived stem cells. *Mol Cell Biol.* Nov 2005; 25(21): 9509–9519.

42. Flaherty MP, Abdel-Latif A, Li Q, Hunt G, Ranjan S, Ou Q, Tang XL, Johnson RK, Bolli R, Dawn B. Noncanonical Wnt11 signaling is sufficient to induce cardiomyogenic differentiation in unfractionated bone marrow mononuclear cells. *Circulation.* Apr 29 2008; 117(17): 2241–2252.

43. Koninckx R, Hensen K, Daniels A, Moreels M, Lambrichts I, Jongen H, Clijsters C, et al. Human bone marrow stem cells co-cultured with neonatal rat cardiomyocytes display limited cardiomyogenic plasticity. *Cytotherapy.* 2009; 11(6): 778–792.

44. Koyanagi M, Bushoven P, Iwasaki M, Urbich C, Zeiher AM, Dimmeler S. Notch signaling contributes to the expression of cardiac markers in human circulating progenitor cells. *Circ Res.* Nov 26 2007; 101(11): 1139–1145.

45. Makino S, Fukuda K, Miyoshi S, Konishi F, Kodama H, Pan J, Sano M, et al. Cardiomyocytes can be generated from marrow stromal cells in vitro. *J Clin Invest.* Mar 1999; 103(5): 697–705.

46. Wang Y, Feng C, Xue J, Sun A, Li J, Wu J. Adenovirus-mediated hypoxia-inducible factor 1alpha double-mutant promotes differentiation of bone marrow stem cells to cardiomyocytes. *J Physiol Sci.* Nov 2009; 59(6): 413–420.

47. Yoon J, Choi SC, Park CY, Choi JH, Kim YI, Shim WJ, Lim DS. Bone marrow-derived side population cells are capable of functional cardiomyogenic differentiation. *Mol Cells.* Apr 30 2008; 25(2): 216–223.

48. Orlic D, Kajstura J, Chimenti S, Jakoniuk I, Anderson SM, Li B, Pickel J, McKay R, Nadal-Ginard B, Bodine DM, Leri A, Anversa P. Bone marrow cells regenerate infarcted myocardium. *Nature.* Apr 5 2001; 410(6829): 701–705.

49. Orlic D, Kajstura J, Chimenti S, Limana F, Jakoniuk I, Quaini F, Nadal-Ginard B, Bodine DM, Leri A, Anversa P. Mobilized bone marrow cells repair the infarcted heart, improving function and survival. *Proc Natl Acad Sci USA.* Aug 28 2001; 98(18): 10344–10349.

50. Kajstura J, Rota M, Whang B, Cascapera S, Hosoda T, Bearzi C, Nurzynska D, et al. Bone marrow cells differentiate in cardiac cell lineages after infarction independently of cell fusion. *Circ Res.* Jan 7 2005; 96(1): 127–137.

51. Rota M, Kajstura J, Hosoda T, Bearzi C, Vitale S, Esposito G, Iaffaldano G, et al. Bone marrow cells adopt the cardiomyogenic fate in vivo. *Proc Natl Acad Sci USA.* Nov 6 2007; 104(45): 17783–17788.

52. Balsam LB, Wagers AJ, Christensen JL, Kofidis T, Weissman IL, Robbins RC. Haematopoietic stem cells adopt mature haematopoietic fates in ischaemic myocardium. *Nature.* Apr 8 2004; 428(6983): 668–673.

53. Murry CE, Soonpaa MH, Reinecke H, Nakajima H, Nakajima HO, Rubart M, Pasumarthi KB, et al. Haematopoietic stem cells do not transdifferentiate into cardiac myocytes in myocardial infarcts. *Nature.* Apr 8 2004; 428(6983): 664–668.

54. Nygren JM, Jovinge S, Breitbach M, Sawen P, Roll W, Hescheler J, Taneera J, Fleischmann BK, Jacobsen SE. Bone marrow-derived hematopoietic cells generate cardiomyocytes at a low frequency through cell fusion, but not transdifferentiation. *Nat Med.* May 2004; 10(5): 494–501.

55. Wei HM, Wong P, Hsu LF, Shim W. Human bone marrow-derived adult stem cells for post-myocardial infarction cardiac repair: Current status and future directions. *Singapore Med J.* Oct 2009; 50(10): 935–942.

56. Schachinger V, Erbs S, Elsasser A, Haberbosch W, Hambrecht R, Holschermann H, Yu J, et al. Improved clinical outcome after intracoronary administration of bone-marrow-derived progenitor cells in acute myocardial infarction: Final 1-year results of the REPAIR-AMI trial. *Eur Heart J.* Dec 2006; 27(23): 2775–2783.

57. Wollert KC, Meyer GP, Lotz J, Ringes-Lichtenberg S, Lippolt P, Breidenbach C, Fichtner S, et al. Intracoronary autologous bone-marrow cell transfer after myocardial infarction: The BOOST randomised controlled clinical trial. *Lancet.* Jul 10–16 2004; 364(9429): 141–148.

58. Meyer GP, Wollert KC, Lotz J, Steffens J, Lippolt P, Fichtner S, Hecker H, et al. Intracoronary bone marrow cell transfer after myocardial infarction: Eighteen months' follow-up data from the randomized, controlled BOOST (BOne marrOw transfer to enhance ST-elevation infarct regeneration) trial. *Circulation.* Mar 14 2006; 113(10): 1287–1294.

59. Abdel-Latif A, Bolli R, Tleyjeh IM, Montori VM, Perin EC, Hornung CA, Zuba-Surma EK, Al-Mallah M, Dawn B. Adult bone marrow-derived cells for cardiac repair: A systematic review and meta-analysis. *Arch Intern Med.* May 28 2007; 167(10): 989–997.

60. Janssens S, Dubois C, Bogaert J, Theunissen K, Deroose C, Desmet W, Kalantzi M, et al. Autologous bone marrow-derived stem-cell transfer in patients with ST-segment elevation myocardial infarction: Double-blind, randomised controlled trial. *Lancet.* Jan 14 2006; 367(9505): 113–121.

61. Pittenger MF, Mackay AM, Beck SC, Jaiswal RK, Douglas R, Mosca JD, Moorman MA, Simonetti DW, Craig S, Marshak DR. Multilineage potential of adult human mesenchymal stem cells. *Science.* Apr 2 1999; 284(5411): 143–147.

62. Zuk PA, Zhu M, Ashjian P, De Ugarte DA, Huang JI, Mizuno H, Alfonso ZC, Fraser JK, Benhaim P, Hedrick MH. Human adipose tissue is a source of multipotent stem cells. *Mol Biol Cell.* Dec 2002; 13(12): 4279–4295.

63. Shiota M, Heike T, Haruyama M, Baba S, Tsuchiya A, Fujino H, Kobayashi H, Kato T, Umeda K, Yoshimoto M, Nakahata T. Isolation and characterization of bone marrow-derived mesenchymal progenitor cells with myogenic and neuronal properties. *Exp Cell Res.* Mar 10 2007; 313(5): 1008–1023.

64. Dai W, Hale SL, Kloner RA. Role of a paracrine action of mesenchymal stem cells in the improvement of left ventricular function after coronary artery occlusion in rats. *Regen Med.* Jan 2007; 2(1): 63–68.

65. Dai W, Hale SL, Martin BJ, Kuang JQ, Dow JS, Wold LE, Kloner RA. Allogeneic mesenchymal stem cell transplantation in postinfarcted rat myocardium: Short- and long-term effects. *Circulation.* Jul 12 2005; 112(2): 214–223.

66. Gnecchi M, He H, Noiseux N, Liang OD, Zhang L, Morello F, Mu H, Melo LG, Pratt RE, Ingwall JS, Dzau VJ. Evidence supporting paracrine hypothesis for Akt-modified mesenchymal stem cell-mediated cardiac protection and functional improvement. *Faseb J.* Apr 2006; 20(6): 661–669.

67. Miyahara Y, Nagaya N, Kataoka M, Yanagawa B, Tanaka K, Hao H, Ishino K, et al. Monolayered mesenchymal stem cells repair scarred myocardium after myocardial infarction. *Nat Med.* Apr 2006; 12(4): 459–465.

68. Noiseux N, Gnecchi M, Lopez-Ilasaca M, Zhang L, Solomon SD, Deb A, Dzau VJ, Pratt RE. Mesenchymal stem cells overexpressing Akt dramatically repair infarcted myocardium and improve cardiac function despite infrequent cellular fusion or differentiation. *Mol Ther.* Dec 2006; 14(6): 840–850.

69. Shake JG, Gruber PJ, Baumgartner WA, Senechal G, Meyers J, Redmond JM, Pittenger MF, Martin BJ. Mesenchymal stem cell implantation in a swine myocardial infarct model: Engraftment and functional effects. *Ann Thorac Surg.* Jun 2002; 73(6): 1919–1925; discussion 1926.

70. Toma C, Pittenger MF, Cahill KS, Byrne BJ, Kessler PD. Human mesenchymal stem cells differentiate to a cardiomyocyte phenotype in the adult murine heart. *Circulation.* Jan 1 2002; 105(1): 93–98.

71. Chen SL, Fang WW, Ye F, Liu YH, Qian J, Shan SJ, Zhang JJ, Chunhua RZ, Liao LM, Lin S, Sun JP. Effect on left ventricular function of intracoronary transplantation of autologous bone marrow mesenchymal stem cell in patients with acute myocardial infarction. *Am J Cardiol.* Jul 1 2004; 94(1): 92–95.

72. Messina E, De Angelis L, Frati G, Morrone S, Chimenti S, Fiordaliso F, Salio M, et al. Isolation and expansion of adult cardiac stem cells from human and murine heart. *Circ Res.* Oct 29 2004; 95(9): 911–921.

73. Smith RR, Barile L, Cho HC, Leppo MK, Hare JM, Messina E, Giacomello A, Abraham MR, Marban E. Regenerative potential of cardiosphere-derived cells expanded from percutaneous endomyocardial biopsy specimens. *Circulation.* Feb 20 2007; 115(7): 896–908.

74. Hierlihy AM, Seale P, Lobe CG, Rudnicki MA, Megeney LA. The post-natal heart contains a myocardial stem cell population. *FEBS Lett.* Oct 23 2002; 530(1–3): 239–243.

75. Martin CM, Meeson AP, Robertson SM, Hawke TJ, Richardson JA, Bates S, Goetsch SC, Gallardo TD, Garry DJ. Persistent expression of the ATP-binding cassette transporter, Abcg2, identifies cardiac SP cells in the developing and adult heart. *Dev Biol.* Jan 1 2004; 265(1): 262–275.

76. Beltrami AP, Barlucchi L, Torella D, Baker M, Limana F, Chimenti S, Kasahara H, et al. Adult cardiac stem cells are multipotent and support myocardial regeneration. *Cell.* Sep 19 2003; 114(6): 763–776.

77. Laugwitz KL, Moretti A, Lam J, Gruber P, Chen Y, Woodard S, Lin LZ, et al. Postnatal isl1+ cardioblasts enter fully differentiated cardiomyocyte lineages. *Nature.* Feb 10 2005; 433(7026): 647–653.

78. Moretti A, Caron L, Nakano A, Lam JT, Bernshausen A, Chen Y, Qyang Y, et al. Multipotent embryonic isl1+ progenitor cells lead to cardiac, smooth muscle, and endothelial cell diversification. *Cell.* Dec 15 2006; 127(6): 1151–1165.

79. Itzhaki-Alfia A, Leor J, Raanani E, Sternik L, Spiegelstein D, Netser S, Holbova R, Pevsner-Fischer M, Lavee J, Barbash IM. Patient characteristics and cell source determine the number of isolated human cardiac progenitor cells. *Circulation.* Dec 22 2009; 120(25): 2559–2566.

80. Liang SX, Tan TY, Gaudry L, Chong B. Differentiation and migration of Sca1+/ CD31- cardiac side population cells in a murine myocardial ischemic model. *Int J Cardiol.* Jan 7; 138(1): 40–49.

81. Oyama T, Nagai T, Wada H, Naito AT, Matsuura K, Iwanaga K, Takahashi T, et al. Cardiac side population cells have a potential to migrate and differentiate into cardiomyocytes in vitro and in vivo. *J Cell Biol.* Jan 29 2007; 176(3): 329–341.

82. Pfister O, Mouquet F, Jain M, Summer R, Helmes M, Fine A, Colucci WS, Liao R. CD31− but Not CD31+ cardiac side population cells exhibit functional cardiomyogenic differentiation. *Circ Res.* Jul 8 2005; 97(1): 52–61.

83. Rota M, Padin-Iruegas ME, Misao Y, De Angelis A, Maestroni S, Ferreira-Martins J, Fiumana E, et al. Local activation or implantation of cardiac progenitor cells rescues scarred infarcted myocardium improving cardiac function. *Circ Res.* Jul 3 2008; 103(1): 107–116.

84. Matsuura K, Nagai T, Nishigaki N, Oyama T, Nishi J, Wada H, Sano M, et al. Adult cardiac Sca-1-positive cells differentiate into beating cardiomyocytes. *J Biol Chem.* Mar 19 2004; 279(12): 11384–11391.

85. Oh H, Bradfute SB, Gallardo TD, Nakamura T, Gaussin V, Mishina Y, Pocius J, et al. Cardiac progenitor cells from adult myocardium: Homing, differentiation, and fusion after infarction. *Proc Natl Acad Sci USA.* Oct 14 2003; 100(21): 12313–12318.

86. Arsic N, Mamaeva D, Lamb NJ, Fernandez A. Muscle-derived stem cells isolated as non-adherent population give rise to cardiac, skeletal muscle and neural lineages. *Exp Cell Res.* Apr 1 2008; 314(6): 1266–1280.

87. Asakura A, Komaki M, Rudnicki M. Muscle satellite cells are multipotential stem cells that exhibit myogenic, osteogenic, and adipogenic differentiation. *Differentiation.* Oct 2001; 68(4–5): 245–253.

88. Reinecke H, Poppa V, Murry CE. Skeletal muscle stem cells do not transdifferentiate into cardiomyocytes after cardiac grafting. *J Mol Cell Cardiol.* Feb 2002; 34(2): 241–249.

89. Murry CE, Wiseman RW, Schwartz SM, Hauschka SD. Skeletal myoblast transplantation for repair of myocardial necrosis. *J Clin Invest*. Dec 1 1996; 98(11): 2512–2523.

90. Taylor DA, Atkins BZ, Hungspreugs P, Jones TR, Reedy MC, Hutcheson KA, Glower DD, Kraus WE. Regenerating functional myocardium: Improved performance after skeletal myoblast transplantation. *Nat Med*. Aug 1998; 4(8): 929–933.

91. Roell W, Lewalter T, Sasse P, Tallini YN, Choi BR, Breitbach M, Doran R, et al. Engraftment of connexin 43-expressing cells prevents post-infarct arrhythmia. *Nature*. Dec 6 2007; 450(7171): 819–824.

92. Murry CE, Field LJ, Menasche P. Cell-based cardiac repair: Reflections at the 10-year point. *Circulation*. Nov 15 2005; 112(20): 3174–3183.

93. Menasche P, Alfieri O, Janssens S, McKenna W, Reichenspurner H, Trinquart L, Vilquin JT, et al. The myoblast autologous grafting in ischemic cardiomyopathy (MAGIC) trial: First randomized placebo-controlled study of myoblast transplantation. *Circulation*. Mar 4 2008; 117(9): 1189–1200.

94. Hollister SJ. Porous scaffold design for tissue engineering. *Nat Mater*. Jul 2005; 4(7): 518–524.

95. Mikos AG, Temenoff JS. Formation of highly porous biodegradable scaffolds for tissue engineering. *Journal of Biotechnology ISSN*. August 15 2000; 3: 114–119.

96. Levenberg S, Huang NF, Lavik E, Rogers AB, Itskovitz-Eldor J, Langer R. Differentiation of human embryonic stem cells on three-dimensional polymer scaffolds. *Proc Natl Acad Sci USA*. Oct 28 2003; 100(22): 12741–12746.

97. Levy-Mishali M, Zoldan J, Levenberg S. Effect of scaffold stiffness on myoblast differentiation. *Tissue Eng Part A*. Apr 2009; 15(4): 935–944.

98. Gao J, Crapo PM, Wang Y. Macroporous elastomeric scaffolds with extensive micropores for soft tissue engineering. *Tissue Eng*. Apr 2006; 12(4): 917–925.

99. Leor J, Aboulafia-Etzion S, Dar A, Shapiro L, Barbash IM, Battler A, Granot Y, Cohen S. Bioengineered cardiac grafts: A new approach to repair the infarcted myocardium? *Circulation*. Nov 7 2000; 102(19 Suppl 3): III56–61.

100. Shapira-Schweitzer K, Habib M, Gepstein L, Seliktar D. A photopolymerizable hydrogel for 3-D culture of human embryonic stem cell-derived cardiomyocytes and rat neonatal cardiac cells. *J Mol Cell Cardiol*. Feb 2009; 46(2): 213–224.

101. Zimmermann WH, Schneiderbanger K, Schubert P, Didie M, Munzel F, Heubach JF, Kostin S, Neuhuber WL, Eschenhagen T. Tissue engineering of a differentiated cardiac muscle construct. *Circ Res*. Feb 8 2002; 90(2): 223–230.

102. Zimmermann WH, Melnychenko I, Wasmeier G, Didie M, Naito H, Nixdorff U, Hess A, et al. Engineered heart tissue grafts improve systolic and diastolic function in infarcted rat hearts. *Nat Med*. Apr 2006; 12(4): 452–458.

103. Guo XM, Zhao YS, Chang HX, Wang CY, Ling-Ling E, Zhang XA, Duan CM, Dong LZ, Jiang H, Li J, Song Y, Yang XJ. Creation of engineered cardiac tissue in vitro from mouse embryonic stem cells. *Circulation*. May 9 2006; 113(18): 2229–2237.

104. Simpson D, Liu H, Fan TH, Nerem R, Dudley SC, Jr. A tissue engineering approach to progenitor cell delivery results in significant cell engraftment and improved myocardial remodeling. *Stem Cells*. Sep 2007; 25(9): 2350–2357.

105. Landa N, Miller L, Feinberg MS, Holbova R, Shachar M, Freeman I, Cohen S, Leor J. Effect of injectable alginate implant on cardiac remodeling and function after recent and old infarcts in rat. *Circulation*. Mar 18 2008; 117(11): 1388–1396.

106. Leor J, Tuvia S, Guetta V, Manczur F, Castel D, Willenz U, Petnehazy O, Landa N, et al. Intracoronary injection of in situ forming alginate hydrogel reverses left ventricular remodeling after myocardial infarction in swine. *J Am Coll Cardiol.* Sep 8 2009; 54(11): 1014–1023.
107. Kofidis T, Lebl DR, Martinez EC, Hoyt G, Tanaka M, Robbins RC. Novel inject-able bioartificial tissue facilitates targeted, less invasive, large-scale tissue res-toration on the beating heart after myocardial injury. *Circulation.* Aug 30 2005; 112(9 Suppl): I173–177.
108. Lu WN, Lu SH, Wang HB, Li DX, Duan CM, Liu ZQ, Hao T, et al. Functional improvement of infarcted heart by co-injection of embryonic stem cells with temperature-responsive chitosan hydrogel. *Tissue Eng Part A.* Jun 2009; 15(6): 1437–1447.
109. Stevens KR, Pabon L, Muskheli V, Murry CE. Scaffold-free human cardiac tis-sue patch created from embryonic stem cells. *Tissue Eng Part A.* Jun 2009; 15(6): 1211–1222.
110. Stevens KR, Kreutziger KL, Dupras SK, Korte FS, Regnier M, Muskheli V, Nourse MB, Bendixen K, Reinecke H, Murry CE. Physiological function and transplantation of scaffold-free and vascularized human cardiac muscle tissue. *Proc Natl Acad Sci USA.* Sep 29 2009; 106(39): 16568–16573.
111. Shimizu T, Yamato M, Isoi Y, Akutsu T, Setomaru T, Abe K, Kikuchi A, Umezu M, Okano T. Fabrication of pulsatile cardiac tissue grafts using a novel 3-dimen-sional cell sheet manipulation technique and temperature-responsive cell cul-ture surfaces. *Circ Res.* Feb 22 2002; 90(3): e40.
112. Shimizu T, Yamato M, Kikuchi A, Okano T. Cell sheet engineering for myocar-dial tissue reconstruction. *Biomaterials.* Jun 2003; 24(13): 2309–2316.
113. Ott HC, Matthiesen TS, Goh SK, Black LD, Kren SM, Netoff TI, Taylor DA. Perfusion-decellularized matrix: Using natures platform to engineer a bioartifi-cial heart. *Nat Med.* Feb 2008; 14(2): 213–221.
114. Tan MY, Zhi W, Wei RQ, Huang YC, Zhou KP, Tan B, Deng L, et al. Repair of infarcted myocardium using mesenchymal stem cell seeded small intestinal submucosa in rabbits. *Biomaterials.* Jul 2009; 30(19): 3234–3240.
115. Singelyn JM, DeQuach JA, Seif-Naraghi SB, Littlefield RB, Schup-Magoffin PJ, Christman KL. Naturally derived myocardial matrix as an injectable scaffold for cardiac tissue engineering. *Biomaterials.* Oct 2009; 30(29): 5409–5416.
116. Wei HJ, Chen CH, Lee WY, Chiu I, Hwang SM, Lin WW, Huang CC, Yeh YC, Chang Y, Sung HW. Bioengineered cardiac patch constructed from multilayered mesenchymal stem cells for myocardial repair. *Biomaterials.* Sep 2008; 29(26): 3547–3556.
117. Romano NH, Sengupta D, Chung C, Heilshorn SC. Protein-engineered bio-materials: Nanoscale mimics of the extracellular matrix. *Biochim Biophys Acta.* Jul 18.
118. Sengupta D, Heilshorn SC. Protein-engineered biomaterials: Highly tunable tis-sue engineering scaffolds. *Tissue Eng Part B Rev.* Jun; 16(3): 285–293.
119. Maskarinec SA, Tirrell DA. Protein engineering approaches to biomaterials design. *Curr Opin Biotechnol.* Aug 2005; 16(4): 422–426.
120. van Hest JC, Tirrell DA. Protein-based materials, toward a new level of struc-tural control. *Chem Commun (Camb).* Oct 7 2001(19): 1897–1904.
121. Liu JC, Tirrell DA. Cell response to RGD density in cross-linked artificial extra-cellular matrix protein films. *Biomacromolecules.* Nov 2008; 9(11): 2984–2988.

122. Nowatzki PJ, Tirrell DA. Physical properties of artificial extracellular matrix protein films prepared by isocyanate crosslinking. *Biomaterials.* Mar–Apr 2004; 25(7–8): 1261–1267.
123. Welsh ER, Tirrell DA. Engineering the extracellular matrix: A novel approach to polymeric biomaterials. I. Control of the physical properties of artificial protein matrices designed to support adhesion of vascular endothelial cells. *Biomacromolecules.* Spring 2000; 1(1): 23–30.
124. Wong Po Foo CT, Lee JS, Mulyasasmita W, Parisi-Amon A, Heilshorn SC. Two-component protein-engineered physical hydrogels for cell encapsulation. *Proc Natl Acad Sci USA.* Dec 29 2009; 106(52): 22067–22072.
125. Heilshorn SC, DiZio KA, Welsh ER, Tirrell DA. Endothelial cell adhesion to the fibronectin CS5 domain in artificial extracellular matrix proteins. *Biomaterials.* Oct 2003; 24(23): 4245–4252.
126. Chin JW, Martin AB, King DS, Wang L, Schultz PG. Addition of a photocross-linking amino acid to the genetic code of *Escherichia coli. Proc Natl Acad Sci USA.* Aug 20 2002; 99(17): 11020–11024.
127. Yoder NC, Kumar K. Fluorinated amino acids in protein design and engineering. *Chem Soc Rev.* Nov 2002; 31(6): 335–341.
128. Jain RK, Au P, Tam J, Duda DG, Fukumura D. Engineering vascularized tissue. *Nat Biotechnol.* Jul 2005; 23(7): 821–823.
129. Nomi M, Atala A, Coppi PD, Soker S. Principals of neovascularization for tissue engineering. *Mol Aspects Med.* Dec 2002; 23(6): 463–483.
130. Kaully T, Kaufman-Francis K, Lesman A, Levenberg S. Vascularization: The conduit to viable engineered tissues. *Tissue Eng Part B Rev.* Mar 20 2009.
131. Levenberg S, Rouwkema J, Macdonald M, Garfein ES, Kohane DS, Darland DC, Marini R, van Blitterswijk CA, Mulligan RC, D'Amore PA, Langer R. Engineering vascularized skeletal muscle tissue. *Nat Biotechnol.* Jul 2005; 23(7): 879–884.
132. Dvir T, Kedem A, Ruvinov E, Levy O, Freeman I, Landa N, Holbova R, Feinberg MS, et al. Prevascularization of cardiac patch on the omentum improves its therapeutic outcome. *Proc Natl Acad Sci U S A.* Sep 1 2009; 106(35): 14990–14995.
133. Polykandriotis E, Arkudas A, Horch RE, Sturzl M, Kneser U. Autonomously vascularized cellular constructs in tissue engineering: Opening a new perspective for biomedical science. *J Cell Mol Med.* Jan–Feb 2007; 11(1): 6–20.
134. Birla RK, Borschel GH, Dennis RG, Brown DL. Myocardial engineering in vivo: Formation and characterization of contractile, vascularized three-dimensional cardiac tissue. *Tissue Eng.* May–Jun 2005; 11(5–6): 803–813.
135. Morritt AN, Bortolotto SK, Dilley RJ, Han X, Kompa AR, McCombe D, Wright CE, Itescu S, Angus JA, Morrison WA. Cardiac tissue engineering in an in vivo vascularized chamber. *Circulation.* Jan 23 2007; 115(3): 353–360.
136. Jain RK. Molecular regulation of vessel maturation. *Nat Med.* Jun 2003; 9(6): 685–693.
137. Richardson TP, Peters MC, Ennett AB, Mooney DJ. Polymeric system for dual growth factor delivery. *Nat Biotechnol.* Nov 2001; 19(11): 1029–1034.
138. Peters MC, Polverini PJ, Mooney DJ. Engineering vascular networks in porous polymer matrices. *J Biomed Mater Res.* Jun 15 2002; 60(4): 668–678.

9

Tissue Engineering: Focus on the Musculoskeletal System

Michael Keeney, Li-Hsin Han, Sheila Onyiah, Fan Yang

CONTENTS

Introduction

The musculoskeletal system defines the major structure of the body, and enables the whole body to move, stabilize itself, and maintain its form and function in load-bearing conditions. The major components of the musculoskeletal system are bone, cartilage, muscle, tendon, and ligaments. The mechanical and biochemical properties of these tissues are vital to the function of the system. Current clinical practice on repairing orthopedic tissue loss relies mostly on metallic implants such as total hip replacement. While effective, these approaches provide only mechanical support and do not mimic biological tissue structure and functions. Successful musculoskeletal tissue engineering requires a good understanding of the structural materials of specific tissues, which are organized in a hierarchical manner. Bones are composed of a central marrow canal surrounded by a bi-layer of trabecular and compact bone encased in an outer periosteum. Trabecular bone contains a tightly woven network of hydroxyapatite and type I collagen fibers, which are composed of tropocollagen units formed by polypeptide helices. All tissues in the musculoskeletal system contain a similar type of hierarchy that defines the diverse tissue structure and functions in our body. The interactions between cells, matrix networks, and biological signals are crucial for the normal tissue structure and function, and disruptions in these processes often lead to musculoskeletal diseases and/or tissue degeneration. Loss of musculoskeletal tissues is common and may be caused by traumatic injury, aging, diseases, or unhealthy lifestyle choices such as alcohol abuse. While our body can regenerate minor injury, guided regeneration is required where the defect size is beyond the self-healing capabilities of the body.

Tissue engineering aims to restore and regenerate the lost tissue structure and function by delivering the right cues to guide tissue regeneration. Collaboration between clinicians and engineers is very important for the advances in this field, and has continued to strengthen over the years. Such interdisciplinary interactions help identify the clinical needs, aid in the design process, and guide surgeons through this transition to tissue engineering technologies. Most tissue engineering strategies are composed of single or a combination of three key components: scaffolds, cells, and instructive signals. Scaffolds can provide a 3D niche for cells to migrate, proliferate, and produce matrices that are needed for tissue repair. Biological or physical cues may also be incorporated into scaffolds to promote the desired cell phenotype and tissue function. Tissues in musculoskeletal system have a broad range of biochemical and mechanical properties; tissue-engineering strategies must therefore be designed and optimized for specific needs of various tissues. There are many excellent sources from which an extensive review on scaffolds, cells, and signals can be obtained [1–6]. This chapter focuses on recent advances in research and technology development that aim to bridge the gaps between the research and clinical applications.

Scaffolds

Scaffolds in tissue engineering refer to three-dimensional carriers that serve as a temporary matrix to support desired cellular-fate processes and tissue regeneration. The scaffold may also serve as a reservoir system that gradually releases biological signals such as bone morphogenetic protein-2 (BMP-2). Materials used for scaffolds are broadly classified into two categories: natural and synthetic. Natural scaffolds are derived from natural origins such as animal tissues and marine algae, while synthetic materials for tissue engineering include polymers and ceramics. The two most studied natural and synthetic polymers are type I collagen and poly-co-glycolic acid (PLGA), respectively. Both materials are FDA approved for specific applications. Table 9.1 summarizes a list of commercially available tissue engineering products for musculoskeletal regeneration. Many of these products utilize calcium phosphate or collagen materials, as both materials are FDA approved. Some products incorporate more complex strategies, such as stem cells or controlled delivery of growth factors. The FDA approval processes of the more complex strategies often take much longer time and are very costly, but are crucial for paving the road of translating more exciting tissue engineering strategies into clinical settings.

TABLE 9.1

Commercially Available Tissue Engineering Products for Musculoskeletal Tissue Repair

Company	Product	Description
Biomet, Inc.	Pro Osteon®	Ha + calcium carbonate
Dentsply Friadent CeraMed	Frios Algipore®	Derived from calcium carbonate in marine algae
	OsteoGraf®	Derived from natural hydroxyapatite
DePuy Inc.	Healos®	HA + collagen type I
Globus Medical, Inc.	MicroFuse ST®	PLGA microspheres
Interpore Cross International	ProOsteon®	Derived from calcium carbonate in marine algae
Medtronic Inc.	Infuse®	BMP-2 + collagen I
Orthovita Inc.	Vitoss®	β-TCP
Osteomed	OsteoVation®	Calcium phosphate
Osteotech Inc.	Xpanse™	Demineralized bone
	Plexur™	Allograft + polylactide-co-glycolide
	Graftech®	Allografts
	Graft Cage®	PEEK cage
Stryker	OP-1®	rhBMP-7 + collagen type I
Synthes, Inc	Norian SRS®	Carbonated apatite
	chronOS®	β-tricalcium phosphate
Wright Medical Technologies Inc.	Osteoset®	Calcium sulphate
Zimmer Inc.	Triosite™	Biphasic calcium phosphate
Zimmer Inc.	Collagraft®	HA/β-TCP + collagen type I

Collagen type I is an example of an extensively used polymer that is derived from natural origins. As these materials are natural to the human body, they are immediately recognized by cell surface receptors and can have significant control over cell functions such as attachment, migration, and proliferation. Collagen type I, in its raw form, can be used to fabricate a broad range of scaffold structures, including hydrogel, sponge, film, or as a composite. Decellularized tissues have also been explored as scaffolds using tissues such as submucosa from the urinary bladder, small intestine, or gallbladder. The advantage of these decellularized tissues is that they have a predefined structure and contain biologically relevant growth factors. However, since these naturally derived materials are often isolated from animal origins, their clinical uses are very limited owing to the potential threat of immunogenicity and batch-to-batch variance.

A new paradigm in scaffold production has arisen recently that utilizes matrix manufactured by the cells themselves in vitro. Choi et al. recently described a cell-derived extracellular matrix composed of cartilage matrix molecules secreted by porcine chondrocytes [7]. The chondrocytes were cultured in monoculture for three weeks followed by a further three weeks in a 3D pellet. The pellet was then freeze dried to remove any cell debris and treated with DNase for purification. The final construct was a sponge-like cartilage material composed mainly of collagen type II and sulfated glycosaminoglycan. Rabbit MSCs were seeded on these scaffolds, and the ability to support chondrogenesis was evaluated both in vitro and in vivo in a mouse model. The chondrocyte-derived ECM was highly efficient at forming cartilage in vitro when seeded with MSCs and delayed cartilage degeneration in vivo. One limitation of this approach for clinical translation is the minimum six-week waiting period required for scaffold manufacture. Lareu et al. propose a strategy to overcome this problem by increasing the production rate of ECM proteins in vitro [8]. Dextran sulfate and neutral dextran were used as macromolecular agents, which resulted in a significantly improved conversion rate of procollagen to collagen, a key component of the cartilage extracellular matrix. This kind of procedure may be used to produce a patient specific cartilage surface in vitro, which could later be transplanted to the patient in place of the degenerated surface.

Protein engineering has also been utilized to make tissue-engineering scaffolds, which allows greater specificity over scaffold composition [9]. Modular peptide domains with various functionalities can be encoded into a plasmid DNA, which is then transfected into an organism of choice to produce proteins with molecular-level sequence specification. The materials can be manufactured to contain functional modules that enhance cell signaling, adhesion, and biodegradability; likewise they can also incorporate domains not normally found in natural ECM, such as DNA-binding sequences [10]. However, one major drawback of many protein-engineered scaffolds is the relatively weak mechanical properties, which limit their uses for musculoskeletal-tissue engineering. In an attempt to overcome this problem, muscle-mimicking protein has been engineered based on the molecular structure of titin [11], a complex molecular spring located within the I-band of muscle tissue and largely responsible for the muscle's elasticity. The polyprotein comprises a composite of GB1 and resilin, both of which were produced by overexpressing DH5α cells containing pQE80L vectors modified with genes representing both proteins. The collected proteins were combined and photochemically crosslinked to produce biomaterial constructs. The resulting constructs were rubber-like and showed high resilience to low strain while acting as a shock-absorber under high strains, hence the material can effectively dissipate energy at high strain levels, much like that of muscle tissue. All the results from this study were based on the evaluation of mechanical properties; however, it would be interesting to quantify cell response, for example, whether stem cells can recognize this as a muscle-like ECM and differentiate down the myogenic lineage, or how cells attach and migrate along the synthesized surface. This research is an excellent example of how molecular engineering can be utilized to fabricate tissue-like materials.

Engineering Functional Materials

Engineering functionality into a synthetic material would better mimic natural extracellular environment and promote desired cell fate and tissue functions. Extensive research effort has been dedicated to design the architecture of scaffolds to influence the cellular fate processes. When plated on electrospun grooved materials, cells will align in the direction of the grooves, and such a technique has been used to direct neuronal cell alignment for spinal cord regeneration. Specific pore sizes have also been engineered to attract certain cell populations, and this technique was explored to attract chondrocytes to the cartilage side of an osteochondral implant [12]. In a recent study, Engelmayr et al. have developed "accordion-like honeycombs" that contract with beating cardiac cells for cardiac tissue engineering (Figure 9.1) [13]. The scaffolds are fabricated using excimer laser microablation to induce an accordion-like architecture in poly(glycerol sebacate) (PGS). The resulting scaffolds were seeded with neonatal rat myocytes, which pulsed the scaffold in a preferred direction according to the design of the honeycomb repeat units. Bi-layered scaffolds were also formed in this study by stacking scaffolds at an oblique angle, which enables interporous connectivity and cell infiltration throughout the network. As with all scaffold for tissue engineering, this design also faces a number of limitations [14]. The overall thickness of the scaffold may hinder oxygen diffusion throughout, hence preventing healing at the injury site. Various techniques have been designed to support vascularization and oxygen diffusion throughout scaffolds, as discussed in the signals section below. This study demonstrates how architecture can be used to functionalize a tissue-engineering scaffold, and more in-vivo work would be important to further validate its efficacy for cardiac tissue engineering.

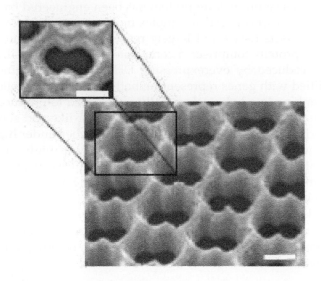

FIGURE 9.1
Accordion-like honeycombs scaffolds that contract with beating cardiac cells for cardiac tissue engineering. Scanning electron micrographs demonstrated the fidelity of excimer laser microablation in rendering an accordion-like honeycomb designs in poly(glycerol-sebacate) (PGS). Scale bars = 200 μm. (From *Nat Mater* 7: 1003–1010, 2008. With permission from the Publisher.)

Recent work has also highlighted the importance of physical cues such as substrate stiffness on stem cell differentiation [15]. Polyacrylamide gels with elastic moduli ranging from 1 to 100 kPa were fabricated as a platform to evaluate the effects on mesenchymal stem cell differentiation. Interestingly, cells cultured on materials with tissue-mimicking stiffness promote cell differentiation towards the corresponding tissue lineages. This study stimulatedion great enthusiasm on elucidating the role of mechano-physical environment on stem cell differentiation. However, it is important to note that results from this study were obtained from a two-dimensional culture, while cells in vivo reside in a three-dimensional environment. Therefore, how mechanical stiffness of scaffold regulates stem cell fate in 3D needs to be elucidated. Not only does the material stiffness play a key role in cell differentiation, small functional chemical groups incorporated into scaffold can also induce lineage differentiation of stem cells [16]. MSC differentiation was shown to be influenced solely by small chemical functional groups while maintaining a constant material stiffness. Benoit et al. used modified PEG hydrogels with tethered functional groups to induce differentiation of MSCs down an osteogenic or adipogenic lineage. Hydrogels were fabricated by mixing a PEG solution with various functional groups and photopolymerized by ultraviolet light. Protein and gene expression were measured along with fluorescent imaging of cell morphology to prove differentiation down each specific lineage. It is important to note that stem cells were cultured in a 3D matrix in this study rather than the conventional 2D culture. The mechanism for differentiation remains unknown; however, the simple translation from 2D to 3D cell culture is a new trend in biomaterial research and is continually outputting new and exciting results that will have a significant impact on the design of future biomaterials.

Hydrogels

Despite the promise of cell-based therapy for tissue engineering, retaining cell viability upon transplantation remains a major hurdle for cell therapies. Studies have shown that upon transplantation in vivo, cell viability is extremely low and cells often get rapidly cleared by the spleen and liver [17, 18]. Possible factors for rapid cell death include immune response and/or environmental stress on the cells. Encapsulating cells in hydrogels may help decrease cell death by providing a protective carrier for transplanted cells. Cell encapsulation was initially demonstrated in the early 1980s for encapsulating islet cells in an effort to prolong their survival in vivo [19]. Late work has expanded to explore engineering 3D hydrogels to mimic the stem cell niche in vivo. Techniques have been developed to crosslink hydrogels while maintaining cell viability using various stimuli such as ions or light [20, 21].

Biodegradable hydrogels that can respond to an external stimulus to facilitate cell migration, and ECM synthesis represents a promising direction for musculoskeletal tissue engineering [22]. Lipase is an enzyme that catalyses the degradation of ester bonds, hence this enzyme can be used to control the degradation of ester-containing polymers [23]. The degradation rate of the polymer can be controlled by the concentration of lipase or even the number of ester-cleavable bonds included in the monomer mix. Peptides that are sensitive to cell-secreted proteases can also be incorporated into 3D scaffolds to fabricate a biodegradable hydrogel network. Inclusion of protease degradable sequences in a polymer will facilitate cell migration, which is critical for many tissue regeneration processes. Lee et al. have developed a collagenase-sensitive poly(ethylene glycol-co-peptide) diacrylate hydrogel that is suitable for cell migration, and a 3D pathway containing cell adhesive peptide was fabricated using photolithography to promote cell migration in specific directions [24]. Human dermal fibroblasts encapsulated within the hydrogel migrated only within the RGDS patterned region. Cell adhesive peptide such as RGDS can also be incorporated into PEG-based hydrogels using a thiol-acrylate mixed-mode polymerization technique [25]. This technique involves the co-polymerization of monomers with thiol-reactive functional groups such as cell recognizable peptide sequences, which are covalently linked to the polymer network. The inclusion of this sequence was required to increase the viability of cells encapsulated within the hydrogel. The hydrogels presented in both these studies study demonstrate a number of exciting functionalities that can be introduced into a scaffold that are useful in the re-creation of cell niches.

Environmentally Responsive Hydrogels

Environmentally responsive materials have always been of interest to tissue engineers because of the range of properties available at various pH and temperatures. Thermally responsive scaffolds are normally designed to impart some functionality upon reaching body temperature, such as by self-hardening or acquiring shape memory, while the majority of research on pH responsive materials is focused on gene delivery, such as by releasing the DNA cargo once a drop in pH is detected within the lysosome.

Temperature-responsive hydrogels are ideal for minimally invasive surgeries, whereby the surgeon can inject solution that can self-harden at body temperature. Li et al. have designed a thermosensitive fast-setting hydrogel for delivering human mesenchymal stem cells that is injectable at 4°C and gels when the temperature rises above 24°C [26]. Gelation was achieved within 7 seconds at 37°C with an elastic modulus of 119 KPa, which is in the range of myocardium tissues. This hydrogel system has also been used to protect the transplanted cells from the body's immune response by delivering enzymes to capture superoxide secreted by inflammatory cells. Thermally responsive materials can also be used for the delivery of therapeutics. Zhang et al. have developed poly(N-isopropylacrylamide-co-acrylamide) hydrogels for targeted delivery of therapeutics to treat tumors in a mouse model [27]. Poly(N-isopropylacrylamide-co-acrylamide) hydrogels remains circulatory at body temperature, and when the temperature is raised above 42°C, the nanogels become hydrophobic and aggregate easily. Zhang et al. proposed to take advantage of this temperature responsiveness to selectively aggregate therapeutic particles in a specific location using a thermal pad in vivo. To show proof of principle, the nanogels were loaded with a fluorescent dye and injected through the tail vein of the mouse. A thermal patch was placed in the vicinity of a tumor and near infrared imaging showed that fluorescent hydrogels aggregated in areas of applied heat. The final example is a shape memory polymer for bone-tissue engineering. Xu et al. have recently designed a thermally sensitive polymer constructed from a well-defined star-branched macromer containing a rigid nanoparticle core [28] (Figure 9.2). This polymer allows covalently conjugating an integrin-binding ligand without compromising the mechanical strength. This material can be molded to a specific shape at approximately 50°C and will retain that shape at body temperature. Such a material would be ideal for maxiofacial-type surgery when a scaffold must be molded to fit an irregularly shaped defect. If the scaffold does not match the defect shape, the materials can be reheated and remolded until a suitable compliance is achieved. Degradation rate of this hydrogel is also tunable by controlling the PLA polymer chain length. Altogether, these studies demonstrate that thermo-responsive materials are a family of smart materials for musculoskeletal tissue engineering applications.

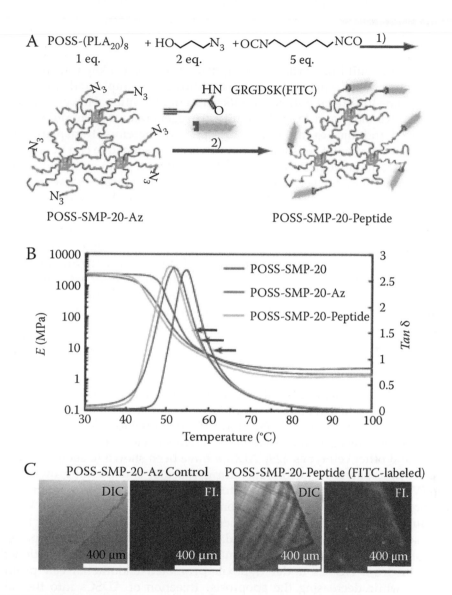

FIGURE 9.2 (See color insert.)
Chemical modification of POSS-SMP with a bioactive peptide: (A) Synthetic scheme illustrating the introduction of azido groups during the covalent cross-linking of POSS-(PLA20)8 and subsequent conjugation of fluorescently labeled integrin-binding peptide to POSS-SMP via "click" chemistry. (B) Storage modulus (E′)-temperature curves and loss angle (Tan δ)-temperature curves (denoted by black arrows) of POSS-SMP-20, POSS-SMP-20-Az, and POSS-SMP-20-Peptide. (C) Differential interference contrast (DIC) and fluorescent (Fl) micrographs confirming the covalent conjugation of the fluorescently labeled peptide via click chemistry. (From *PNAS* 107:7652–7657, 2010. With permission from the Publisher.)

Cells

Cells are the building blocks of living tissues and play a key role in tissue repair processes. Tissue engineering scaffolds can be implanted acellularly and become infiltrated with local cells upon implantation. However, as the body grows older and loses its ability to regenerate, stem population is decreased and injuries can occur more readily. Therefore, transplanting a vibrant population of cells with a new ability to regenerate can be used to aid this regenerative process. Stem cells represent a promising cell source for tissue engineering because of their unique ability to self-renew and differentiate down multiple lineages. There are two major classes of stem cells: adult stem cells and embryonic stem cells. Adult stem cells are multipotent cells that can be isolated from many adult tissues, such as bone marrow and adipose tissues. Embryonic stem cells are derived from embryos, that is, fertilized female eggs, which are normally fertilized in vitro and donated to research with the consent of the donors. This section focuses on recent advances in cell-based therapy for musculoskeletal tissue engineering, with a focus on the use of adipose-derived stem cells (ADSCs), induced pluripotent stem (IPs) cells, and stem-cell homing.

Adipose-Derived Stem Cells

Adipose-derived stem cells (ADSCs) are isolated from fat tissue and have advantages that include relative abundance and ease of isolation via a minimally invasive procedure. ADSCs have the potential to differentiate along multiple cell lineages into osteoblasts, chondrocytes, endothelials, myocytes, and other cell types. [29]. ADSCs have been shown to secret a broad spectrum of paracrine factors, including vascular endothelial growth factor (VEGF), hepatocyte growth factor (HGF), and transforming growth factor-β (TGF-β). Under hypoxia conditions, ADSCs up-regulate VEGF production, which makes them particularly attractive candidates for treating ischemic diseases. Rehman et al. investigated the angiogenic potential of ADSCs in a murine hind-limb ischemia model [30]. Conditioned medium from ADSCs under hypoxia promoted endothelial cell proliferation while decreasing the apoptosis. Injection of ADSCs into the tail vein of mice suffering from hind limb ischemia restored blood perfusion in the ischemic limb to approximately 60% (relative to non-ischemic) after 10 days, while the non-treated groups showed a base line increase of only 25%. After direct injection of labeled ADSCs into the tibialis anterior muscle, only 28% of implanted cells were detectable after one week, suggesting that cell survival after transplantation needs to be improved. In another study, by Bhang et al., ADSCs were delivered into a hind-limb ischemia model via a heparin-containing PLGA nanosphere/fibrin carrier loaded with fibroblast growth factor-2 (FGF2). The FGF2 was shown to promote ADSC proliferation and inhibit apoptosis under hypoxia conditions in vitro. Furthermore, the presence of FGF2 increased VEGF and HGF mRNA expression under hypoxia conditions, indicating that FGF2 promotes ADSC survival and also up-regulates their angiogenic capability. When injected into the ischemic hind limb, limb perfusion was increased to approximately 55% (relative to non-ischemic), while ADSCs without FGF remained at approximately 35% (non-cell carriers did not increase above 20%). Groups treated with FGF/ADSC also led to significant enhanced limb salvage (4 out of 10) compared to the control (0 out of 10) (Figure 9.3). This work demonstrates the promise of using a combination of ADSC and biomaterials therapy for promoting tissue regeneration.

Induced Pluripotent Stem Cells

For repairing large tissue defects, it is often a challenge to obtain a large number of cells that can differentiate into the desired cell phenotype without immunogenicity. The recent discovery of induced pluripotent stem cells (iPS) will likely provide a solution to this problem by reprogramming the patient's own differentiated cells back to a pluripotent state. Takahashi and Yamanaka first proved in 2006 that embryonic-like stem cells could be derived from adult fibroblasts, which can then be used to generate any type of differentiated cells for repairing tissues [31]. Re-programming was achieved by virally introducing four factors, Oct3/4, Sox2, c-Myc, and Klf4, into adult fibroblasts under embryonic stem cell culture conditions. When injected subcutaneously into a mouse model, tumors formed containing a variety of tissues from all three germ lines, which confirmed the pluripotency of these cells. While the iPSCs demonstrated a lot of similarity to ESCs in terms of cell morphology, proliferation, pluripotency and tumorigenicity, they express a global gene expression pattern different from that of embryonic stem cells [32]. In a follow-up study by the same research group, germline transmission was achieved by using more specific markers for iPS selection, and adult chimaeras were successfully obtained from iPS cell clones [33]. Within this study, 20% of the offspring developed tumors attributed to reactivation of the c-myc transgene. While the work with iPSC is still relatively new, it holds great promise for wide applications for repairing musculoskeletal tissues, where large defects are common and iPSCs would offer a source for generating all the desired cell types, including osteoblasts, chondrocytes, and endothelial cells. More work needs to be done to control the specific differentiation of iPSCs towards these lineages by defining the right induction cues.

Cell Homing

Localizing the transplanted cells to the disease or defect site in vivo is important for successful tissue repair, and biological signals may be used to guide such homing processes. Upon intravenous injection, hematopoietic stem

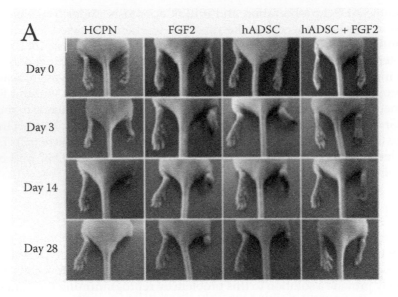

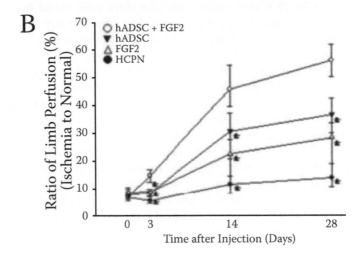

FIGURE 9.3

Improvement of ischemic limb salvage by hADSC transplantation with local FGF2 delivery. (A) Representative photographs of HCPN-, FGF2-, hADSC-, and hADSC þ FGF2-treated ischemic hind limbs on days 0, 3, 14, and 28 after treatment. (B) Blood perfusion ratio of ischemic limbs measured by laser Doppler imaging 0, 3, 14, and 28 days after treatment. The ratio of ischemic to normal limb blood perfusion was significantly improved by combined therapy at all time points (*, $p < .05$, compared with hADSC+ FGF2 group). Abbreviations: FGF2, fibroblast growth factor-2; hADSC, human adipose-derived stromal cell; HCPN, heparin-conjugated poly(lactide-co-glycolide) nanosphere. (From *Stem Cells*, 27:1976–1986, 2009. With permission from the Publisher.)

cells can engraft into the marrow with over 90% efficiency [34]. The mechanism for stem cell homing is highly complex and multi-factorial, and stromal derived factor-1 (SDF-1) is a key player in this process. SDF-1 is part of the chemokine family and is expressed by bone marrow stromal cells and endothelial cells. Its receptor, CXCR4, is expressed on human stem and progenitor cells [35]. Of particular interest to bone tissue engineering, parathyroid hormone treatment led to an increase in SDF-1 production at the growth plate in mice and a decrease in the serum [36]. This suggests a SDF-1 gradient for the localization of stem cells to the growth front in mice. Aiuti et al. demonstrated that human hematopoietic progenitor cells migrated in vitro and in vivo towards a gradient of SDF-1 produced by bone marrow stromal cells [37]. Schantz et al. used this to their advantage in a bone tissue engineering model via the delivery of three separate proteins [38] in an experiment designed to promote vascularization, stem cell recruitment, and bone formation, respectively. A polycaprolactone scaffold was connected to an external microneedle reservoir system that was programmed to sequentially deliver VEGF, SDF-1, and BMP-6. In-vitro findings showed that MSCs expressed CXCR4 and migrated in response to stimulation by SDF-1; in-vivo results showed enhanced vascularization. This study is a good example of delivering biological signals to guide stem cell homing for the purpose of musculoskeletal tissue engineering.

Instructive Signals

Biological signals play an important role during normal tissue development process and mediate the tissue remodeling during injury. Musculoskeletal diseases such as osteoporosis and arthritis are often accompanied by a change in local biochemical cues, which set off the degenerative process. To improve endogenous regeneration, we can restore the local microenvironment by delivering the right signals. For bone tissue repair, active clinical trials are going on to assess the efficacy of delivering recombinant bone morphogenetic protein-2 (rhBMP-2) from 3D collagen sponge implants [39]. Signals coordinate physiological responses on different length scales; soluble signals can act locally or systemically depending on their diffusion limit, access to the circulatory system, and stability in the body. Among the factors that play a role in this process, BMP-2 received premarket approval from the Food and Drug Administration (FDA) for spinal fusion in the case of disk generation from L4 to S1, acute tibial fractures, and sinus augmentations [40]. BMP-7 has received a Humanitarian Device Exemption (HDE) from the FDA for spinal fusion and as an alternative bone autograft when other treatments have failed [41]. Vaccaroet et al. performed a one-year study on the effect of BMP-7 as an iliac crest autograft for non-instrumented posterolateral fusions in 12 (6 female, 6 male) patients with degenerative spondylolisthesis [42]. The Oswestry Disability Index (ODI) and dynamic radiographs showed solid bone fusion. Clinical trials are also investigating parathyroid hormone–related peptide (PTHrP), which is marketed as a treatment for osteoporotic fracture [43]. The clinical uses of BMPs include spinal fusion, treatment of long-bone defects and non-unions, and osteointegration with metallic implants [44]. Parathyoid hormone (PTH) is an anabolic agent whose recombinant human (rhPTH (1-34)) form is approved by the FDA [45]. This 34-amino-acid fragment of PTH is currently used by post-menopausal women with osteoporosis and men with advanced osteoporosis. Antagonists such as sclerostin are also being incorporated into clinical practices [46]. These clinical issues drive scientist to continue to find solutions through tissue engineering strategies.

Synergistic Delivery

Synergistic delivery has emerged as an effective approach for delivering therapeutic growth factors and genes in a physiological relevant manner to mimic the natural musculoskeletal tissue healing process [47]. The bone remodeling process is characterized by an early upregulation of angiogenic growth factors such as VEGF, which is followed by expression of bone morphogenic protein, insulin-like growth factor (IGF), and fibroblast growth factor (FGF). Such a multi-phasic factor expression induces angiogenesis, which creates highly vascular tissue optimal for bone regeneration later on [48]. Kumar et al. reported that bone formation was enhanced in a mouse model of segmental bone defect using genetically engineered mesenchymal stem cells (MSC) that expressed osteogenic and angiogenic factors [47]. The MSCs were either mock-transduced or transduced with recombinant adeno-associated virus 6 (rAAV)-BMP2:VEGF or rAAV-GFP. The groups that received dual delivery of growth factors VEGF and BMP-2 resulted in prosperous bone growth and consequently an increase in bone mineral density, as reported by the DXA analysis. Micro-computed tomography (μCT) analysis also confirmed enhanced tissue repair in the group receiving simultaneous release of both growth factors. The control group that received VEGF alone did not exhibit sufficient tibiae bone correction, compared to the dual-delivery group. After 16 weeks of physical therapy, the tibia from the dual-delivery treated groups showed a substantial improvement in peak load, stiffness, and toughness compared to the control. This study suggests the promise of using genetically modified stem cells to synergistically deliver multiple signals to promote musculoskeletal tissue repair. Biological signals such as growth factors can also be controlled-released from a 3D depot in situ to achieve synergistic delivery [49]. A dual growth factor–releasing scaffold consisting of $VEGF_{165}$ and BMP2 in PDLLA was developed to guide MSC differentiation and bone repair in a murine femur defect model. The results showed that groups receiving synergistic delivery of both angiogenic and osteogenic growth factors experience the strongest bone regenerative response. It is important to note that the technology platform for synergistic delivery is versatile and can be applied to deliver growth factors of any type. Other examples include sequential delivery of FGF-2 and PDGF-BB [50] or by coexpressing FGF2 and cyr61 [51] to promote angiogenesis and muscle regeneration.

DNA Delivery

Delivering genetic signals such as DNA into target cells holds great promise for directing stem cell differentiation by turning on activator signaling pathways. Current approaches to gene delivery are two-fold: viral and non-viral. While highly efficient, viral vectors are limited by safety concerns such as immunogenicity and toxicity. Non-viral gene delivery is safer, but suffers from low-transfection efficiency and short-term expression. Extensive effort has been dedicated towards overcoming the many obstacles such as extracellular transport, intracellular transport, unpacking the DNA-vector complex, and the nuclear transport [52]. Yang and colleagues have recently reported poly (β-amino esters) (PBAEs) as a promising biodegradable polymeric vectors for efficient gene delivery to adult or embryo-derived human stem cells with minimal toxicity compared with other physical methods such as electroporation [53]. PBAE/VEGF nanoparticle-treated embryonic stem cells demonstrated significantly upregulated endothelial markers such as Tie 2, von Willebrand factor, and platelet endothelial cell adhesion molecules (PECAM). These results highlight the potential of using PBAE as biodegradable non-viral gene delivery system to direct stem cell differentiation via non-viral gene delivery.

DNA delivery is a complex process that involves many barriers at the extracellular and intracellular level. Developing methods to overcome such trafficking barriers would greatly facilitate their wide applications in tissue engineering applications. Hama et al. performed a quantitative comparison on intracellular trafficking and transcription efficiency between adenoviral and lipoplex systems. It was shown that the lipoplex vector was more efficient at crossing the cell membrane, while the adenovirus was more efficient at endosomal escape and nuclear uptake. The major difference observed is that the lipoplex system required three orders of magnitude more intranuclear gene copies to achieve transfection efficiency on the level of the adenovirus. Nuclear localization signals (NLSs) act like a chauffeur for plasmid DNA and may aid DNA transport across the nuclear membrane. NLSs are like tags that present on the surface of a protein; the tag is identified intracellularly by importins, which attach to the NLS and transport the protein to the nucleus. NLSs were discovered in the 1980s but have only recently been applied to gene therapy. The most commonly used nuclear import signal is the simian cancer virus large T antigen–derived sequence (SV40). Plasmids containing the SV40 sequence have been shown to promote nuclear import in all cell lines tested from a variety of species, such as human, rat, monkey, mouse, hamster, and chicken [54, 55]. The SV40-derived sequence in itself is not an NLS. However, when a plasmid containing this sequence enters the cytoplasm, it quickly binds to transcription factors that contain NLSs. These NLSs are bound by importin and transported to the cell nucleus. Another sequence used to induce nuclear import is the multiple nuclear factor-κB binding sites. It was further demonstrated that transfection could be further increased by treating cells with tumor necrosis factor-α, an NF-κB activator [56]. In a separate work, Breuzard et al. have shown that up to 60 times more plasmid DNA can enter the nucleus using multiple NF-κB binding sites, which proves a case for increasing nuclear up-take of plasmid DNA [57]. While searching for similar sequences that could be combined with plasmid DNA, cell-type specific sequences were identified. As each cell type contains specific transcription factors and each transcription factor has specific binding sites, it was hypothesized that the binding sites could be incorporated into plasmid DNA for cell specific uptake. One example is the smooth muscle specific DNA-targeting sequence, of which only a 176-bp sequence of the smooth muscle γ-actin promoter can promote plasmid DNA nuclear import in smooth muscle cells but not in other cell types [58]. It is proposed that NLS-containing transcription factors also bind to this promoter and facilitate nuclear import. In another study, Shen et al. have used a high-mobility group box 1 (HMGB1) protein, which contains lysine and arginine groups that can facilitate DNA condensing and also present NLSs for nuclear import [59]. When this system was further condensed with polyethyleneimine, transfection efficiency was increased from approximately 21% to 62% in HeLa cells. This system is bi-functional and achieves both DNA condensation and presentation of NLSs.

siRNA Delivery

Advances in gene therapy provide a powerful tool to promote lineage-specific differentiation via directly regulating the intrinsic signals of stem cells. Today, technology is being developed with the potential to either "turn on" a target gene, through DNA delivery, or "turn off" a gene by small interference RNA (siRNA) delivery. Unlike DNA delivery, the siRNA mechanism was a discovery within the past decade or so and was initially demonstrated in the nematode worm *Caenorhabditis elegans* [60]. When long pieces of double-stranded siRNA are present in the cytoplasm, cleavage occurs by interaction with an enzyme named dicer. The smaller fragment is then incorporated into a protein complex called the RNA-induced silencing complex (RISC). This protein complex causes unwinding of the siRNA and cleavage of the sense strand. The antisense strand is then transported to mRNA that contains the complimentary RNA sequence [61]. The mRNA strand is sliced and degraded by the cell while the RISC moves to seek out the next mRNA. In dividing cells, siRNA is diluted out within approximately 7 days, whereas in non-dividing cells it can remain present for several weeks. siRNA can be delivered either locally to areas such as the eyes and lungs, or intravenously, which requires transport through the blood system to the target site. Intravenous injection has the obvious limitations of being cleared by the liver and distributed throughout non-target tissues, but inhalation has been used to deliver siRNA to the lungs for the treatment of respiratory syncytial virus (RSV) [62]. This study used a siRNA called ALN-RSV01, which acts against the mRNA of N-protein of RSV to down-regulate its expression. This was a safety and tolerability study, but testing has now entered phase II clinical trials. Another example of siRNA delivery is the intratumoral injection of siRNA condensed by using a PEI carrier [63]. Injection of siRNA complexes into tumors within the cranium successfully decreased the expression of pleiotrophin, a known promoter of U87 glioblastoma cell proliferation. While this is a tumor model, this technique can be easily adapted to treat other diseases such as osteoarthritis, by down-regulating tumor necrosis factor-α in mice suffering from collagen-induced arthritis [64]. Although applying siRNA delivery for tissue engineering applications is a relatively young field, it holds great promise for the future and we expect to see rapidly growing research efforts in this area in the near future.

Technology for Manufacturing Tissue Engineering Scaffolds

Tissue-engineering scaffolds aim to induce tissue regenerations by engineering the behaviors of individual cells. To reach this goal, a properly designed scaffold architecture should be developed to trigger desirable cellular fates for biological functions of specified organs [65]. This requires not only controlling the biochemical and physical properties of scaffold materials, as previously mentioned, but also developing microstructures within the scaffolds. These microstructures should have dimensions comparable to the size of cells (1 to 100 micrometers); they facilitate tissue regeneration by physically directing cell morphology, positioning, and alignment that resemble those of tissues in vivo. The aforementioned honeycomb microchannels, which direct the polarization and self-alignment of seeded cardiomyocytes, are a good example. Likewise, microchannels of poly(lactide-co-glycolide) (PLG) were used to guide the unidirectional extension and growth of neurons in three-dimensional space [66].

In addition to guiding cell morphology, microstructures also take important roles in sustaining cell proliferation. For example, interconnected microporosity is commonly used to facilitate the diffusion of nutrients, wastes, and signaling molecules in a tissue-engineering scaffold [67].

In fabricating a scaffolding microstructure, one needs first to decide which type of scaffold precursor should be used; this precursor is usually a fluid that can be solidified by certain mechanism.

Polymerization

Free-radical polymerization is widely used to develop scaffolding microstructures. Materials for free-radical polymerization are composed of monomers and initiators. As the building blocks for microstructures, monomers have one or more crosslinking groups, which can be activated by free radicals and forms covalent binding between the monomers. The monomers can be either synthesized materials or derivatives from the aforementioned natural materials. For example, gelatin and hyaluronan were modified by methacrylate groups and became crosslinkable gels [68, 69]. On the other hand, initiators generate the free radicals in response to specific stimulations, such as pH value, heat, and irradiation. For example, azobisisobutyronitrile (AIBN) decomposes and produces free radical 2-cyanoprop-2-yl at elevated temperature (above 50°C) [70]. As well, 4-(2-hydroxyethoxy)phenyl-(2-hydroxy-2-propyl)ketone (Irgacure 2959), which is a commonly used photoinitiator for tissue engineering, produces free radicals upon irradiation with ultraviolet light [71]. Among the various initiators, photoinitiators are mostly used for making three-dimensional tissue-engineering scaffolds because of the convenience and non-invasive process by irradiation [72].

Enzyme-catalyzed polymerization of proteins is another method for creating solid microstructures from biomaterials. In the mechanism to form a blood clot, for example, fibrinogen molecules crosslink and become solid fibrin gel in the present of thrombin [73]. This type of gelation has been used for cell-culture study and was applied to clinical application as "tissue glue" [74, 75]. Protein polymerization also takes place in the present of synthetic catalysts. For example, in a light-activated state, ruthenium trisbipyridyl chloride [RuII(bpy$_3$)$^{2+}$] catalyzes the formation of dityrosine bonds between tyrosine-abundant proteins, such as fibrinogen, and converts the peptides into a solid gel [76, 77].

Charged bio-polymers, such as chondroitin sulfate, alginate, and hyaluronan, can be electrostatically crosslinked by oppositely charged ions [78]; to crosslink alginate, for example, calcium ions are used. Ionic crosslinking can also become photo-controllable. Photoliable chelators for calcium, such as DM-nitrophen (a brand name for 1- (2-nitro-4,5-dimethoxyphenyl)- N,N,N′,N′-tetrakis [(oxycarbonyl) methyl]-1,2-ethanediamine), were used as "calcium cages" to temporarily bind to calcium ions and neutralize the ions' charges; upon irradiation, however, these chelators degrade and release the calcium ions, which then crosslink the charged polymers [79].

Microfabrication Platforms

Scanning Laser Stereolithography (SLS)

SLS by Continuous-Wave (CW) Laser

The main components of a SLS microfabrication platform include a collimated laser and an optical lens; the lens focuses the laser beam in a bath of photocrosslinkable monomer; at the focal point the monomer solidifies and becomes a volumic pixel (or *voxel*) [80]. In fabricating a scaffolding micro-structure, a SLS platform scans the monomer with the focused laser beam, moving the focal point along a three-dimensional route and creating continuous voxels for the microstructure. The minimum feature that a CW-SLS platform can develop equals the size of a voxel, which is about 100–500 microns. The scanning of the focal point, relative to the monomer bath, can be created by either moving the monomer bath by a three-dimensional stage or moving the focal point by motorized optics.

SLS by Femtosecond Pulsed (fs) Laser

The minimum features created by using SLS with fs-laser are drastically finer than with CW laser; the typical size of voxels created by fs-SLS is 200 nm (0.0002 mm) [81]. The mechanism behind the performance of fs-laser is named by "two-photon absorption," a non-linear effect that takes place under sufficiently strong irradiation, as in the case of using fs-laser [82]. In two-photon absorption, a light-sensitive molecule, such as photoinitiators, becomes sensitive to photons half the energy level at which the absorption of a single photon takes place. For example, given that a molecule absorbs a wavelength only around 380 nm in single-photon absorption, in two-photon absorption it absorbs 760 nm light. Red to near-infrared (640–800 nm) femtosecond lasers are frequently used to create voxels in monomers that polymerize in ultraviolet (320–400 nm) light. Because the probability that two-photon absorption takes place in a molecule is in proportion to the square of light intensity [83], under focused fs irradiation the monomers polymerize in a very narrow region, in the order of 100 nm, around the focal point.

Projection-Printing Stereolithography

Projection-printing stereolithography (PPS) develops a microstructure by forming a sequence of cross-sectional slices in a photocurable monomer [84]. A typical PPS platform includes a sequence of photomasks to pattern the cross-sections of microstructure, a light source to illuminate the photomask patterns, and an optical lens to project the illuminated patterns onto the photocurable monomer and cure the monomer according to the patterns; the surface of monomer turns into a thin cross-sectional slice upon the projection. After the formation of each cross-sectional slice, the microstructure is repositioned (downward) by the stage to prepare for creating another slice; this procedure is repeated until every cross-sectional slice for a microstructure is built. PPS differs from SLS in that the former generates a one-dimensional scanning (to stack the cross-sectional slices) to fabricate a three-dimensional structure, while the latter generates a three-dimensional scanning (by the 3-D motion of a motorized optics) to create continuous, small voxels in three-dimensional space. This gives PPS a much higher fabrication speed, especially for making larger tissue-engineering scaffolds. The resolution of PPS microfabrication depends on the quality of the projecting lens; normally, the minimum feature that PPS can fabricate is thinner than 10 micrometers, or at a cellular scale.

The photomasks for PPS can be fabricated like the ones for electronics photolithography [85]; however, it is more convenient and cost-efficient to use a dynamic photomask instead. The DLP chipsets (from Texas Instruments) for digital micromirror display are widely used digital photomasks [86, 87]. A DLP chip has on its surface more than a million digitally controllable micromirrors; each micromirror can be electrostatically tilted to an "on" or an "off" state. Upon an illumination, the micromirrors at the "on" state reflect the incident light toward the projecting lens and become the bright pixels at the projected curing image. Using this dynamic mask, therefore, the photomask pattern for each cross-sectional slice of a microstructure becomes programmable, making scaffold microfabrication faster and easier.

Syringe-Pump Microfabrication

In syringe-pump microfabrication (SPM), monomers for microstructure are extruded at a stable rate from a microsyringe, which is actuated by a syringe pump. The monomer squeezed from the needle is immediately cured to form a continuous, micron-scaled line to construct microstructures; the method to cure the monomer depends on which type of crosslinkable material the monomer is. To cure alginic acid, for example, the needle tip is submersed in a calcium chloride solution; the extruded alginic acid is thus crosslinked by calcium ions through electrostatic binding [88]. As well, to cure a photocrosslinkable monomer, the tip of the needle can be continuously exposed to a curing wavelength [89]. The minimum feature (diameter of the line) created by this method is about 1 micron. Similar to SLS microfabrication, SPM is a 3-D scanning method and includes a three-dimensional motorized system to continuously change the relative position between the syringe tip and the fabricated microstructure.

Selecting a Microfabrication Platform

In making microstructures for tissue engineering, it is important to select a fabrication platform to comply with the material properties of monomers, the solidifying mechanism, and the geometry of fabricated scaffolds. SLS and PPS platforms, for instance, are limited for photopolymerizable monomers. On the other hand, microsyringe stereolithography has the advantage that it is applicable to every type of the aforementioned crosslinking mechanism, including ionic and catalyzed crosslinking. Manufacturing resolution is another important issue to consider. For example, SLS with femtosecond laser offers sub-micron resolution and is suitable for patterning subcellular-scaled microstructures, while the other platforms are more suitable for making cellular-scaled microstructures. The time for creating scaffolds is another important issue. The speed of scaffold manufacturing impacts fabrication cost and the feasibility for clinical applications. Longer manufacturing time also impacts cell viability when making cell-encapsulating scaffolds. Because CW-SLS, fs-SLS, and microsyringe stereolithography are 3-D volumic scanning methods, the time for making one scaffold using these platforms is proportional to the scaffold volume, or the 3rd order of scaffold dimension; these platforms become very inefficient for building scaffolds of centimeter size. Compared to other platforms, PPS-developed microstructures through the aforementioned 1-D scanning have the advantage of building large scaffolds at a much higher fabrication rate. Table 9.2 summarizes the different aspects of the aforementioned platforms: their working principles, fabrication speed, structure resolutions, and suitable monomer types.

Supercritical CO_2 and Gas-foamed Scaffolds

Gas foaming is a technique that allows for processing polymers such as PLGA into highly porous scaffolds without using organic solvents or high temperature, which facilitates incorporating sensitive biological signals such as growth factors or nucleic acids. Supercritical CO_2 has been employed to make polymeric scaffolds at approximately 31°C and 1500 psi. Under these conditions, the biomolecules remain intact while the polymer transforms into a liquid stage, and therefore encapsulation of biomolecules within the polymer can be achieved. When the pressure is decreased to ambient conditions, the polymer reverts to a solid state and swells during the removal of CO_2; a sponge-like scaffold is formed that is directly related to the rate of CO_2 removal (Figure 9.4). Howdle and colleagues have performed extensive research on this topic using a variety of polymers, biomolecules, and even cells to form porous scaffolds for non-viral gene delivery and tissue engineering applications [90–93]. Plasmid DNA was complexed with a cationic polymer and deposited on the PLA powder before freeze drying. The mixture was then exposed to supercritical CO_2 at 35°C and 2500 psi. The pressure was gradually released to form porous constructs encapsulating the polyplexes. The resulting scaffolds showed a sustained release of functional polyplexes from the scaffolds, with a moderate level of transfection over 60 days. This study demonstrates the potential of using supercritical CO_2-sustained delivery of functional polyplexes. Ginty et al. have evaluated viability of multiple cell lines (C2C12 cell line, 3T3 fibroblasts, chondrocytes, and hepatocytes) under supercritical conditions (35°C and 1070 psi) [94]. Cell/polymer composites were exposed to supercritical CO^2 for 30 seconds with an additional 80 seconds for pressurization and depressurization of the cylinder. Cell viability was confirmed and osteogenic differentiation of C2C12 cells was shown by alkaline phosphatase staining. Combining the supercritical technique of scaffold fabrication with biomolecule encapsulation provides a versatile platform for delivering biological cues from 3D depot to guide tissue regeneration. High-pressure CO_2 (800 psi) has also been demonstrated

TABLE 9.2

Microfabrication Platforms for Tissue Engineering Scaffolds

	cw-SLS	fs-SLS	PPS	SPM
Working Principle	Photocrosslinking	Photocrosslinking by two-photon absorption	Photocrosslinking	Extrusion and curing
Method to Create 3-D Microstructures	3-D Scanning	3-D Scanning	1-D Scanning	3-D Scanning
Suitable Monomers	Photocrosslinkable	Photocrosslinkable	Photocrosslinkable	Versatile
Structure Resolution (micron)	100	0.1	10	1
Fabrication Time	Proportional to volume (αL^3)	Proportional to volume (αL^3)	Proportional to the number of cross-sectional slices (αL)	Proportional to volume (αL^3)

Abbreviations:
Cw-SLS: Scanning laser stereolithography by continuous-wave laser
fs-SLS: Scanning laser stereolithography by femtosecond pulsed laser
PPS: Projection-printing stereolithography
SPM: Syringe-pump microfabrication
L: Dimension of scaffold such as a length, width, or height

(a)   (b)

FIGURE 9.4

Micro-CT images of a) the 30 and b) 60 min vented scaffolds. Changes in porosity, pore size and interconnectivity are clearly visible from the images (From *J Mater Sci* 41: 4197–4204, 2006. With permission from the Publisher.)

to fabricate porous PLA scaffolds for bone tissue engineering applications [95–97]. The polymer discs are pre-formed by compressing ground particles in a hydraulic press and subsequently exposing the disc to high pressure for 48–72 hours. A porogen can also be used during this process to increase the porosity of the final scaffold. These scaffolds have been used for the delivery of polyplexes, naked plasmid, and growth factors for bone tissue engineering with excellent outcomes.

Self-Assembly from Amphiphilic Peptide

Amphiphilic peptides are molecular building blocks for forming highly porous scaffold by self-assembly [98]. A typical amphiphilic peptide contains four domains: (i) a hydrophobic site, (ii) a self-assembly site to form one-dimensional alignment with other peptides, (iii) a charged peptide to promote hydrophilicity, and (iv) a bioactive peptide [99] (Figure 9.5). In an acidic environment (pH < 5), which protonates carboxylic residues, these peptides form hydrogen bonds between each other and then form one-dimensional (1-D) β-sheet structures. Driven by thermodynamic law, the hydrophobic sides of the peptides aggregate and cause the β-sheets to self-assemble further, forming cylindrical micelles. These cylindrical micelles form long, interconnected nanofibers (50–200 nm in diameter), which turn the amphiphilic peptide solution into a water-insoluble hydrogel network. These peptide hydrogels are reported to promote cell adhesion and proliferation by incorporating relevant biological sequence [99]. A hydrophobic site other than alkane chains [98, 99] may further promote the forming of nanofibers. Ma and coworkers [100] used aromatic compounds, such as pyrene and fluorene, for the hydrophobic site; their report suggests that the π-π stacking among the aromatic compounds adds to the 1-D alignment among the peptides. Low mechanical strength is one technical challenge of using self-assembling peptides.

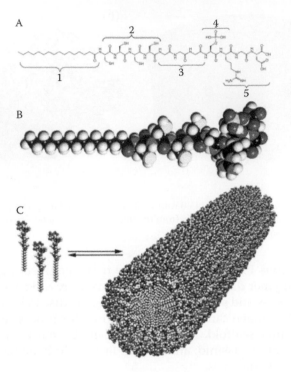

FIGURE 9.5

(A) Chemical structure of the peptide amphiphile, highlighting five key structural features. Region 1 is a long alkyl tail that conveys hydrophobic character to the molecule and, when combined with the peptide region, makes the molecule amphiphilic. Region 2 is composed of four consecutive cysteine residues that when oxidized may form disulfide bonds to polymerize the self-assembled structure. Region 3 is a flexible linker region of three glycine residues to provide the hydrophilic head group flexibility from the more rigid cross-linked region. Region 4 is a single phosphorylated serine residue that is designed to interact strongly with calcium ions and help direct mineralization of hydroxyapatite. Region 5 displays the cell adhesion ligand RGD. (B) Molecular model of the PA showing the overall conical shape of the molecule going from the narrow hydrophobic tail to the bulkier peptide region. Color scheme: C, black; H, white; O, red; N, blue; P, cyan; S, yellow. (C) Schematic showing the self-assembly of PA molecules into a cylindrical micelle. (From *Science* 294: 1684–1688, 2001. With permission from the Publisher.)

To address this problem, Cui and coworkers [101] synthesized an amphiphilic peptide with diacetylene moiety at the hydrophobic site. After self-assembly, the as-formed nanofibers were crosslinked using UV light and reinforced the nanofiber network. The photocrosslinking among peptides, however, may lead to non-biodegradable scaffolds. The need for acidic solution to induce self-assembly is also an issue for encapsulating cells. Alternatively, Greenfield et al. [102] used calcium ions to induce nanofiber assembly by forming salt bridges between the peptides. Rheological testing showed that these calcium-bound nanofibers have a higher rigidity compared to nanofibers assembled by hydrogen bonds.

Electrospun Nanofibers

Electrospinning is a method of fabricating unlimitedly long nanofibers (100–200 nm in diameter) from a viscous polymer [103]. An electrospinning platform consists of a syringe, a direct current (DC) power source connected to the needle of the syringe, a syringe pump to extrude polymers, and a grounded plate to collect polymers. During fabrication, the polymer extruded from the syringe needle is highly charged (10–100 kV) by the DC power and is stretched by electrostatic repulsion between the polymer molecules; the electrostatic force induces a jet of polymer toward the grounded plate. Before reaching the grounded plate, as a result of charge repulsion the polymer jet performs a whipping motion, which leads to elongated, thin polymer fibers of submicron diameters. For most applications, the fibrous polymer should be solidified after being deposited onto the grounded plate; this can be achieved by using either a molten polymer that turns solid at room temperature, a dissolved polymer that rapidly dries, or a pre-polymer that solidifies in a post-crosslinking. Compared to other scaffolding materials, electrospun nanofibers have an advantage of higher mechanical strength; the electrospinning process allows creating nanofibers from organic or inorganic materials of high Young's modulus. Nanofibers woven from strong, synthetic materials, such as poly(lactic-co-glycolic acid) (PLGA) and polycaprolactone (PCL), have been intensively used to create electrospun scaffolds for bone regeneration [103].

In addition to synthetic polymers, electrospinning has also been applied to fabricate scaffolds from natural materials. Rnjak and coworkers [104] created a synthetic elastin scaffold from human tropoelastin and tropocollagen. Human dermal fibroblasts adhered to and proliferated across the electrospun nanofibers and deposited extracellular matrix proteins, such as type I collagen and fibronectin. The mechanical modulus of the created elastin scaffold was 0.3–1 MPa, which is comparable to the strength of human skin. The application of electrospun nanofibers is normally for 2-dimensional cell cultures; sheets of nanofibers could be applied to suturing surface tissues, such as at artery walls. To explore the application of electrospun sheets in 3D, Panseri and coworkers [105] created a cylindrical grounded collector to weave PLGA/PCL nanofibers into tubular sheets or tubes. Their report shows that the electrospun tubes effectively induced nervous regeneration of the severed nerve tracts in a rat model.

Conclusions

While significant progress has been made in musculoskeletal tissue engineering research, translating these technologies into tissue engineering products has been slow, and a huge lag remains between research findings and commercially available products. Further research is needed to address the challenges, including vascularizing tissue engineered constructs, tissue heterogeneity, host-tissue integration, overcoming immunogenicity, and improved mechanical properties. When designing a tissue engineering strategy, it is important to also consider the ease of use by clinicians for specific applications. While the FDA approval process is often long and costly for tissue engineering products, recent approval of musculoskeletal tissue engineering products releasing biologics or cells suggest more acceptance of advanced technology. Cell-based therapies that utilize differentiated cells or adult stem cells will likely be the first ones that get translated into clinical settings. Future efforts on better understanding of cell–niche interactions is also crucial, and recent advances in combinatorial approach for studying cell–materials interactions will likely provide a useful tool for further advances in this direction. In vivo testing of musculoskeletal tissue engineering products in appropriate animal models is crucial and will warrant the successful development of functional tissue substitutes for repairing musculoskeletal tissue injuries.

References

1. Lanza, R.P., R.S. Langer, and J. Vacanti. 2007. *Principles of tissue engineering*. 3rd ed. Boston: Elsevier/Academic Press. xxvii, 1307 p.
2. Hutmacher, D.W. 2000. Scaffolds in tissue engineering bone and cartilage. *Biomaterials*, 21(24): 2529–43.
3. Ratner, B.D. 2004. *Biomaterials science: An introduction to materials in medicine*. 2nd ed. Boston: Elsevier/Academic Press. xii, 851 p.
4. Atala A. and D.J. Mooney. 1997. Synthetic biodegradable polymer scaffolds. *Tissue Engineering*. Boston: Birkhäuser. xii, 258 p.
5. Elçin, Y.M. 2003. *Tissue engineering, stem cells, and gene therapies*. Advances in experimental medicine and biology. New York: Kluwer Academic/Plenum Publishers. ix, 340 p.
6. Guilak, F. 2003. *Functional tissue engineering*. New York: Springer. xvi, 426 p.
7. Choi, K.H., et al. 2010. The chondrogenic differentiation of mesenchymal stem cells on an extracellular matrix scaffold derived from porcine chondrocytes. *Biomaterials* 31(20): 5355–65.
8. Lareu, R.R., et al. 2007. In vitro enhancement of collagen matrix formation and crosslinking for applications in tissue engineering: A preliminary study. *Tissue Eng* 13(2): 385–91.
9. Sengupta, D. and S.C. Heilshorn. 2010. Protein-engineered biomaterials: Highly tunable tissue engineering scaffolds. *Tissue Eng Part B Rev* 16(3): 285–93.
10. Dai, H., et al. 2005. Nonequilibrium synthesis and assembly of hybrid inorganic–protein nanostructures using an engineered DNA binding protein. *J Am Chem Soc* 127(44): 15637–43.
11. Lv, S., et al. 2010. Designed biomaterials to mimic the mechanical properties of muscles. *Nature* 465(7294): 69–73.

12. Sherwood, J.K., et al. 2002. A three-dimensional osteochondral composite scaffold for articular cartilage repair. *Biomaterials* 23(24): 4739–51.
13. Engelmayr, G.C., Jr., et al. 2008. Accordion-like honeycombs for tissue engineering of cardiac anisotropy. *Nat Mater* 7(12): 1003–10.
14. Zimmermann, W.H. 2008. Tissue engineering: Polymers flex their muscles. *Nat Mater* 7(12): 932–33.
15. Engler, A.J., et al. 2006. Matrix elasticity directs stem cell lineage specification. *Cell* 126(4): 677–89.
16. Benoit, D.S., et al. 2008. Small functional groups for controlled differentiation of hydrogel-encapsulated human mesenchymal stem cells. *Nat Mater* 7(10): 816–23.
17. Hofmann, M., et al. 2005. Monitoring of bone marrow cell homing into the infarcted human myocardium. *Circulation* 111(17): 2198–202.
18. Toma, C., et al. 2002. Human mesenchymal stem cells differentiate to a cardiomyocyte phenotype in the adult murine heart. *Circulation* 105(1): 93–8.
19. Lim, F. and A.M. Sun. 1980. Microencapsulated islets as bioartificial endocrine pancreas. *Science* 210(4472): 908–10.
20. Mazumder, M.A., et al. 2009. Core-cross-linked alginate microcapsules for cell encapsulation. *Biomacromolecules* 10(6): 1365–73.
21. Nuttelman, C.R., M.C. Tripodi, and K.S. Anseth. 2005. Synthetic hydrogel niches that promote hMSC viability. *Matrix Biol* 24(3): 208–18.
22. Nuttelman, C.R., et al. 2008. Macromolecular monomers for the synthesis of hydrogel niches and their application in cell encapsulation and tissue engineering. *Prog Polym Sci* 33(2): 167–179.
23. Rice, M.A., J. Sanchez-Adams, and K.S. Anseth. 2006. Exogenously triggered, enzymatic degradation of photopolymerized hydrogels with polycaprolactone subunits: Experimental observation and modeling of mass loss behavior. *Biomacromolecules* 7(6): 1968–75.
24. Lee, S.H., J.J. Moon, and J.L. West. 2008. Three-dimensional micropatterning of bioactive hydrogels via two-photon laser scanning photolithography for guided 3D cell migration. *Biomaterials* 29(20): 2962–68.
25. Salinas, C.N., et al. 2007. Chondrogenic differentiation potential of human mesenchymal stem cells photoencapsulated within poly(ethylene glycol)-arginine-glycine-aspartic acid-serine thiol-methacrylate mixed-mode networks. *Tissue Eng* 13(5): 1025–34.
26. Li, Z., et al. 2009. Injectable, highly flexible, and thermosensitive hydrogels capable of delivering superoxide dismutase. *Biomacromolecules* 10(12): 3306–16.
27. Zhang, J., et al. 2008. The targeted behavior of thermally responsive nanohydrogel evaluated by NIR system in mouse model. *J Control Release* 131(1): 34–40.
28. Xu, J. and J. Song. 2010. High performance shape memory polymer networks based on rigid nanoparticle cores. *Proc Natl Acad Sci USA* 107(17): 7652–57.
29. Gimble, J.M., A.J. Katz, and B.A. Bunnell. 2007. Adipose-derived stem cells for regenerative medicine. *Circ Res* 100(9): 1249–60.
30. Rehman, J., et al. 2004. Secretion of angiogenic and antiapoptotic factors by human adipose stromal cells. *Circulation* 109(10): 1292–98.
31. Takahashi, K. and S. Yamanaka. 2006. Induction of pluripotent stem cells from mouse embryonic and adult fibroblast cultures by defined factors. *Cell* 126(4): 663–76.
32. Yamanaka, S. 2009. A fresh look at iPS cells. *Cell* 137(1): 13–17.

33. Okita, K., T. Ichisaka, and S. Yamanaka. 2007. Generation of germline-competent induced pluripotent stem cells. *Nature* 448(7151): 313–17.
34. Benveniste, P., et al. 2003. Hematopoietic stem cells engraft in mice with absolute efficiency. *Nat Immunol* 4(7): 708–13.
35. Lapidot, T., A. Dar, and O. Kollet. 2005. How do stem cells find their way home? *Blood* 106(6): 1901–10.
36. Jung, Y., et al. 2006. Regulation of SDF-1 (CXCL12) production by osteoblasts: A possible mechanism for stem cell homing. *Bone* 38(4): 497–508.
37. Aiuti, A., et al. 1997. The chemokine SDF-1 is a chemoattractant for human CD34+ hematopoietic progenitor cells and provides a new mechanism to explain the mobilization of CD34+ progenitors to peripheral blood. *J Exp Med* 185(1): 111–120.
38. Schantz, J.T., H. Chim, and M. Whiteman. 2007. Cell guidance in tissue engineering: SDF-1 mediates site-directed homing of mesenchymal stem cells within three-dimensional polycaprolactone scaffolds. *Tissue Eng* 13(11): 2615–24.
39. NIH. 2009. rhBMP-2 versus autograft in critical size tibial defects. [cited 2010 Sept. 6]; Available from: http://clinicaltrials.gov/ct2/show/NCT00853489.
40. Axelrad, T.W. and T.A. Einhorn. 2009. Bone morphogenetic proteins in orthopaedic surgery. *Cytokine Growth Factor Rev* 20(5–6): 481–88.
41. Moghaddam, A., et al. 2010. Clinical application of BMP 7 in long bone nonunions. *Arch Orthop Trauma Surg* 130(1): 71–76.
42. Johnsson, R., B. Stromqvist, and P. Aspenberg. 2002. Randomized radiostereometric study comparing osteogenic protein-1 (BMP-7) and autograft bone in human noninstrumented posterolateral lumbar fusion: 2002 Volvo Award in clinical studies. *Spine* (Philadelphia, Pa 1976), 27(23): 2654–61.
43. Aspenberg, P., et al. 2010. Teriparatide for acceleration of fracture repair in humans: A prospective, randomized, double-blind study of 102 postmenopausal women with distal radial fractures. *J Bone Miner Res* 25(2): 404–14.
44. Bessa, P.C., M. Casal, and R.L. Reis. 2008. Bone morphogenetic proteins in tissue engineering: The road from laboratory to clinic, part II (BMP delivery). *J Tissue Eng Regen Med* 2(2–3): 81–96.
45. Lane, N.E. and S.L. Silverman. 2010. Anabolic therapies. *Curr Osteoporos Rep* 8(1): 23–27.
46. Padhi, D., et al. 2010. Single-dose, placebo-controlled, randomized study of AMG 785, a sclerostin monoclonal antibody. *J Bone Miner Res.* 26(1): 16–26.
47. Kumar, S., et al. 2010. Mesenchymal stem cells expressing osteogenic and angiogenic factors synergistically enhance bone formation in a mouse model of segmental bone defect. *Mol Ther* 18(5): 1026–34.
48. Gerber, H.P., et al. 1999. VEGF couples hypertrophic cartilage remodeling, ossification and angiogenesis during endochondral bone formation. *Nat Med* 5(6): 623–28.
49. Kanczler, J.M., et al. 2010. The effect of the delivery of vascular endothelial growth factor and bone morphogenic protein-2 to osteoprogenitor cell populations on bone formation. *Biomaterials* 31(6): 1242–50.
50. Li, J., et al. 2010. Synergistic effects of FGF-2 and PDGF-BB on angiogenesis and muscle regeneration in rabbit hindlimb ischemia model. *Microvasc Res* 80(1): 10–17.
51. Rayssac, A., et al. 2009. IRES-based vector coexpressing FGF2 and Cyr61 provides synergistic and safe therapeutics of lower limb ischemia. *Mol Ther* 17(12): 2010–9.
52. Luo, D. and W.M. Saltzman. 2000. Synthetic DNA delivery systems. *Nat Biotechnol* 18(1): 33–37.

53. Yang, F., et al. 2009. Gene delivery to human adult and embryonic cell-derived stem cells using biodegradable nanoparticulate polymeric vectors. *Gene Ther* 16(4): 533–46.

54. Miller, A.M. and D.A. Dean. 2009. Tissue-specific and transcription factor-mediated nuclear entry of DNA. *Adv Drug Deliv Rev* 61(7–8): 603–13.

55. Lam, A.P. and D.A. Dean. 2010. Progress and prospects: Nuclear import of nonviral vectors. *Gene Ther* 17(4): 439–47.

56. Reich, Z. 2005. Nucleic acid constructs capable of high efficiency delivery of polynucleotides into DNA containing organelles and methods of utilizing same. United States Patent Application.

57. Breuzard, G., et al. 2008. Nuclear delivery of NFkappaB-assisted DNA/polymer complexes: Plasmid DNA quantitation by confocal laser scanning microscopy and evidence of nuclear polyplexes by FRET imaging. *Nucleic Acids Res* 36(12): e71.

58. Miller, A.M. and D.A. Dean. 2008. Cell-specific nuclear import of plasmid DNA in smooth muscle requires tissue-specific transcription factors and DNA sequences. *Gene Ther* 15(15): 1107–15.

59. Shen, Y., et al. 2009. High mobility group box 1 protein enhances polyethylenimine mediated gene delivery in vitro. *Int J Pharm* 375(1–2): 140–47.

60. Fire, A., et al. 1998. Potent and specific genetic interference by double-stranded RNA in *Caenorhabditis elegans*. *Nature* 391(6669): 806–11.

61. Whitehead, K.A., R. Langer, and D.G. Anderson. 2009. Knocking down barriers: Advances in siRNA delivery. *Nat Rev Drug Discov* 8(2): 129–38.

62. DeVincenzo, J., et al. 2008. Evaluation of the safety, tolerability and pharmacokinetics of ALN-RSV01, a novel RNAi antiviral therapeutic directed against respiratory syncytial virus (RSV). *Antiviral Res* 77(3): 225–31.

63. Grzelinski, M., et al. RNA interference-mediated gene silencing of pleiotrophin through polyethylenimine-complexed small interfering RNAs in vivo exerts antitumoral effects in glioblastoma xenografts. *Hum Gene Ther* 2006. 17(7): 751–66.

64. Schiffelers, R.M., et al. 2005. Effects of treatment with small interfering RNA on joint inflammation in mice with collagen-induced arthritis. *Arthritis Rheum* 52(4): 1314–18.

65. Palsson, B.O. and S.N. Bhatia. 2004. Tailoring biomaterials, in *Tissue Engineering*, pp. 270–287. Upper Saddle River, NJ: Pearson Prentice Hall.

66. Houchin-Ray, T., et al. 2007. Patterned PLG substrates for localized DNA delivery and directed neurite extension. *Biomaterials* 28(16): 2603–11.

67. Osathanon, T., C.M. Giachelli, and M.J. Somerman. 2009. Immobilization of alkaline phosphatase on microporous nanofibrous fibrin scaffolds for bone tissue engineering. *Biomaterials* 30(27): 4513–21.

68. Suri, S. and C.E. Schmidt. 2010. Cell-laden hydrogel constructs of hyaluronic acid, collagen, and laminin for neural tissue engineering. *Tissue Eng Part A* 16(5): 1703–16.

69. Van Den Bulcke, A.I., et al. 2000. Structural and rheological properties of methacrylamide modified gelatin hydrogels. *Biomacromolecules* 1(1): 31–38.

70. Lee, S.H., et al. 2009. Rapid formation of acrylated microstructures by microwave-induced thermal crosslinking. *Macromol Rapid Commun* 30(16): 1382–1386.

71. Fedorovich, N.E., et al. 2009. The effect of photopolymerization on stem cells embedded in hydrogels. *Biomaterials* 30(3): 344–53.

72. Nguyen, K.T. and J.L. West. 2002. Photopolymerizable hydrogels for tissue engineering applications. *Biomaterials* 23(22): 4307–14.

73. Ahmed, T.A., E.V. Dare, and M. Hincke. 2008. Fibrin: A versatile scaffold for tissue engineering applications. *Tissue Eng Part B Rev* 14(2): 199–215.
74. Berdajs, D., et al. 2010. Seal properties of TachoSil: In vitro hemodynamic measurements. *Interact Cardiovasc Thorac Surg* 10(6): 910–13.
75. Miyamoto, H., et al. 2010. The effects of sheet-type absorbable topical collagen hemostat used to prevent pulmonary fistula after lung surgery. *Ann Thorac Cardiovasc Surg* 16(1): 16–20.
76. Karpel, R., G. Marx, and M. Chevion. 1991. Free radical-induced fibrinogen coagulation: modulation of neofibe formation by concentration, pH and temperature. *Isr J Med Sci* 27(2): 61–66.
77. Elvin, C.M., et al. 2009. The development of photochemically crosslinked native fibrinogen as a rapidly formed and mechanically strong surgical tissue sealant. *Biomaterials* 30(11): 2059–65.
78. Chu, C., et al. 2009. Three-dimensional synthetic niche components to control germ cell proliferation. *Tissue Eng Part A* 15(2): 255–62.
79. Chueh, B.H., et al. 2010. Patterning alginate hydrogels using light-directed release of caged calcium in a microfluidic device. *Biomed Microdevices* 12(1): 145–51.
80. Lee, J.W., et al. 2008. Fabrication and characteristic analysis of a poly(propylene fumarate) scaffold using micro-stereolithography technology. *J Biomed Mater Res B Appl Biomater* 87(1): 1–9.
81. Ovsianikov, A., et al. 2007. Two-photon polymerization technique for microfabrication of CAD-designed 3D scaffolds from commercially available photosensitive materials. *J Tissue Eng Regen Med* 1(6): 443–49.
82. Li, L., et al. 2009. Achieving lambda/20 resolution by one-color initiation and deactivation of polymerization. *Science* 324(5929): 910–13.
83. Oheim, M., et al. 2006. Principles of two-photon excitation fluorescence microscopy and other nonlinear imaging approaches. *Adv Drug Deliv Rev* 58(7): 788–808.
84. Han, L.H., et al. 2010. Fabrication of three-dimensional scaffolds for heterogeneous tissue engineering. *Biomed Microdevices* 12(4): 721–25.
85. Madou, M.J. 2002. *Fundamentals of Microfabrication: The Science of Miniaturization.* 2nd ed., Boca Raton, FL: CRC Press.
86. Singh-Gasson, S., et al. 1999. Maskless fabrication of light-directed oligonucleotide microarrays using a digital micromirror array. *Nat Biotechnol* 17(10): 974–78.
87. Han, L.H., et al. 2010. Fluorinated colloidal emulsion of photochangeable rheological behavior as a sacrificial agent to fabricate organic, three-dimensional microstructures. *Langmuir* 26(9): 6108–6110.
88. Khalil, S. and W. Sun. 2009. Bioprinting endothelial cells with alginate for 3D tissue constructs. *J Biomech Eng* 131(11): 111,002–111,008.
89. Barry, R.A., et al. 2009. Direct-write assembly of 3D hydrogel scaffolds for guided cell growth. *Advanced Materials* 21(23): 2407–10.
90. Heyde, M., et al. 2007. Development of a slow non-viral DNA release system from PDLLA scaffolds fabricated using a supercritical CO_2 technique. *Biotechnol Bioeng* 98(3): 679–93.
91. Tai, H., et al. 2007. Putting the fizz into chemistry: Applications of supercritical carbon dioxide in tissue engineering, drug delivery and synthesis of novel block copolymers. *Biochem Soc Trans* 35(Pt 3): 516–21.
92. Tai, H., et al. 2007. Control of pore size and structure of tissue engineering scaffolds produced by supercritical fluid processing. *Eur Cell Mater* 14: 64–77.

93. Barry, J.J., et al. 2006. Porous methacrylate tissue engineering scaffolds: Using carbon dioxide to control porosity and interconnectivity. *J Mater Sci* 41: 4197–4204.

94. Ginty, P.J., et al. 2006. Mammalian cell survival and processing in supercritical CO_2. *Proc Natl Acad Sci U S A* 103(19): 7426–31.

95. Huang, Y.C., et al. 2005. Bone regeneration in a rat cranial defect with delivery of PEI-condensed plasmid DNA encoding for bone morphogenetic protein-4 (BMP-4). *Gene Ther* 12(5): 418–26.

96. Huang, Y.C., et al. 2005. Combined angiogenic and osteogenic factor delivery enhances bone marrow stromal cell-driven bone regeneration. *J Bone Miner Res* 20(5): 848–57.

97. Kaigler, D., et al. 2006. VEGF scaffolds enhance angiogenesis and bone regeneration in irradiated osseous defects. *J Bone Miner Res* 21(5): 735–44.

98. Semino, C.E. 2008. Self-assembling peptides: From bio-inspired materials to bone regeneration. *J Dent Res* 87(7): 606–16.

99. Hartgerink, J.D., E. Beniash, and S.I. Stupp. 2001. Self-assembly and mineralization of peptide-amphiphile nanofibers. *Science* 294(5547): 1684–88.

100. Ma, M., et al. 2010. Aromatic-aromatic interactions induce the self-assembly of pentapeptidic derivatives in water to form nanofibers and supramolecular hydrogels. *J Am Chem Soc* 132(8): 2719–28.

101. Cui, H., M.J. Webber, and S.I. Stupp. 2010. Self-assembly of peptide amphiphiles: From molecules to nanostructures to biomaterials. *Biopolymers* 94(1): 1–18.

102. Greenfield, M.A., et al. 2010. Tunable mechanics of peptide nanofiber gels. *Langmuir* 26(5): 3641–7.

103. Jang, J.H., O. Castano, and H.W. Kim. 2009. Electrospun materials as potential platforms for bone tissue engineering. *Adv Drug Deliv Rev* 61(12): 1065–83.

104. Rnjak, J., et al. 2009. Primary human dermal fibroblast interactions with open weave three–dimensional scaffolds prepared from synthetic human elastin. *Biomaterials* 30(32): 6469–77.

105. Panseri, S., et al. 2008. Electrospun micro- and nanofiber tubes for functional nervous regeneration in sciatic nerve transections. *BMC Biotechnol* 8: 39.

92. Barry JJ, et al. 2006. Porous methacrylate tissue engineering scaffolds: Using carbon dioxide to control porosity and interconnectivity. Biomat 27, 41. 4196-4204.

94. Curtis FL, et al. 2004. Mammalian cell survival and proliferation in supercritical CO2. Biol. Eng 10 U.S.A 10179-A-51.

96. Huang YC, et al. 2005. Bone regeneration in a rat cranial defect with endogenously of FH overexpressed plasmid DNA encoding for bone morphogenetic protein-4 (BMP-4). Gene Ther 12(6), 418-426.

98. Huang YC, et al. 2005. Combined angiogenic and osteogenic factor delivery enhances bone marrow stromal cell-driven bone regeneration. J Bone Miner Res 20(5), 848-857.

96. Kaigler D, et al. 2006. VEGF scaffolds enhance angiogenesis and bone regeneration in irradiated osseous defects. J Bone Miner Res 21(5), 735-44.

98. Serraton C, et al. 2006. Controlled drug delivery properties from bio-inspired materials in bone regeneration. Biomaterials 27, 40-55.

99. Hartgerink JD, Beniash E, and Stupp SI. 2001. Self-assembly and mineralization of peptide-amphiphile nanofibers. Science 294(5547), 1684-88.

100. Niu SI, et al. 2006. A simple approach towards to induce the self-assembly of peptide-amphiphiles in water to form 3D nanofibers and supramolecular hydrogels. J Am Chem Soc 135(8), 219-56.

101. Guo H, Li J, Witten and SI Stupp. 2006. Self assembly of peptide amphiphiles from pharmaceuticals to drug delivery. Biomaterials 99(1), 1-14.

102. Greenfield MA, et al. 2010. Tunable mechanics of peptide nanofiber gels. Langmuir 26(5), 3641-3647.

103. Silva LI, Czeisler C, Kam JW Kim, SI. Electrospun nanofibrils as potential platforms for tissue engineering. J Biomater Sci Polym Ed 16(1), 104-11.

104. Silva GI et al. 2004. Selective human neuronal differentiation with open water 3D-three-dimensional scaffolds prepared from amino acid human elastin. Biomaterials 20(14), 54-64.

105. Panseri S, et al. 2005. Electrospun micro- and nanofiber tubes for functional nervous regeneration in sciatic nerve transactions. BMC Biotechnol 8, 39.

10

Regulatory Challenges in Biomaterials: Focus on Medical Devices*

Pablo Gurman, Orit Rabinovitz-Harison, Tim B. Hunter

CONTENTS

---

* Information in this chapter is used with permission from Argonne National Laboratory, operated by UChicago Argonne, LLC, for the U.S. Department of Energy, under contract No. DE-AC02-06CH11357.

Preface

The field of biomaterials has grown significantly over the last years. The conceptual limits of the discipline have thus become confusing and a matter of debate. Besides, since it is a multidisciplinary field, a biomaterial may have many diverse definitions. One of these definitions states that biomaterials are "materials that are used in medical devices." This definition is supported by the fact that biomaterials are being currently used in as many as 8,000 medical devices [1]. In addition, this multidisciplinary concept has an important practical significance because biomaterials are being used with multiple applications in a variety of fields such as stem cells and tissue engineering, gene therapy, and micro-electro-mechanical/nano-electro-mechanical systems (MEMS/NEMS) [1, 2]. Moreover, biomaterials are incorporated into a variety of medical technologies and play a critical role in the failure or success of these technologies. These technologies include cardiovascular and gastrointestinal stents, defibrillators, artificial hips, and pacemakers, all of which are appearing in increasing numbers in the market. These devices, which save the lives of millions of people in the world every year, must be manufactured with appropriate/functional biomaterials for optimum, robust, long-life performance when implanted in the human body.

The federal government of the United States has created a number of agencies in charge of ensuring that, once they enter the market, these products will perform as expected, without causing harm to the general population. The Food and Drug Administration (FDA) is the federal regulatory agency involved in the regulation of biomedical products, including those made of or using biomaterials. The regulation of biomaterials involves several kinds of products, each of which is under the jurisdiction of one of the FDA divisions empowered to control their commercialization by enforcing manufacturers to comply with standards that ensure the safety and effectiveness of such products [3].

A major objective of this chapter is to focus on some of the regulatory challenges faced by medical devices containing biomaterials. In addition, the chapter aims to provide a general picture of the regulatory process for readers not involved in this subject. Finally, it is the intention of the author to use this chapter as a reminder of the great importance of these regulatory agencies, as they are critical for maintaining public health while allowing the introduction of innovative approaches and technologies in health care.

The views and conclusions expressed in this work are those of the author and do not necessarily represent those of or imply endorsement from the Food and Drug Administration.

Introduction: Biomaterials and Regulation of Biomedical Research—A Historical Overview

A Brief History of the Development of Biomaterials

If we take a brief look at our history, from the Paleolithic to the Industrial Revolution and the modern era of nuclear energy, we will see that humankind has dealt with drugs, food, and devices since very early times. By experimentation with techniques and tools developed throughout our history, like that of knives made of stones, humans have learned how to use tools for medical purposes. The same concept applies to the use of medicines obtained from plants and food obtained by bioprospecting the natural resources of our world by the first hunters. We can track many of these facts by archeological research and other sources. For example, a prosthetic used to replace an amputated toe was found in Theban tombs dated 1065–740 BC [4].

With the advent of new technologies, the science of biomaterials has evolved rapidly. This evolution has brought about several consequences, including a more intimate contact of materials with the human body, more complexity, and better performance. To better understand this technological evolution, we can divide the history of biomaterial development in three stages. The first stage was characterized by a demand of materials capable of being implanted inside the body. At the beginning, these materials were taken from industrial raw materials and used in applications ranging from orthopedics to cardiovascular surgery and ophthalmology. However, they elicited unacceptable toxicity. Many materials were not biodegradable where biodegradability was desired. Of these classes of materials, we can mention pure metals that elicited toxicity by corrosion when exposed to the biological media, polymers such as cellulose acetate (originally used for dialysis tubes), rubber, and ceramics such as zirconia. The second stage in biomaterials development took place during the 1970s, which provided materials that were not only biocompatible and biofunctional, but also biodegradable or bioactive. Among this "second generation" of biomaterials we can mention synthetic polymers such as absorbable sutures made of polylactic and polyglycolic acid or chitosan, bioactive ceramics such as calcium phosphate, and biodegradable ceramics such as hydroxyapatite, used in metallic prostheses to improve fixation to the bone. The third stage encompasses the development of biomaterials since 2000 to the present. These materials are both bioactive and biodegradable and take advantage of new microfabrication and nanofabrication techniques that make these materials very complex by introducing sensing and actuating elements. Examples of third-generation biomaterials are the bio-micro-electro mechanical systems (BioMEMS) [1].

The Role of Ethics in Biomedical Research

As more complex devices were developed as a consequence of human needs and technological evolution, many ethical issues emerged. In many cases, the development and testing of new medical products has been carried out while ignoring these ethical implications, following only economic profits as the main goal and bringing catastrophic consequences to the patients (Figure 10.7). In the best scenario, these practices were developed without weighing the risks against the benefits for the patient [5, 6]. It was not until the end of World War II, after the atrocity of the Nazi genocide, which included human experimentation, that the first code of conduct for research on humans was written and approved, in Nuremberg in 1947. The Nuremberg code stated three principles for conducting research on human subjects, namely: voluntary and informed consent, favorable risk-benefit analysis, and the right of the subject to withdrawn from the research at any time without further consequences [6]. Later on, in 1964, the Helsinki Declaration was written by the World Medical Association. This declaration stated two fundamental principles in addition to the Nuremberg principles: a) the subject of research will be provided with the best treatment available, and b) care over the subject will be the first priority over the benefits for society. Then, in 1978, the Belmont Report was written, establishing three fundamental pillars for conducting biomedical research in human subjects: 1) respect for the human subject, 2) beneficence, and 3) justice [6]. Currently, the Helsinki Declaration (the latest amendment of which was approved in Seoul, Korea, in 2008) is being used as one of the main guidelines to conduct clinical research ethically [7].

Regulation of Medical Devices in the United States, Europe, and Japan

In order to assess the safety, performance, and effectiveness of biomedical devices, to control their commercialization, and to protect the public health, several national agencies and worldwide private organizations are responsible for overseeing the manufacture and market access of medical products. These agencies and organizations have different policies, according to national requirements. In Japan, the agency in charge is the Pharmaceutical and Medical Devices Agency (PMDA), although final authority falls under the jurisdiction of the Japan Ministry of Health, Labor and Welfare (MHLW). Within the MHLW, the Pharmaceuticals and Food Safety Bureau reviews the regulation of pharmaceuticals, food, and medical devices. In Europe, there operates a decentralized system based on Notified Bodies (NB), which are private organizations with authority to bestow the "CE" marking, which provides clearance for commercialization to medical device companies within European countries.

Notified Bodies are audited by the national agencies of each country (e.g., the Medicines and Healthcare Products Regulatory Agency, in the U.K.). As we shall discuss later during this chapter, within the United States, the Food and Drug Administration (FDA) takes responsibility for regulating medical devices and ensuring their safety and effectiveness for the general population [3, 8–11].

But why is this chapter oriented solely on the FDA? The reason to have chosen the FDA for an overview of biomaterial-based medical device regulations is that it is not possible to describe in detail every aspect of the European, Japanese, and U.S. regulatory agencies. Accomplishing this task would require more than one chapter; otherwise there is the risk of confusing readers with a plethora of policies and names. Instead, briefing a few useful concepts in biomaterial-medical device regulation is the scope of this chapter. Moreover, the FDA is one of the most important agencies worldwide, playing a very significant role, since United States leads the pharmaceutical and medical device market. Finally, the FDA has launched innovative initiatives such as the regulation of nanotechnology products and the Critical Path Initiative, which provides an oversight of innovative initiatives in regulation of biomaterials and medical devices. However, since both European and Japanese regulatory systems are also key players in the global scenario of regulation of biomedical products, some aspects of these as well are summarized here (Table 10.1). The list is by no means complete, and interested readers are referred to the websites of the regulatory agencies, private organizations, and the referred literature. We suggest scanning the table and returning to it after finishing this chapter in order to get a better understanding of its contents.

Brief History of the FDA

The Food and Drug Administration (FDA), a federal agency within the United States Department of Health and Human Services, is "responsible for protecting the public health by ensuring the safety, efficacy and security of human and veterinary drugs, biological products, medical devices, food supply, cosmetics and products that emit radiation" [3]. The FDA was created as a result of several laws that were approved by the Congress of the United States to protect public health. The first of these laws was the Food and Drug Act of 1906, launched to monitor meat and food quality to protect against any possible adulteration and to control medicines that were not properly labeled or that had questionable compounds (called "patent medicines," but which, ironically, were never patented, to avoid having to disclose critical information about the ingredients needed to produce them). Later on, in 1938, after a tragic episode where more than 100 patients died as a result of the presence of diethylene glycol in sulfonamide products, which had been used to create a liquid form of this antibiotic, a comprehensive law called the Federal Food and Drug Cosmetic Act was created in the United States to improve the safety of foods, drugs and cosmetics. With the tragedy of thalidomide (contergan) in Europe (in the U.S. the drug had not been approved)

TABLE 10.1

Differences between United States, European, and Japanese Regulatory Agencies

Regulatory Agency	United States	Europe	Japan	References
History	1976 Medical Device Amendment 1990 Safe Medical Device Act 1997 Food and Drug Administration Act 2002 Medical Device User Fee and Modernization Act	In 1998 the 93/42/EEC Medical Device Directive (MDD) became fully operational that together with the Active Implantable Medical Device Directive 90/385/EEC and the In Vitro Medical Device Directive regulates medical device manufacturing and sale within Europe. The MDD demands either self-declaration of conformity with established standards or accreditation of standards compliance by a third party (Notified Bodies). These standards are based on "essential requirements" that the product must fulfill, including safety, due performance, and technical issues regarding design and manufacturing of medical devices.	1997 Creation of the Pharmaceuticals and Drugs Evaluation Center (PMDC) 2004 creation of the Pharmaceutical Medical Device Agency (PMDA) 2005 Implementation of the Pharmaceutical Affairs Law: Approval and licensing systems for biomedical products under revision The Marketing Authorization Holder becomes operative: a new system for foreign medical device manufacturers that intends to improve the safety and efficacy of medical devices by separating manufacturing responsibilities from marketing responsibilities.	[12] [13] [14] [10] [3] [15]
Authority	U.S. FDA – Center for Devices and Radiological Health	National Agencies – Notified Bodies (NB), private organizations that certify medical device manufacturers with CE mark under National Agencies Supervision	Ministry of Health, Labor and Welfare (MHLW) – Pharmaceutical Medical Device Agency (PMDA)	

Classification of medical devices (a risk-based approached is common for all the classification systems).	Class I general controls (surgical gloves) Class II special controls (endoscopes) Class III high-risk devices (pacemakers) Classification is based on intended used of the devices, indication of use of the devices, and risk (from lower risk to higher risk for the patient).	Class I low-risk devices (e.g., stethoscopes) Class I sterile Class I measure Class IIa medium- to low-risk devices (e.g., electrocardiographs) Class IIb medium- to high-risk devices (intensive-care monitoring system) Class III high-risk devices (e.g., prosthetic heart valves)	General medical devices (class I) (e.g., dental accessories) only premarket submission (self-declaration of conformity) required , not certification required by PMDA Designated Controlled Medical devices (class II) Premarket Certification required through third parties (similar to CE mark in Europe) (e.g., tracheotomy tubes) Controlled medical devices (class II devices other than designated devices) and highly controlled medical devices (class III: orthopedic implants, catheters; class IV: pacemakers, stents) Pre-market approval required through PMDA
Clinical trials for high-risk devices Examples: drug-eluting stents, coronary guidewires	Clinical trials designed to prove safety and effectiveness (comparative, large group trials with longer approval times)	Clinical trials designed to prove safety and performance (small groups with faster approval times)	Clinical trials designed to prove safety and effectiveness

(Continued)

TABLE 10.1 (CONTINUED)

Differences between United States, European, and Japanese Regulatory Agencies

Regulatory Agency	United States	Europe	Japan	References
Review and certification times	Premarket notification (510k) (class II devices) Review time from CDRH could take 3 months. Premarket approval (class III devices) could take 6 months. Process to get FDA approval to initiate first clinical studies 3–6 months. Review time for an Investigational Device Exemption (IDE) could take 1 month. Institutional Review Board review times (IRB) could take 3 months or longer.	Fast track reviews from some NB could take 45 days for class III medical devices and less than 45 days for Class II devices	Highly controlled medical devices. Premarket approval from PMDA could take 6–18 months (up to 90% of medical devices could be approved in one year and some cases could take up to 3 years). Controlled medical devices (class II) could take 1–3 months (certification time).	
Time to the market	⇧⇧	⇧	⇧⇧⇧	

in 1962, which caused severe birth defects, including heart defects and missing limbs, a new law known as the Kefauver-Harris Drug Amendments Act was passed to strengthen the power of the FDA to monitor drug safety in the United States [46]. In 1976, the creation of the Medical Device Amendment Act came to be the first medical device regulation in the United States. This amendment called for quality control procedures and the registering of manufacturers of medical devices with the FDA [16, 17]. In 1990, the Safe Medical Device Act required safety studies and reporting from public institutions of any adverse event that could be related to a particular medical device. In addition, device manufacturers were required to perform safety monitoring of implantable medical devices in the form of post-marketing safety studies, including reports and tracking of the patients using their products. In 1997, the Food and Drug Administration Act was approved. By this act, the FDA reaffirmed its commitment to protect public health by ensuring the safety of foods and the effectiveness and safety of drugs, medical devices, and radiation-emitting products. This act sought also to improve the labeling of these products, consult with experts of every field about every aspect involved in public health protection, and decrease the level of bureaucracy and improve harmonization with other regulatory agencies worldwide. In 2002, the Medical Device User Fee and Modernization Act was incorporated. In addition, the Office of Combination Products was created [16, 17]. The 2002 act incorporated fees for the premarket approval of devices, allowed for the inspection at the manufacturer site by third parties, and introduced new regulations in reprocessed single-use devices [16, 17].

FDA Organization and Operation

FDA Organization

To address the question of how the FDA is able to regulate almost one quarter of the products sold in the United States, it is important to understand its organization. To achieve its goals, the FDA is organized into six product centers, one research center, and two offices: the Center for Biologics and Evaluation Research (CBER), the Center for Drug and Evaluation Research (CDER), the Center for Food Safety and Applied Nutrition (CFSAN), the Center for Tobacco Products (CTP), the Center for Veterinary Medicine (CVM), the Center for Devices and Radiological Health (CDRH), the National Center for Toxicological Research (NCTR), the Office of Regulatory Affairs (ORA), and the Office of the Commissioner.

> Center for Biologics and Evaluation Research (CBER): The CBER is in charge of the regulation of a wide variety of products, such as gene therapy, blood and blood components, vaccines, tissues (such as

biological heart valves) and tissue-engineering products. Some of the issues where the FDA takes regulatory action are product and manufacturing establishment inspections, licensing and safety of blood supplies (blood and blood components), and post-marketing activities, including the post-marketing surveillance of biological products where many adverse events related to biological products could be detected.

Center for Drug and Evaluation Research (CDER): The CDER evaluates every drug intended to be marketed.

Center for Food Safety and Applied Nutrition (CFSAN): The CFSAN regulates the labeling and safety of food and cosmetics.

Center for Tobacco Products (CTP): The CTP evaluates tobacco products and inspects manufacturers of tobacco products.

Center for Veterinary Medicine (CVM): The CVM is responsible for assessing the safety and effectiveness of animal food, drugs, and devices. It is important to note that some of the products, such as implantable identification devices, demand the use of biomaterials to achieve biocompatibility since they are implanted inside the animal body.

Center for Devices and Radiological Health (CDRH): The CDRH is the center responsible for evaluating medical devices (and therefore the biomaterials that are used in such devices), premarket approval of new devices, manufacturing and performance standards, and tracking reports of device malfunctioning and serious adverse reactions. In addition, the CDRH reviews radiation safety performance standards, including those products that emit X-rays, microwaves, radiofrequencies (RF), and ultraviolet light. The relevance of radiation-emitting products in biomaterials and medical devices could be better understood if we think of the use of magnetic resonance imaging (MRI) in patients carrying pacemakers, where electromagnetic interactions between the pacemaker and the magnetic and RF fields produced by the MRI equipment may take place and interfere with the correct functioning of the pacemaker. One can envision novel biomaterials capable of shielding the device against unwanted magnetic fields. Another example is the use of magnetic nanoparticles made of iron oxides (magnetite), which are used as heating elements for cancer therapy. Here, the magnetic, biocompatible, and biodegradable properties of the magnetic nanoparticles are exploited as implantable antennas that produce heat when they are excited by an external magnetic field and in this manner are used as therapeutic tools to destroy tumor cells. The CDRH also has a center for science and research that supports the regulatory decisions on a scientific basis. This center is divided into several areas, including biology, physics, and electrical engineering. One of these areas is focused on chemistry and materials science, supporting the necessary knowledge to evaluate the diverse repertoire of biomaterials that are found in many medical devices.

National Center for Toxicological Research (NCTR): The NCTR plays a critical role in assessing the safety of new technologies. In addition, the NTCR assesses the toxicology of certain biomaterials, such as titanium oxide and zinc oxide nanoparticles. These compounds are commonly used in sunscreens and other products.

Office of Regulatory Affairs (ORA): Among other relevant tasks, the ORA is responsible for FDA inspections of manufacturers' facilities and products as well as for the control of imported products.

Office of the Commissioner: This office takes responsibility for effectively conducting the FDA's mission. Within this office, the Office of Combination Products (OCP), created in 2002, addresses the regulatory gap of the increasing number of products made by more than one technology (devices with biologics, drugs with devices, or drugs with biologics), such as drug-eluting stents, drug-delivery polymer scaffolds, and antibiotic bone cements, which have been under development in the last years and are intended to obtain FDA clearance to be commercialized. The OCP has a jurisdictional and classification authority that allows it to determine which center will be responsible for performing a premarket review of a combination product. This is accomplished by defining the primary mode of action of the product that will contribute to the desired therapeutic effect to a larger extent. For instance, in a drug-delivery device, where the drug is the main determinant in the mode of action of the product to produce the desired therapeutic effect, the CDER will have primary jurisdiction over the product review, whereas if the device itself determines the primary mode of action to produce the desired therapeutic effect, then the CDRH will be the final reviewer of the documents submitted by the manufacturer [3].

With a few exceptions, there is usually a series of technical requirements called standards that must be followed by any medical product containing biomaterials (from a stent to a polymer suture) in order to ensure its quality and safety before obtaining FDA approval to enter the market. Thus, we will next briefly describe some aspects of standards, to give a better understanding of the field before going deeper into the existing approval pathways for medical devices.

FDA Standards

To ensure optimal performance and safety of medical devices and the biomaterials they are made of, a comprehensive technical assessment is needed. This technical assessment is based on obtaining reliable information of the material or device performance and contrasting it against parameters that are adopted as "standards," which will determine whether the product being assessed complies with the necessary parameters to be commercialized. According to the International Standard Organization (ISO), a standard could be defined as "a document established by consensus approved by a recognized body, which, for common and repeated use, provides rules, guidelines and/or characteristics for activities or their results, aimed to achieve the optimum degree of order in a given context" [18].

Voluntary Standards

Standards are usually developed by nongovernmental voluntary organizations called standard-developing organizations by consensus among experts and later adopted by the industry, sometimes as a competitive advantage, and otherwise because they are mandatory. In the latter case, these standards adopted by the regulatory agency enforcing their use must be followed by the industry as a requirement for obtaining permission to commercialize their products. In many cases, however, these standards remain voluntary. Standard-developing organizations are important because a) regulatory agencies cannot afford the economic cost and human resources necessary to set standards for any product or process that is or will be in the market, and b) voluntary standards allow industry to adopt the standard that fulfills its objectives without being enforced to comply with standards that are either not relevant or not related to the industry's activities.

There are several standard-developing organizations, such as the International Organization for Standardization (ISO) and the American Society for Testing and Materials (now known as ASTM International). In addition, many other Developing Standard Organizations are working with biomaterial standards, many of which have been adopted by the FDA.

International Organization for Standardization (ISO)

The ISO is devoted to developing standards across a wide variety of disciplines. To this end, it is divided into committees, each of which takes part in writing guidance documents, which become standards after being subjected to a general vote among all the groups participating around the world. Some examples of ISO committees focusing on medical devices are ISO/TC 194 (which focuses on the biological evaluation of medical devices) and, although still in a nascent stage, the ISO/TC 229 WG3 (Nanotechnologies in Health). Standards related to medical devices currently in use are the ISO 10993, for the evaluation of biological performance of medical devices, and ISO 13485, a quality standard for medical devices [18].

American Society for Testing and Materials (ASTM)

ASTM International (formerly known as the American Society for Testing and Materials) is an internationally recognized voluntary organization that develops standards for products, processes, and services that are used by industry. Within the ASTM, the F04, or Medical and Surgical Material and Device Committee, sets the standards in issues regarding biomaterials and medical devices. The ASTM has developed several tests, standard specifications, and standard terminology for a wide variety of biomaterials that have been adopted by the FDA. Among these tests are the corrosion and fatigue test for metallic implants, an *in vitro* test to evaluate the hydrolytic degradation properties of polymers used in surgical implants, the standard test for measuring magnetically induced torque in medical devices in a magnetic resonance environment, the standard specification for unalloyed titanium for surgical implants applications, and the standard specification for calcium phosphate coatings for implantable materials.

With the aim to improve the safety of medical devices, the FDA has created the Standard Management Staff (SMS) within the CDRH. The SMS is responsible for the FDA Standards Program, which is in charge of adopting the standards developed by the above-mentioned organizations. The SMS has developed a standard database where several of the FDA recognized standards are listed divided by specific areas, including materials and devices, thus facilitating the selection of appropriate standards by manufacturers. The use of the recognized standards that are listed in this database could benefit manufacturers during the regulatory process [19, 20].

Mandatory Standards

Although some standards are voluntary, other standards are adopted by regulatory agencies and become mandatory. GMP, GLP, and GCP (good manufacturing practices, good laboratory practices, and good clinical practices, respectively) are examples of mandatory standards that have been agreed upon by the International Conferences on Harmonization (ICH), an international effort carried out by Europe, Japan, and the United States to establish scientifically consensuated guidelines in biomedical research and development, and adopted by regulatory agencies all over the world to ensure that standards are followed by manufacturers during the development, laboratory testing, and clinical evaluation of medical products.

Good manufacturing practices (GMP) are a set of requirements including facilities, equipment, personnel, packaging, and labeling of medical devices that should comply with the FDA regulations according to the Code of Federal Regulations (the Code of Federal Regulations, or CFR, is a set of rules in a number of volumes produced by departments and agencies of the federal government) to ensure that device manufacturing is safe and that any chance of defects in the final product or adulteration is minimized or avoided. FDA personnel have the authority to perform inspections at any of the manufacturer facilities where devices are developed and inspect all the records and documents. If the FDA finds a violation to GMP, it can halt the production of the device for up to 30 days [16].

Good Laboratory Practices (GLP) are standards adopted by the FDA and registered in the Code of Federal Regulations that ensure validation of the data arising from preclinical studies (but not clinical data) of medical products, including biomaterial-based medical devices. Examples of areas covered by GLP are organization and personnel, facilities, equipment, records and reports, and protocols for conducting a non-clinical laboratory study. For instance, a manufacturer developing new bone cement would have to conduct animal studies according to GLP. It is important to note that GLP are regulations and not just standards, and therefore are mandatory. If at an early stage of development a manufacturer wants to use a standard other than those described in GLP, he or she should discuss with an FDA representative the use of voluntary standards instead of standards specified in GLP [21, 22].

Good clinical practices (GCP) are international standards that provide the scientific and ethical validity of the clinical data originating in clinical trials and submitted to the FDA by the manufacturer of a medical product. GCP include ethical principles, scientific procedures, including how a clinical trial should be conducted according to a clinical protocol, reporting of adverse events that could be associated with a medical product, organization and responsibilities of an Institutional Review Board, audits, informed consent of human subjects participating in a clinical trial, and all the relevant information that is required to ensure protection of any human subject participating in a clinical study [23, 24].

Now that we have an overview of some of the technical requirements that are necessary to market a medical product, we will review in more detail what the pathways are that a medical device should follow to obtain FDA approval. To review the regulatory aspects of every product and the center responsible for it, however, goes beyond the scope of this chapter. Instead, a general description of medical device regulation will be presented. The importance of understanding medical device regulations in biomaterial research is evident, since there are presently biomaterials in some 8000 medical devices. This approach will provide a general idea of how the FDA regulates biomaterials (or at least those that are used in medical devices) [3, 16].

FDA Pathways for the Approval of Medical Devices

Medical Devices

Before we start reviewing the FDA pathways for the approval of medical devices, it might be useful to try to answer the following questions: What is considered a medical device? Why are medical devices important in healthcare?

According to the FDA, a medical device is

> an instrument, apparatus, implement, machine, contrivance, implant, *in vitro* reagent, which is recognized in the United States Pharmacopoeia, intended for use in the diagnosis of disease or other conditions, or in the

cure, treatment, or prevention of disease, or does not achieve its primary intended purposes through chemical action within the body and which is not dependent upon being metabolized for the achievement of any of its primary intended purposes. (Adapted from [16]).

To address the second question, we should mention that medical devices are used in so many people, assisting them in the workforce and improving their quality of life, that they have become a cornerstone of patient care. Medical devices are used for a variety of purposes. Some of them, such as ECG leads, temperature-sensing bladder catheters, and pulse oximeters, are used to monitor a patient, whereas others, such as prosthetic heart valves, abscess drainage catheters, and fracture fixation plates, are used for therapeutic purposes. Thus, their proper functioning is requisite for patient care and evaluation. Unfortunately, some problems and complications related to medical devices are sometimes overlooked or misinterpreted. Besides, the complications may become far more problematic to the patient than the conditions for which they were intended to treat or monitor. Examples of these complications are a contaminated orthopedic fixation plate leading to a limb-threatening bone infection, and a misbehaving cardiac pacemaker causing cardiac arrest. For these reasons, patients and/or their families need to be familiar with their own medical apparatus, its presence, its use, its complications, and plans for its removal or deactivation. Furthermore, the personnel rendering care to the patient need to be familiar with any medical apparatus involved in the patient's care. Physicians not only need to know about the devices being used for their patients, but should also be familiar with the uses, complications, and contraindications for the medical devices they use as part of their practice. Physicians should also have an overview of medical device regulations and the reporting requirements for medical complications related to device use. Engineers working in biomaterials should also be aware of the complexity of medical device regulations, since they play a fundamental role in the development process of each device.

Another reason why medical devices are important is that, from an economic point of view, medical device manufacture and sale are important business operations employing thousands of people, with a staggering number of medical devices currently on the market. A large number of medical devices using many kinds of biomaterials are launched into the market every year. It is estimated that 4% of the American population uses at least one implantable medical device. Before 2000, the FDA had approved almost 500,000 medical devices developed by 23,000 manufacturers [25]. In 2008, there were nearly 350,000 pacemakers, 140,000 implantable defibrillators, and 1,230,000 stents implanted in the United States [26]. According to the Advanced Medical Technology Association, the medical devices market reached $77 billion in 2002 [27], whereas a WHO report estimated a 260-billion-dollar market in 2006 [28]. According to a recent study that took into account different scenarios, the cost to produce a medical device was estimated to be between $322 million and $522 million [29].

Medical Device Regulation

How a medical device is regulated depends on how the device is classified in the eyes of the FDA, a consideration that needs a description of the approval pathways of the FDA. The FDA has established three categories of medical devices according to the risk they represent to the patient: Class I, Class II and Class III. Class I devices are considered low-risk devices and therefore require low control levels. An example of a Class I device is examination gloves. Class II devices are of intermediate risk and have more control levels than Class I devices. Examples of Class II devices are CT scanners. Finally, Class III devices are high-risk devices that require high-level controls. Examples of Class III devices include defibrillators and pacemakers. A high-risk device represents a life-supporting device or a device that could have life-threatening consequences if it fails. A low-risk device is any device other than a high-risk device [3]. Once a device is categorized in one of the classes described above, specific requirements should be followed. The requirements that apply for each class are "pathways" for obtaining FDA approval. The possible "regulatory routes" a device could follow are described below.

Premarket Notification (510k)   The premarket notification, called a 510k, is the pathway followed by most biomaterial-based medical devices in order to get clearance from the FDA and enter the market in the United States. By requesting a 510k, the FDA establishes a level of equivalence to an existing device. The level of equivalence is based on the intended use of the device or the technological characteristics that could affect the effectiveness and safety of the new device. Most of the evidence submitted in a 510k is based on bench and sometimes pre-clinical data. Only 10 to 15% of the 510k notifications are required to submit clinical data. If the device has a degree of equivalence to one preexisting in the market that fulfills FDA requirements, a premarket notification is sufficient to market the product within the United States. If the device does not meet FDA criteria for equivalence to an existing device, a premarket approval is required. The reason for this difference is that if there is enough similarity to an existing device in the market, it will be expected that the new device will perform similarly in terms of safety and effectiveness to the preexisting one.

Premarket Notification (510k) Exemptions   Class I and Class II devices can be exempted from a premarket notification if

- They are pre-amendment devices, which means that they existed in the market before 1976 and have not significantly changed since then.
- They are exempted by regulation.

Examples of Class I and Class II devices exempted from 510k because the material used for these kinds of devices existed before 1976 are neurological devices such as two-point discriminators made of stainless steel, dental materials such as carboxymethylcellulose sodium denture adhesive, and a hydrogel wound dressing used in plastic surgery.

Exemptions to the regulations occur when the FDA decides to refocus the resources used to assess 510k notifications onto public health issues of higher priority.

A complete list of Class I and Class II devices exempted from 510k is available from the FDA site [30].

Even when a medical device is exempted from submitting a 510k, the following information should still be submitted:

- Registration and listing
- Labeling
- Good manufacturing practices (GMPs). (It is worth noting the relevance of standards in that devices not subjected to premarket notification are still required to present evidence of compliance with manufacturing standards.)

Premarket Approval (PMA)   The premarket approval is the way in which the FDA assesses Class III devices. The PMA is required when it is not possible to prove substantial equivalence to an existing device, and therefore, in addition to bench and pre-clinical data, additional clinical data must be provided to demonstrate the effectiveness and safety of the investigational device. This is clearly the hardest and more expensive way of bringing medical devices to the market. High-risk, innovative devices are more likely to follow this pathway.

Two other important categories or pathways are the Investigational Device Exemption (IDE) and the Humanitarian Device Exemption (HDE).

Investigational Device Exemption (IDE)   If a device is intended to be used for the first time in a clinical trial (see further), an IDE must be submitted to the FDA by the manufacturer. An IDE could be required by the CDRH to support a 510k or PMA application when clinical data are needed. If the device is Class III, there is also a need for an Institutional Review Board (IRB) approval. As previously described in this chapter, as a reaction to the atrocities committed during World War II, a code of conduct in clinical research was created by consensus and formally written in the form of declarations. An IRB ensures that the principles behind these declarations will be followed. Therefore, the aim of an IRB is to ensure that the rights of the subjects enrolled in a clinical study will be protected [3, 6]. An IDE should also be submitted when the sponsor wants to submit a new use for an existing device.

Humanitarian Device Exemption (HDE)   Medical devices that will be used for rare diseases affecting fewer than 4,000 patients a year in the United States belong to the HDE category. One example of this category is fetal bladder stents. This category was developed to help manufacturers in commercializing medical devices that have small markets and therefore small financial incentives [31].

FDA Fast-Track Approvals   Although the FDA could take actions that may delay the biomaterial-medical device approval process (such as requesting additional laboratory tests and/or clinical data), it could also, under some circumstances, accelerate it. The FDA created fast-track approvals as a pathway to facilitate the access to the market of certain products that fulfill an unmet medical need. One type of fast-track are some devices (not available to the general population) that are exempted from FDA review and approval, including registry, standard conformity or premarket approval, when they are considered custom devices and are ordered from some healthcare professionals to treat special populations or individuals, such as the case of dental and orthopedic devices [16, 32].

CASE STUDY: HIP REPLACEMENT

A medical device company developed a new hip system that claimed to be better than the current systems in the market. The technology innovation was based on the combination between a ceramic femoral head and the metallic acetabular insert, although both systems already existed as separate products in the market, having both been cleared through 510k. The better performance of the device claimed by the company was in terms of efficacy (less friction, longer durability) and safety (less complications) over their competitors.

FDA rejected the 510k claiming the lack of evidence to ensure that an equivalent product was already present in the market although the same company had been commercializing the ceramic femoral head and the metallic acetabular insert as two separate products. The company had to withdraw the 510k and submit a Premarket Approval (the FDA pathway for class III devices).

The company performed a pivotal trial, a non-inferiority study to demonstrate that the invention was non-inferior in terms of performance over the existing products. This study, in addition to data provided by the products already in the market and a planned post-market study, allowed the company to submit a PMA in order to obtain FDA approval (Ref 49).

Clinical Trials

Critical Steps in Medical Device Development

The medical device development process is extremely complex and takes place in several stages, as depicted in Table 10.2. From the initial idea to the design and fabrication stages, the medical device development process is

TABLE 10.2

Critical Steps in Medical Device Development

Stages	Comments
Initial Stage	
Familiarize oneself with applicable regulations, including requirements of Design Controls.	As discussed on the FDA Device Advice web page, Design Control requirements begin at the initial stages of device development. You cannot go back later!
Preclinical Stages	
Establish an animal model to evaluate biocompatibility and device functioning.	Animal model should be scientifically supported based on device design. Good laboratory practices requirements apply.
Meet with FDA to discuss biocompatibility testing and device design.	The types of biocompatibility tests rather than detailed data would be discussed.
Submit pre-IDE.	Provides opportunity for comments.
Submit IDE for FDA review.	Local IRB submission may precede this.
Submit study application to local IRB.	Studies in U.S. require both IRB and IDE approval.
Initial Clinical Stages (Phase 1)	
Conduct initial phase of human study and analyze data.	Clinical data would be submitted at least yearly in IDE annual progress report.
Modify device and clinical protocol as appropriate.	Consider outcomes of feasibility phase in future plans.
Later Clinical Stages (Phases 2 and 3)	
Submit IDE supplement to FDA for next phase of clinical testing (two or more phases may be necessary).	Discussion with FDA advised, particularly if device is re-designed.
Obtain IRB approval as necessary.	More than one IRB approval may be needed, depending on institutions involved.
Meet with FDA for pre-PMA submission meeting.	Provides opportunity to discuss format and content of PMA.
Submit PMA to FDA for review of marketing application.	Database will have been closed and audited prior to PMA submission.
Post Marketing Study (Phase 4)	
If required as a condition of PMA approval.	A post-approval study requirement is dependent upon many factors.

Reprinted with permission from IOP Publishing Ltd.
Courtesy of Saviola, J. 2005. The FDA role in medical devices clinical studies of human subjects. *J Neural Eng* 2: S1–S4.

intimately tied to regulatory requirements, including materials being used, animal testing, and finally, in some cases, clinical trials.

Simply stated, a clinical trial is a research performed on human subjects. This research is a scientifically and ethically controlled experiment performed on a limited population that intends to statistically represent the real population to assess the performance of the medical product (drug, device, biologic) after being tested during the laboratory and pre-clinical stages of development. As stated previously, clinical trials are required in PMA submissions that apply for high-risk devices (Class III) and in some cases for 510k submissions as well. The reason behind this is that high-risk devices or Class II devices submitting a 510k require more evidence that supports what the manufacturer alleges about the product. Preclinical data may not be enough for the FDA to classify a product as safe and effective, and therefore, in addition to preclinical and bench data, the FDA requires clinical data. A recent search in the public clinical trials database clinical trials.gov showed that there are currently more than 1,000 ongoing clinical trials involving medical devices [34].

Clinical Trials Protocol

In order to perform a clinical trial, a clinical protocol should be written to provide the following information:

a) The idea that justifies the clinical trial: In clinical research as well as in other fields of science, researchers want to test ideas. These ideas could be expressed in terms of a question or research hypothesis that the researchers aim to demonstrate (as in any other experiment). There are different kinds of hypotheses:

- *Explanatory vs. Pragmatic*: The study may be *explanatory* of some physiopathological hypotheses (such as the biofouling of implantable devices related to macrophage activation and inflammatory response around the implant) or be *pragmatic* (biofouling could be avoided by bio-inert materials that avoid protein adhesion) [35, 36].

- *Efficacy vs. Effectiveness*: The material was effective in decreasing biofouling (efficacy), or the material achieved the goal of reducing biofouling by decreasing the thickness of the fibrous capsule around the implant by 50% against the current best available material (*effectiveness*).

- *Superiority vs. Equivalence*: Treatment A is superior to treatment B (*superiority*) or treatment A is as effective as treatment B (*equivalence*). Equivalence studies are common in drug trials assessing generic drugs vs. brand name drugs.

b) The purpose of the study: The protocol should state in a single phrase what the condition the researchers intend to treat is (what the patient population target is) and what the clinical outcomes are that they expect to achieve through the proposed intervention.

c) Exclusion/inclusion criteria for patient enrollment: When enrolling patients in the clinical trial, the researcher will select those patient conditions that will allow the assessment of the effectiveness of the proposed treatment (*inclusion criteria*) and will exclude others that could interfere in the assessment (*exclusion criteria*).

d) Clinical end points: The clinical outcomes should be measured to evaluate how good the proposed intervention is against a placebo or the current best management approach. This is done by using some clinical condition, laboratory values, or other measurable parameter, which are called *end points*. End points could be either direct (mortality rates among patients undergoing a new device for a cardiac condition such as arrhythmia) or indirect (when we cannot use a direct measurement for ethical or technical issues; an example of this is the measurement of biomarkers that indirectly relates to patient mortality).

e) Study design: The main issues that are usually found in the literature when describing trial design are:

1. *Controlled vs. non-controlled trials*: There is a group that will be receiving the experimental device in contrast to a group that will be receiving the best standard current management or in some cases a placebo [35, 36].

2. *Randomized vs. non-randomized trials*: Randomized trials are used to define the random nature of assignment of the patient into one of the groups taking part in the clinical study and implies that any patient has the same chances to receive either the standard device or the experimental device [35, 36].

3. *Blindness*: This concept refers to the persons (the patient, the clinical investigator, and/or external evaluators) who are prevented from knowing which treatment is being assigned to each group. According to this, trials are classified as "open" (when all the participants, including the patient, know the treatment they are receiving), "single blind" (when only the patient is blinded), "double blind" (when both the patient and the clinical investigator are blinded), and other more complex methodologies such as "double dummy", "triple blind," and "quadruple blind." The aim of blindness is to avoid the so-called blindness bias that occurs either in the subject (the patient improves his/her condition because he/she knows that he/she is assigned to the new treatment) or in the physician (he/she sees improvement in the patient that is using the new treatment as he/she expects that this treatment will work out). In this regard, some precautions to reduce blindness bias could be taken. Some of the precautions that may be used include allocation concealment of the site where the operator will be performing the procedure on the patient and masking the outcome assessors after the medical device has been implanted [37].

4. *Parallel, cross-over, or factorial*: This is related to how the groups receive the treatment. In parallel trials, each group receives either the best management treatment or placebo and the other

group receives the treatment under study. In cross-over trials, the same individuals receives the standard treatment and the treatment under study at different times, and the clinically relevant outcomes (end points) are assessed in the same subjects. In factorial trials, more than one treatment is evaluated in the same trial by assigning different groups to receive different treatments that are assessed against the standard treatment or placebo.

5. *Number of patients taking part of the study*: The trial may involve either only one patient, a fixed number of patients (from smaller trials up to multicenter trials), or a sequential number of patients (the number of patients increases as the trial is being run).

d) Trial development: During the trial development, patients start to take the study medication or are subject to a procedure such as device implantation, and they are followed up to see the effect of the new intervention. Data regarding their medical status along the study are collected and their quality confirmed before the final data analysis is performed.

e) Analysis of the results: Statistical analysis of the results should be carried out in order to know whether the differences found between the experimental device and its comparator are statistically significant (and therefore could be attributed to the new intervention) or are a random difference [35, 36].

Drug Trials vs. Medical Device Trials

We will now describe briefly the differences between trials performed to assess a drug's effectiveness and safety, and trials performed to assess biomaterial-based devices' effectiveness and safety, because biomaterials are being used in both areas. Examples range from a drug delivery system that uses biopolymers surrounding a core where the drug (payload) is embedded to a less sophisticated titanium femoral head for hip replacement. Therefore, it could be useful for those interested in the biomaterial field to become familiar with both regulatory pathways.

As we have previously stated, drug regulation is older than device regulation. Drugs have been regulated since 1906 (Food and Drug Act), whereas medical devices started to be regulated more recently (after the 1976 Amendment Act). There are some differences in the phases of the trial, the number of patients being enrolled in each phase, the time to the market for each product, and the training of the investigator.

1) *Phases:* Drug development and device development are conducted in several stages or phases, each designed with a specific aim.

In drug development, there are the following phases:

Phase I, which is usually carried out in human volunteers, and is aimed at studying pharmacokinetic parameters such as metabolism, clearance, maximum dose tolerated, and safety issues. Phase I trials are the first stage of testing in human subjects, and, normally, a small number (20 to 100) of healthy volunteers are selected.

Phase II, which is designed to assess therapeutic efficacy, and is thus carried out with volunteer patients with the condition to be treated. Some pharmacokinetic parameters such as dose range also are studied in this phase in a selected group of patients. Phase II trials are performed in larger populations (100 to 300 patients) than Phase I trials. Phase II studies are sometimes divided into *Phase IIA* and *Phase IIB:*

- *Phase IIA* is specifically designed to address dosing requirements (i.e., how much drug should be given).

- *Phase IIB* is specifically designed to study efficacy (i.e., how well the drug works at the prescribed dose(s)).

Some trials, however, combine *Phase I* and *Phase II*, and test both efficacy and toxicity in the same study.

Phase III, which is the phase during which the drug of interest is compared against the best management treatment currently available, or in some situations, and if ethically approved, against a placebo. These are usually randomized controlled multicenter trials in a large population of patients (300–3000 or more) to mimic the real population. Phase III trials are intended to obtain an additional information about drug efficacy and to evaluate the overall benefit-risk ratio of the investigational drug.

Phase IV, which is carried out to collect safety information once the drug is in the market and is being used in the general population, although, in some cases, used solely for marketing purposes [38].

On the other hand, in medical device development, trials are conducted in the following phases:

Phase I, or the feasibility trial, during which the design and operating features of the devices are evaluated. This stage is carried out in a patient population of 20–50 (here we can try to see some analogy with Phase I drug trials, where the operating features of the device could be thought as equivalent to pharmacokinetic optimization of drugs).

Phase II, or the pivotal trial, which is usually a multicenter randomized controlled trial in which the effectiveness and safety of the device is being assessed. Usually, large populations (300–800) of patients are enrolled to mimic the real population. Still, as the number of patients that will be using a device is significantly smaller than the population that will be using drugs, the number of patients enrolled in a device trial is usually smaller than the number of patients enrolled in a drug trial.

Phase III, or post-marketing trials, which are performed to assess the long-term safety of the device up to 5 years after it has been launched onto the market. Post-marketing trials of medical devices represent a very important part of the device development pathway, mainly because many medical devices are approved without being tested in clinical trials. Moreover, even when clinical trials are performed, there are some adverse events that could not be detected owing to the small number of subjects enrolled (300 to 3,000) or the limited time during which these clinical trials are run, which preclude the detection of rare or long-term adverse events. As an example, we can mention the need of device post-market surveillance in drug-eluting stents for detection of very late stent thrombosis (which has been associated with the polymer used in these stents) [27, 39, 40].

To achieve this goal, FDA has developed special programs for post-market surveillance of medical devices. This program is called the Medical Device Report (for manufacturers that are required to submit reports of adverse events that could be related to a medical device produced by them) or Med Watch, which is a voluntary adverse-event report system that the FDA has developed for consumers who want to report an adverse event that they think might be caused by a medical device they are using. In spite of these efforts, adverse events are still sub-reported. In 1986, a report from the General Accounting Office of the FDA said that the number of adverse events caused by medical devices reported in hospitals was less than 1% and that, what was even worse, the more serious the adverse event, the less likely it was to be reported [41].

2) *Number of patients*: Drug trials could involve between 1,000 and 3,000 patients, whereas medical device trials could involve between 500 and 1,000 patients (since more people will be using the drugs, as already explained above).

3) *Time to the market*: While it could take from 10 to 15 years for a drug to enter the market, it could take only between 5 to 10 years for a medical device to be commercialized [14, 42, 43]. Some of the reasons behind these differences are based on the fact that medical devices are dynamically improved during the development process, and the regulatory process to approve such incremental modifications needs to be less stringent. Moreover, medical device trials do not need to evaluate pharmacokinetics, which represents an important component of drugs trials. It is relevant to note that there are exceptions in both cases. For instance, HIV protease inhibitors, which are drugs developed for the treatment of AIDS, were developed at a

fast rate, perhaps an effect of the enormous pressure made by advocates of patients, as well as to the economic interest of pharmaceutical companies in having a product that would fulfill the enormous demand created by the increasing number of patients infected with HIV. In addition, HIV protease inhibitors were developed by rational computer design rather than by combinatory chemistry or high-throughput screening, both time-consuming processes [44]. In the case of medical devices, we have the case of highly innovative and complex micro devices such as bio-micro-electro-mechanical systems (BioMEMS), e.g., microchip drug-delivery systems, which could take a considerable time for research and development (owing to the several iterative cycles to optimize the device) before entering a clinical phase and thereafter the market [45].

4) *The control, or comparative, group*: Many medical device trials use historical controls instead of a placebo or the current best management standards (BMS) as a comparative group. This could be attributed to the fact that there are many problems in designing a control group in device trials. An implantable device that needs a placebo surgery against which it could be compared may not be ethically accepted. The testing of some devices used to treat low-incidence diseases where patient enrollment could be difficult to achieve would benefit from using historical control groups. It could also become difficult to provide a control group with an alternative device because such a device might not exist.

5) *The environment*: The environment in which a trial is being run differs in drug and in medical device trials. Whereas medical device trials are performed in highly specialized centers with highly trained operators, drug trials are conducted through visits to the clinical investigator office where the patient takes a simple pill, without any inconvenience and without further need from any operator or trained professional [37].

Table 10.3 sums up all the above regarding FDA pathways for the approval of medical devices. We will describe a case study related to hip replacement.

Challenges in Regulatory Approval of Medical Devices

In spite of the many efforts made by the FDA, many challenges remain to be solved in the regulatory process of biomaterial medical devices. In the following section, we will describe some of these challenges that in the coming years will demand innovative solutions.

TABLE 10.3

Key Differences between Drug Trials and Device Trials

	Drug Development	Medical Device Development	References
FDA experience	From 1906	From 1976	www.fda.gov [3]
Phases	I, IIa, IIb, III, IV (pharmacosurveillance)	I (feasibility II (pivotal)) III post-marketing	FDA Handbook [16] Kaplan et al. [14]
Number of patients enrolled (average)	3000 (phase III)	500 (pivotal trial)	Dhruva [27]
Time to the market (from concept design and development to regulatory approval)	10–15 years	5–10 years	Reed et al. [42] Kaplan et al. [14] Kaitin et al. [47]
Operator	Clinical investigator	Highly trained operator	Sedrakyan et al. [37]
Clinical trial design	Multicenter randomized controlled trials Control: placebo or current best management available	Not multicenter, not randomized historical Control	Dhruva [27], Abded-Aleem [35]
Environment where the clinical trial is developed	Clinical investigator office	Highly specialized centers	Sedrakyan et al. [37]
FDA stringency	Very high	Variable	Dhruva [27]
Control group	Placebo or BMS	Historical	Dhruva [27]
Follow-up	Compliance	Tracking (high-risk devices)	Schoonmaker [48]
Average timeframes in FDA approval pathways (in days)	ANDA*: 510 NDA**: 231 (exceptions) / 357 (standard)	PMA: 151 510(k): 76	Schoonmaker [48]

* ANDA: Abbreviated New Drug Approval
** NDA: New Drug Approval

Technological Challenges

Engineering and Medical Science Integration

Engineering and medical sciences are two disciplines needed in medical device development. For instance, in designing micro-electro-mechanical system (MEMS)–based drug delivery devices, mechanical engineers must consider mechanical parameters that could affect the diffusion of the drug or release time from a reservoir once the system is activated. On the other hand, physicians should advise engineers regarding the biological environment in which the system will be implanted, as the local response to the implant will affect drug diffusion because of the inflammatory response towards the implant (a fibrous capsule that surrounds the implant). Therefore, an alternative approach for drug release that could either elicit more pressure to overcome biological barriers or take advantage of bio-inert MEMS-compatible biomaterials such as ultra-nanocrystalline diamond (UNCD) coatings, which inhibit fouling by impeding cell and protein adhesion, must be considered. These and other technical issues could affect device performance, and as a consequence might not comply with technical standards demanded by the FDA to ensure safety of the device. Therefore, to overcome many of the existing barriers to commercialize a device, a multidisciplinary team is essential. Another example where multidisciplinary work could be of critical importance is through the reporting of adverse events in the patient history records called Case Report Forms, or CRFs, during a clinical trial. Although CRFs are valuable for gathering information for regulatory purposes, they could also be a valuable source of information for bioengineers, since an adverse event could be the first step of a bioengineering problem that has not been considered so far. It would thus be desirable that clinical trials of biomaterial-based medical devices could operate with a feedback loop (currently known as *translational research*), thus improving device performance dynamically. In order to achieve this goal, bioengineers must work coordinately with clinical investigators endeavoring to bring clinical solutions from the bench to the bedside in a timely manner [50–52].

Customization of Implantable Medical Devices

Although general standards in biomaterials and medical devices are of extreme importance for medical device development, it is being increasingly recognized that personalized medicine will play a critical role in medicine in the coming years. Drug and device customization will allow for more effective therapies by adapting them to the patient's unique profile. Customization also means that some procedures are dependent on operator experience. Surgery, for example, provides plenty of examples in which the implant is blamed for the device infection and rejection, although the infection may have been produced as a result of the surgical procedure or a particular patient condition. These conditions include physiological (age, genetic susceptibility, immunological status), psychological and social (emotional factors, family environment), and technical aspects (type of procedure and operator experience). Customization of medical devices thus represents a daunting task in an attempt to design medical devices that could fit individual needs or target specific populations. The following examples will illustrate the complexity of this issue.

Newborns and Children

Congenital cardiac defects are found in a small percentage of newborns. From an economic standpoint, this implies a small market for device industry, resulting in small financial incentives that do not stimulate the development of cardiac devices for congenital heart defects. As a result, adaptation of adult cardiovascular devices to pediatric population is a cost-effective alternative. However, the drawbacks of this approach are evident if one considers the obvious differences between the adult cardiovascular system and that of the newborn, including significant anatomical and physiological differences such as heart valve size, ventricular volume, and the presence of a different circuit in the newborn, such as the ductus arteriosus. Therefore, there is an urgent need in developing customized cardiovascular devices in pediatric populations. Another illustrative example is the early exposure during infancy to isocyanates found in polyurethanes, a material used in medical devices that has been associated with the development of asthma in childhood. The customization approach here relies on replacing this kind of

CASE STUDY: MEDICAL DEVICES IN PEDIATRIC PATIENTS

A bioresorbable material made of polylactic and polyglycolic acid (REPEL-CV Bioabsorbable Adhesion Barrier™) was developed for the treatment of adhesions developed after repeated cardiac surgery. The manufacturer claimed that the product could improve cardiac surgery outcomes by decreasing the number of adhesions as well as their extension and severity, a common complication that arises after repeated cardiac surgeries. This, in turn, would result in decreasing surgery time, hemorrhages and thus morbidity and mortality rates. To demonstrate these claims, the product went through four clinical trials in patients undergoing repeated sternotomies. Although the manufacturer claimed that the product could be used to decrease the extension, severity and incidence of adhesions in repeat cardiac surgeries in adult and pediatric populations, FDA found evidence for approval of the product under only one of the claims, namely, that the product could decrease the severity of postoperative adhesions in pediatric populations undergoing repeat sternotomies. None of the other claims including incidence, extension of adhesion, and different age groups demonstrated to be equal or superior to current therapies in the clinical trials that were performed. In addition, FDA required a non-inferiority post-approval study, providing that the product was as safe as the best management technique available, in the incidence of mediastinitis, rebleeding during exploratory surgery and cardiac tamponade. This was required because post-approval studies are carried out in the general population. This means that events with, say, a frequency in the order of 1 in 100,000 in a population of 10,000,000 could be captured. In a phase III clinical trial that is being run with 3,000 patients, the same event would probably not be detected because the probability of having the case in the population of patients would be extremely low.

material for a suitable one to be used in children, promoting an early intervention that prevents the development of asthma in later stages of life [53].

Elderly People

Elderly people are more prone to complications during cardiac surgery than younger people because of co-morbidities, including impairment in their coagulation system, wound healing capacity, and immunological status, as well as their deteriorated vital organ function capacity as a consequence of aging. Moreover, in the case of heart valve surgery, such as aortic valve replacement, the selection of the material for an artificial heart valve is critical when deciding on a heart valve replacement in an octogenarian person. A representative example of this occurs when the surgeon chooses between mechanical heart valves (metallic) and biological heart valves before operating. This is related to the fact that mechanical valves require anticoagulation during the entire lifetime of the person, which could be associated with bleeding or thromboembolic events, in addition to the psychological impact of this kind of treatment [54].

Pregnant Women and Newborns during Breastfeeding

Pregnancy is a unique condition, in which many physiological parameters are either below or above those considered standards. Pregnant women have a hyper-coagulation state that increases the risk of thrombosis and serious events, including stroke and acute myocardial infarction. If mechanical valves are implanted, an additional risk of valve thrombosis occurs. On the other hand, if bioprosthetic valves are used, the need for an earlier valve replacement becomes imminent. Therefore, careful consideration must be taken when deciding on a heart valve replacement during pregnancy. The problem becomes even more complex in that the medication that given to avoid thrombosis (cumarinic oral anticoagulants) could have serious adverse effects on the fetus (teratogenicity), and alternatives to cumarinic drugs, such as heparin and low molecular weight heparin (LMWH), are less effective. Therefore, a common approach in heart valve replacement during pregnancy is to implant a bioprosthetic valve to avoid the use of anticoagulants. The issue of teratogenicity is of extreme importance, as denoted in other relevant clinical situations, such as the use of anticonvulsants in the management of epilepsy in pregnancy, which at high doses could have teratogenic effects. It is the opinion of the author that this issue calls for therapeutic alternatives, including biomaterial-based devices. Last but not least is the issue of the passage of drugs to the newborn during breastfeeding. Again, creative ideas in the field of medical devices and biomaterials could bring new approaches for these particular conditions [55, 56].

The Psychological Status of the Patient

Some implantable medical devices could have a profound effect on the lifestyle of patients by affecting their psychological status such as defibrillators, with the feeling of dependence, fear of malfunction, and memory of pain during defibrillation possibly indirectly affecting device performance [57].

The Operator's Experience

In contrast to drug therapies, medical devices are dependent not only on the device but also on the operator's experience (e.g., it is not necessary to be trained in administering a drug pill for heart failure, but it is necessary to have extensive experience in order to replace a heart valve that is no longer working). The learning curve is the term used to describe the period in which an operator gains experience in mastering a technique. The learning curve is dependent on several parameters, including patient volume per year and device complexity. The learning curve has important implications in medical device development because the longer an operator needs to manage a procedure (e.g., stent implantation), the longer it will take to market the device.

In conclusion, innovation in medical products is deemed without doubt necessary. However, FDA assessments are extremely stringent in ensuring that an innovative product has unquestionable superiority over the best management product available in the market. Moreover, an innovative product could represent a significant improvement in the patient's quality of life, but it would also need to achieve this without impairing safety. Finally, this case provides evidence that special populations such as pediatric patients (three out of four of the clinical trials were performed to assess efficacy and safety in newborns aged between 3 and 54 days depending on the study) suffering from cardiac congenital malformations have unique needs that will be greatly benefited by innovative approaches such as the one discussed above [60, 61].

Radiological Imaging of Implantable Medical Devices

A significant percent of all radiographic studies, whether they are standard radiographs ("plain films"), CT, ultrasound, MRI, or nuclear medicine exams, show evidence of a medical device in a patient. Sometimes, the overlying medical apparatus can render a study suboptimal to the point of being useless (Figure 10.1). The bewildering array of apparatuses that confronts the radiologist interpreting an ICU portable chest radiograph on a postoperative cardiac patient can be truly staggering (Figure 10.1). Some devices are easily recognized and evaluated, such as a nasogastric tube or a feeding tube, while other critical devices are easily overlooked, such as an intra-aortic balloon pump (IAPB) (Figure 10.2). Additional challenges in identifying medical devices are foreign bodies (bullets, shrapnel, swallowed coins, and so forth) that may simulate medical devices. Foreign bodies are not common but may cause considerable patient harm and be difficult to diagnose if not properly recognized (Figure 10.3). What starts out as a useful device

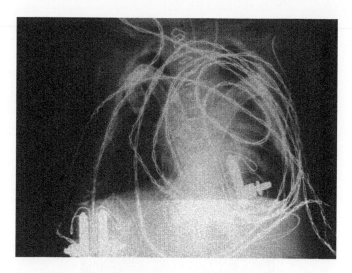

FIGURE 10.1

Portable chest radiograph of an infant after surgery for a congenital cardiac anomaly. The overlying medical apparatus obscures the underlying chest findings so as to make the study non-diagnostic. Some of the apparatus pictured include skin staples, surgical drains, overlying monitoring lead wires, a central venous catheter, and an endotracheal tube. (Courtesy of Hunter, T.)

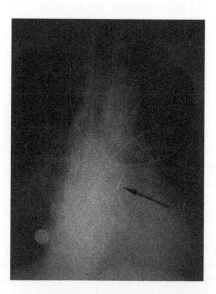

FIGURE 10.2

(left) Intra-aortic balloon pump (IABP) in the descending aorta (arrow). The only visible portion of the pump is the metallic marker on its end. This is often confused with other chest apparatus or obscured by overlying lines, tubes, and leads. (Courtesy of Hunter, T.)

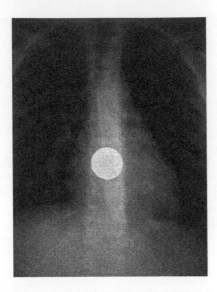

FIGURE 10.3
(right) Frontal view of the chest showing a quarter stuck in the distal portion of the esophagus in a young child. (Courtesy of Hunter, T.)

(e.g., a surgical sponge) can cause problems later when its intended purpose goes awry (e.g., a sponge left in the abdomen after surgery, a chest tube collapsing the lung instead of removing the air collection in a pneumothorax) (Figures 10.4 and 10.5).

Despite the extensive regulatory procedures for introducing new medical devices for medical applications and for reporting death or injury caused by such devices, there seems to be no specific national standards or laws that define or ever require an expected radiologic appearance for a given medical device. This can be very problematic for physicians and healthcare workers using diagnostic radiologic studies to care for patients. For example, it is often difficult to discern the difference between a nasogastric tube and an endotracheal tube on a chest radiograph. They usually lie on top of each other in an anterior-posterior direction on a portable chest radiograph and can have similar appearances (Figure 10.6). It would be ideal for the endotracheal tube tip to be outlined with a prominent radiographic density, allowing for the tube's presence and tip position to be easily noted.

The days of totally radiolucent central catheters seem to be behind us, but many of today's catheters and tubes still remain poorly marked and difficult to appreciate. A particularly bad offender is the intra-aortic balloon pump (IABP) (Figure 10.2), which is typically totally radiolucent, except for a tiny metallic marker on its tip. Even though proper positioning of an IABP is critical for patients with severe cardiac problems, it is nearly impossible to discern

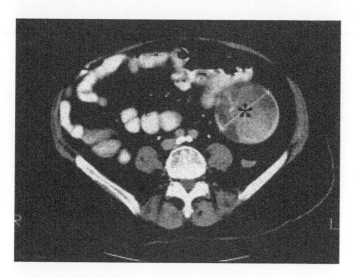

FIGURE 10.4
CT image of the abdomen shows a large granulomatous mass (gossypiboma) resulting from a retained surgical sponge. (Courtesy of Hunter, T.)

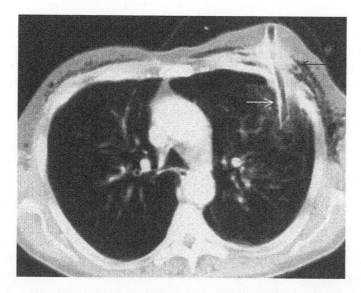

FIGURE 10.5
Chest CT image showing a chest tube (white arrow) piercing the left lung instead of removing the air collection (pneumothorax) surrounding and partially collapsing the left lung. The back arrow shows air that has leaked from the lungs into the chest wall. (Courtesy of Hunter, T.)

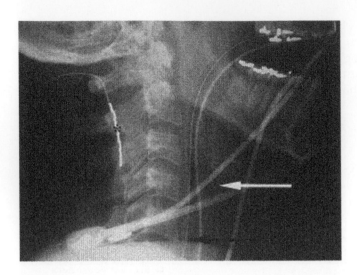

FIGURE 10.6
Lateral view of the neck. An endotracheal tube (white arrow) lies directly anterior to a nasogastric tube (black arrow). Note also the dental fillings and overlying connecting wires. A neurostimulator device is present in the posterior aspect of the upper cervical spinal canal. (Courtesy of Hunter, T.)

the presence of such a pump on portable "chest films." It is not uncommon to see partially radiolucent ECG leads that simulate lung nodules.

In 1986, Wheeler and Scott pointed out the lack of standardization and consistency in medical device radiologic visibility. This remains an important issue today. Hopefully, the radiologic community, along with other physicians and health care workers, will be able to persuade the CDRH to formulate reasonable guidelines for the radiologic appearance of medical devices [62–68].

Nanotechnology

Nanotechnology could be defined as a group of enabling technologies capable of developing structures, systems, and devices in the range scale between 1 and 100 nm. In order to have an idea of this scale, the diameter of a hair is about 80,000 nm and the size of red blood cell is in the order of 7,000 nm. For a broader definition of nanotechnology, refer to the National Nanotechnology Initiative website [69].

Although nanotechnology has been around for many years, it was not until recently that governments and the industry have started to realize the great potential of small technologies for biomedical sciences, food, and cosmetics. In the United States, the National Nanotechnology Initiative (NNI) was launched in 2000. Currently, many federal agencies, private research and development centers, and universities are including nanotechnology in their agendas. Nanotechnology is a growing market representing now almost 1,300 products already being commercialized [80]. The market of nanotechnology-based products reached $254 billion in 2009 and is expected to rise to $1 trillion worldwide by 2015 [70]. Within the nanotechnology market, nanomaterials and bionanotechnology are two very important sectors. In this regard, in recent years, nanoparticle technology, a field that encompasses both areas, has drawn a lot of attention from the industry not only because of its potential market, but also because of its technological maturity. One of the existing definitions states that nanoparticles are particles with at least one dimension between 1 and 100 nm and that are specifically engineered in the laboratory with a desired shape, size, porosity and coating. Nanoparticles could be made of a wide variety of biomaterials, including core metal particles surrounded by a ceramic shell and a biointerface where the target moiety (a monoclonal antibody) is covalently attached, metallic cores with metallic shells (iron oxide nanoparticles surrounded by a gold thin layer), polymer matrixes (nanospheres), polymer shells (nanocapsules), lipid particles such as liposomes, dendrimer-like structures, carbon-based structures such as adamantine, ultra-nanocrystalline diamond (UNCD) particles, fullerenes, and carbon nanotubes (although the last are being subjected to intense scrutiny because of their toxicological profile [71–73].

One of the most promising biomedical applications of nanoparticles is in the field of drug delivery systems. Nanoparticle-based drug delivery systems or devices (we extend the term *device* to nanoparticle systems since complex nanoplatforms including sensing and actuating mechanisms are currently being developed, thus fulfilling the concept of "device") are becoming a very attractive field for the industry [74]. The reason why industry is pursuing these amazing technologies is that nanoparticles represent excellent platforms for drug delivery systems owing to their extraordinary properties. Some of these properties are that:

1. Nanoparticles have a high surface-to-volume ratio (increase biological active surface).
2. Nanoparticles could be modified ("functionalized") at their surface with molecular moieties such as antibodies, adding extreme versatility for various therapeutic applications, allowing the creation of many different kinds of nanoplatforms, which is an attractive technology to increase the portfolio of pharmaceutical products by performing minor modifications.
3. Nanoparticles are so small that they could pass biological barriers and gain access to some parts of the human body otherwise not accessible (e.g., the blood–brain barrier).
4. Nanoparticles, such as magnetic nanoparticles, could actively transport a precise amount of payload (drug) to the target tissue and release it only in the affected area, improving drug availability and decreasing systemic toxicity.

TABLE 10.4

Some of the Studies Performed by the Nanotechnology Characterization Laboratory to Address Critical Issues Regarding the Toxicity of Nanoparticles

Assay Category	Questions to Address
In vitro	
Hemolysis	Do nanoparticles change integrity of red blood cells?
Platelet aggregation	Do nanoparticles interfere with cellular components of the blood coagulation cascade?
Coagulation time	Do nanoparticles cause changes in coagulation factors' function?
Complement activation	Do nanoparticles activate the complement system?
CFU-GM	Do nanoparticles cause myelosuppression (toxicity to bone marrow precursors)?
Leukocyte proliferation	Do nanoparticles have adverse effects on leukocyte proliferative responses?
Uptake by macrophages	Are nanoparticles internalized by specialized phagocytes?
Cytokine induction	Do nanoparticles activate immune cells to elicit cytokine production or interfere with that caused by known immunogens?
Nitric oxide production	Do nanoparticles induce oxidative stress? Indirect test for potential endotoxin contamination.
Cytotoxicity of natural killer cells	Do nanoparticles interfere with the ability of natural killer cells to recognize and kill tumor target cells?
Endotoxin contamination	Pyrogen contamination test
Microbial contamination	Sterility test
Viral/mycoplasma contamination	Sterility test
In vivo	
Single-dose toxicity studies	
Standard toxicity tests	
Blood chemistry	Do nanoparticles cause toxicity to immune cells and organs?
Hematology	Are there any indications for additional toxicity studies?
Histopathology	Additional toxicity studies are conducted on a case-by-case basis using weight-of-
Gross pathology	evidence approach.
TDAR	The test is receded for Is high predictability for human models.
Host-resistance studies	
Evaluation of cell-mediated immunity	These tests are recommended for (1) testing the potential effects that particles might have on host resistance towards pathogens and tumor cells and (2) to check for contact sensitization and delayed type hypersensitivity reactions.
Repeated dose toxicity study	
Immunogenicity	Do nanoparticles elicit particle specific immune response?

Reprinted with permission from Nature Publishing Group, 2007. (Drobovolskaia, M.A and McNeil, S. 2007. Immunological properties of nanoscale materials. *Nat. Nanotechnol.* 2: 469–478.)

Nanotechnology Regulation

In spite of this tremendous potential of nanoparticles for drug delivery (among other applications), there has been a call for a tight control (including prohibition, moratorium, and the use of the precautionary principle) over nanotechnology products. This could be explained by the increasing concerns about the toxicological and environmental fate of nanoparticles and their social and ethical consequences. Some of these concerns are that the same features that make nanoparticles an attractive platform for drug delivery systems (e.g., passage through the blood–brain barrier) could promote toxicity issues, and that more studies should be carried out before any product enters the market [73, 75, 76].

In order to afford these challenges, the FDA has embarked on a series of public-private collaborations, including the Nanotechnology Characterization Laboratory (NCL), which was created in collaboration with the National Institute of Standards and Technology (NIST) and the National Cancer Institute (NIC), in order to perform preclinical assessment of nanotechnology products for cancer therapy, including physical, chemical, and biological characterization of nanoparticles. Companies are encouraged to request the from NCL a complete characterization process for their products. In this manner, there is a mutual benefit between regulators and the industry. In addition to the NCL, the National Center for Toxicological Research (NCTR) is carrying out several projects to better understand the toxicological implications of nanoparticle-based products. Among these projects are the following:

a) Nanotube-based technologies: their implications in public health

b) Skin penetration of quantum dots (nanocrystal semiconductor particles)

c) Tumorigenicity of titanium dioxide nanoparticles

d) Neurotoxicity of manganese nanoparticles

e) Neurotoxicity of silver nanoparticles

f) Application of electron spin resonance spectroscopy to gain more insights into the interaction of nanoparticles with biological systems

The FDA has established other external collaborations, including with John Hopkins Hospital and the Houston Nanohealth Initiative. In addition to external collaborations, the FDA has been engaged in developing an internal expert group in nanotechnology. In August 2006, the FDA created the Nanotechnology Task Force, which, among its duties, has to find the adequate pathways to regulate nanotechnology products. The creation of a Nanotechnology Task Force at the FDA allowed the agency to have the tools for reviewing and/or characterizing nanotechnology-based products, with the authority to request information from nanotechnology manufacturers about formulations, including the use of nanoscale material, and proof of safety and effectiveness if deemed necessary. This was also applied to medical devices. For instance, current Class I or Class II medical devices not otherwise subjected to 510k could require a 510k when using a nanomaterial if the added change surpasses the limit established by the Code of Federal Regulations (CFR) regarding exemptions for devices that are similar to ones existing in the market (21 CFR 878.9). In addition, a new 510k application could be required from the FDA to an already cleared Class I or Class II product when incorporating nanoscale materials [77, 78].

Concerns Regarding New Technologies among the General Population

As new technologies are created, new regulatory gaps in the regulation arise as a consequence of the increasing complexity and uncertainty of such technologies. We mentioned as an example micro and nanotechnologies, but the same trend could be seen in other disciplines using biomaterials such as tissue engineering and gene therapy. The finding of a genetically modified organism (GMO), the finding of lead in toys, the alarming discovery of contaminated peanuts, and the presence of *Escherichia coli* bacteria in meat has raised increasing outrage within the general population in terms of how federal agencies will afford the challenge of new technology regulation. It is important to point out that the main beneficiary of new technologies is the society. A recent report from the Woodrow Wilson Center Project on Emerging Nanotechnologies mentioned that the public trust in the manner in which regulatory agencies are dealing with new technologies has declined over the last years. This can be explained by the fact that public confidence is based on a benefit/risk ratio where high benefits are expected from high risk or undetermined risk products. Because of the incremental benefit of nanotechnologies at this time of development, and the great amount of uncertainty existing among toxicological issues of nanotechnology products, it is legitimate to wonder whether the public will accept such a balance [80, 81].

Ethical Challenges

Ethics vs. Economics

Why do we need to write about ethics in a book about biomaterials? As we have seen during the course of this chapter, biomaterial and device development is a process that takes place through several steps, including design and fabrication, *in vitro* testing, preclinical or animal testing, and ultimately clinical trials. It is important to point out that clinical trials where human subject research was carried out without considering the ethical aspects involved in conducting experimental work in human subjects, some historical tragedies took place (Figure 10.7). Therefore, it should be realized that biomaterial development is a process where ethical considerations must be applied with the same rigor as with any other technical requirement.

1932: The Tuskegee Scandal

The Tuskegee study was carried out to study the evolution of patients with syphilis. Although by that time it was known that penicillin was effective against syphilis, the researchers decided not to give penicillin to patients in order to continue with the study. This study is still today a matter of discussion and a critical call to conduct clinical research ethically [6].

1950: The Jewish Chronic Disease Hospital Study

In studying the rejection process that takes place in transplantation, cancer cells were inoculated in immundepressed patients to observe the effect of cancer spreading with a depleted immunological system (nowadays a method used in animal models to recreate tumorigenesis). Most of the subjects of the study were elderly people with an impaired mental status. No informed consent was required from the helpless patients and when they asked about the study they were told that "the study was a simple skin test" [6].

1960: The Thalidomide Tragedy

Thalidomide is a drug that was prescribed to pregnant women due to its potential use in morning sickness caused during pregnancy. Soon it was realized that thalidomide was related to deformities in newborns (embryopathies) such as focomelia, heart defects and cranial nerve palsies, among others. Almost 8,000 cases of abnormalities in newborns were reported in Europe whereas only 17 cases were reported in the United States (in clinical trials).This was because the drug was not approved in the United States due to concerns regarding potential neuropathy related to the drug (but not because of the potential embryopathies) (Rice, 2008). This tragedy resulted in an amendment of the Food and Drug Cosmetic Act by including the informed consent of the patient as a primary document to conduct human subject research [46].

1970s: The Dalkon Shield Case and Intrauterine Devices

One of the most important scandals in the history of medical devices was that of the Dalkon Shield, an intrauterine contraceptive, designed with nylon-6 and with a multifilament tail, which was withdrawn from the market in 1974 after several cases of Pelvic Inflammatory Disease and Sepsis were reported. The problem was blamed on the design of the device where multifilament tails were incorporated to allow the physician to have access

FIGURE 10.7
Historical tragedies involving biomedical research in human subjects.

to the device from the vagina and thus modify the position of the device if necessary. These tails were made of nylon-6, which was colonized by bacteria and biodegraded, producing holes that allowed the passage of bacteria from the vagina to the uterus. The case had a tremendous impact on the general population, particularly in women who started to see the intrauterine devices as a hazardous method for contraception. As a consequence, after the Dalkon Shield case, the use of contraceptive devices has steadily decreased in the United States, although they are currently considered one of the most common strategies for family planning worldwide, with an estimated unintended pregnancy rate of 0.1%.The Dalkon Shield produced such an impact on the public opinion that the case triggered the FDA amendment act of 1976 [82, 83, 84].

1996: Breast implants were withdrawn from the market

Several reports of people complaining of problems at the site of the implant and even cancer pushed FDA to withdrawn breast implants from the market, taking the manufacturer to the bankrupt. Breast implants were re-incorporated to the market by FDA 14 years later [85].

1999: The Vioxx™ Scandal [86]

Vioxx™ (Rofecoxib), a non-steroidal anti-inflammatory drug designed to treat osteoarthritis without the gastrointestinal side effects of non-steroidal anti-inflammatory drugs was launched by Merck in 1999.

It was known that because of its mode of action, Vioxx could affect the cardiovascular system, increasing the risk of thrombosis. In spite of this information, clinical trials were developed in such a way that potential cardiovascular events (end-points) could not be detected. Some of these studies were written by companies hired by Merck and published in prestigious journals that did not acknowledge the bias present in these studies, allowing these articles to be published.

FIGURE 10.7 (CONTINUED)

Unfortunately, in our days, this kind of misconduct still persists. In this regard, there is a first issue regarding economic influences. It would be striking to know that a federal agency that regulates one quarter of the American product market could be under the influence of economic lobbies, and that this could in turn have profound influences on how decisions are made within the FDA. For instance, recent papers have called attention to increasing concerns regarding conflict of interest within FDA committees, particularly in the pharmaceutical field. These papers advocated for more transparency in the way that members of these committees are elected, pointing out that many of these members, who are in charge of deciding whether a product will be approved or not, have important positions in the biomedical industry. The effects of such influence are seen in how regulatory barriers become less rigorous than they should be and products of unclear quality might be allowed to enter the market. This is clearly shown in medical devices (many of them using biomaterials), where it seems to be a growing concern regarding their assessment, particularly in the quality of medical device clinical trials. For example, a recent meta-analysis has shown that several premarket approvals submitted to the FDA between 2000 and 2007 lacked the minimum methodological requirements to be considered reliable, claiming that only 27% of the studies supporting premarket approvals were randomized and only 17% of them were blind, and 88% of all end points were surrogate (surrogate end points are indirect forms for assessing the effectiveness of a treatment, as opposed to more reliable direct end points) [26]. This is of particular concern since it would be expected that medical devices would be submitted to a more rigorous assessment than drugs, because, as opposed to a drug that could be "washed out" from the body, once a medical device is implanted, any failure such as malfunctioning, contamination, electrical breakdown, heating, degradation, and corrosion could have profound and catastrophic consequences in a patient. In addition, the FDA has more recently started to deal with medical device regulation (from 1976 to date) than with drugs (from 1906 to date), which ideally should make the regulatory process for biomaterial-medical devices much more stringent. Finally, it is obvious that from an industrial standpoint, a less rigorous FDA approval for medical devices would imply faster translation to the market of biomaterial medical-device developments.

The issue of potential conflict of interest within FDA brought about the advent of independent committee assessment groups that would collaborate with FDA experts in monitoring medical device technology by performing critical reviews of clinical studies in order to assess the strength of evidence necessary to launch a product to the market without being under any economic influence. Among these groups, we can mention the Institute of Medicine (IOM), Medicare and professional associations, and the California Technology Assessment Forum (CATF). The last has been assessing medical devices including capsule endoscopy (a biomaterial-based capsule that is ingested and allows the visualizing of the gastrointestinal tract) and mammography. This organization works by performing evidence-based reports based on reviews from medical databases such as the Cochrane Library and Pub Med. These reports made by CAFT consultants are later discussed with a multidisciplinary panel composed of scientists, manufacturers, and consumers and subjected to a general vote. In this way, strong evidence is used to assess the safety of medical technology beyond FDA [25, 87–89].

Ethics vs. Time

Another ethical issue of increasing concern is how the approval times are managed by the FDA. For example, in some populations, such as European countries, some medical devices reach the market earlier than in the United States. The drug-eluting stent, for example, was available in Europe a year earlier than in the U.S., and 75% of cardiovascular devices find their first clinical testing outside the United States, looking for faster approval times [14].

This issue is very complex. On the one hand, there is the necessity to bring new solutions for unmet needs in health care as soon as possible. On the other hand, advocate and political groups and the public opinion entrust the FDA with the task of ensuring that any product entering the market will be safe for the population. This, in turn, leads to a regulatory delay that could have deleterious effects on public health by denying effective treatments in a timely manner. To illustrate this example, we can think about neurological diseases and cancer. These are usually devastating entities that cause suffering to both the patient and his or her family. Some neurodegenerative diseases and many types of cancer have low survival rates, which demands earlier interventions, thus patient advocate groups are created trying to get the attention of the government on these issues. Therefore, the question remains about how to promote a dynamic regulatory process without hindering safety issues. One approach that has been developed by the FDA in order to afford this problem is a "regulator accelerator" or Critical Path Initiative (CPI). Although not focused solely on cancer and neurological diseases, the Critical Path Initiative represents a strategic plan to accelerate the way products are regulated at FDA. The Critical Path Initiative has some specific aims in critical areas where the FDA pipeline demands urgent solutions for unmet critical needs in public health. Some of these aims related to medical devices include modernization of clinical trials to improve the manner in which medical device trials are designed. This initiative has been called the Clinical Trial Transformative Initiative. Another approach was presented in the 2009 Critical Path Initiative report titled *Key Achievements in 2009*, where cardiovascular devices were selected as an example on how innovative solutions could accelerate the development of medical devices. This report showed how computer simulation methods, integrated with medical imaging, are used to improve cardiovascular devices design performance and safety. A collaborative effort, called the DAPT Initiative, is a post-marketing clinical trial among medical device firms, drug firms, and the FDA to improve stent safety by assessing the optimal time that patients carrying a stent must be under antiplatelet therapy. Last but not least is an ongoing project to develop an artificial pancreas at the FDA Critical Path Initiative. This project involves a device that could mimic the pancreas by sensing and releasing insulin on demand. This is an example of how new technologies and biomaterials converge in a device, and how these devices are being considered a priority within the FDA. As it has been described previously in this chapter, nanotechnology represents a very active area of research with enormous technological and economic potential. The Critical Path Initiative has considered nanotechnology a critical tool for improving manufacturing capabilities in the 21st century [85]. Finally, in an unprecedented effort, the CPI will bring together several sectors, including academia, regulatory experts, and industry, forming a multidisciplinary, multitask group to provide outstanding capabilities for the forthcoming enormous challenges that complex biomaterial medical devices are going to represent for the FDA.

The Importance of Helsinki Declaration in Current Clinical Research

At this point of the chapter it is important to highlight how important the Helsinki Declaration is for conducting clinical trials ethically. A few statements mentioned in this declaration are now cornerstones in any clinical trial being performed, regardless of what the clinical trial involves. We will show the validity that this declaration still has to date by describing some of these cornerstones. *Informed consent*: current clinical trials involving biomaterials are required to provide the patient with an informed consent with which the patient decides voluntarily whether he or she wants to take part in the research or not. *Operator experience*: a medical device clinical trial demands the intervention of an operator trained to perform any procedure such as the implantation of a pacemaker. *Well being of the patient*: the well being of the subject during a medical device research takes precedent over any other interest, implying that any procedure that threatens the subject's well being is not justified even if proving to be beneficial for society. *Scientific principles*: any biomaterial-device clinical trial must be based on a comprehensive scientific basis that has been consensuated and validated by the scientific community through the literature, such as peer review journals or other relevant publications. As stated previously in this chapter, institutional review boards are responsible for ensuring the compliance with these principles.

Conclusions and Future Perspectives

Hopefully in the near future, we will witness the arrival of a new generation of medical devices in the hospitals and at the patient's home, with fewer reports of adverse events and better outcomes that will improve the quality and extension of human life. Micro and nanotechnologies will be part of this change, bringing implantable devices to the market with new functionalities and better performance. As the trend to personalized medicine is expected to impact drug therapies by applying pharmacogenetic and pharmacogenomic tools, the same trend towards customization of medical devices taking into account special populations with particular needs will improve the ways medical devices are adapted to the host, and will therefore improve the effectiveness and safety of implantable devices.

However, the current status in medical device technology regulation seems to favor technological improvements that are incremental in nature (improvements of devices already existing in the market) rather than disruptive technologies, and the reason is that breakthrough medical devices are subjected to a much more comprehensive review process by the FDA. This favors the development of conservative approaches that hinder any innovation. Therefore, innovation in medical devices will be possible only by the modernization of regulatory science. In order to achieve this goal, interdisciplinary groups working together at the FDA and at the manufacturer worksite will enable a more dynamic regulatory process, bringing medical devices in a faster and safer way to the population. Hopefully, in years to come, programs such as the Critical Path Initiative will impact the regulatory science and policy by adopting new approaches in clinical trials of medical devices, incorporating new technologies such as computer simulations for studying implantation performance and other advanced technological tools now available. Furthermore, this amazing repertoire of techniques will provide the FDA personnel with more accurate information, thus decreasing the risk of device failure and increasing the chances of success, and paving the way to bring new disruptive medical device technologies to the patients in a timely fashion without hindering safety performance.

From an ethical point of view and based on recent criticism made to the FDA about potential conflicts of interest, the advent of independent committee assessment groups that would collaborate with FDA experts in monitoring the strength of the evidence in clinical trials of medical device technology will help make the regulatory process more transparent and less prone to economic bias.

Acknowledgments

The author thanks Dr. Yitzhak Rosen for his invaluable comments, critiques, and editing of this chapter and is grateful also to Dr. Orlando Auciello for his continuous support during the writing of this chapter. The author would also like to acknowledge the radiological pictures kindly provided by Dr. Tim Hunter. The author also thanks María Victoria Gonzalez Eusevi for all her technical support in the review of this manuscript.

References

1. Woo, R., Jenkins, D.D., Greco, R.S. 2005. Biomaterials, and historical overview and current directions, in *Nanotechnology in Nanoscale Technology in Biological Systems* Ralph Greco, ed. Boca Raton, FL: Taylor and Francis.
2. Williams, D.F. 2009. On the nature of biomaterials. *Biomaterials* 30: 5897–5909.
3. FDA website: http://www.fda.gov (accessed December 2010).
4. Huebsch, N., and Mooney, D.J. 2009. Inspiration and applications in the evolution of biomaterials. *Nature* 462: 426–432.
5. Constantz B. 2005. Crossing the chasm: Adoption of new medical device nanotechnology. In *Nanoscale Technology in Biological Systems*, Ralph Greco, ed., Chapter 20. Boca Raton, FL: CRC Press.
6. Rice, T.W. 2008. The historical, ethical and legal background of human subjects research. *Respir Care* 53(10): 1325–1329.
7. Helsinki Declaration. World Medical Association website: http://www.wma.net (accessed December 4, 2010).
8. European Commission Enterprise and Industry website http://ec.europa.eu/atoz_en.htm.
9. Ministry of Health Labor and Welfare, Japan website: http://www.mhlw.go.jp/english/.
10. Pharmaceutical and Medical Device Agency, Japan website: http://www.pmda.go.jp/english/index.html.
11. Medicines and Health Care Products Regulatory Agency, United Kingdom website http://www.mhra.gov.uk/index.htm.
12. Schuh, J.C.L. 2008. Medical device regulation and testing for toxicological pathologists. *Toxicologic Pathology* 36: 63–69.
13. Jeffreys, D.B. 2001. The regulation of medical devices and the role of the medical device agency. *Br J Clin Pharm* 52: 229–235.
14. Kaplan, A.V., et al. 2004. Medical device regulation: From prototype to regulatory approval. *Circulation* 109: 3068–3072.
15. Nagasaka, E.T., et al. 2008 An overview pharmaceutical and medical device regulation in Japan. Available at http://www.morganlewis.com/pubs/Overview_Pharma_device_reg.pdf.
16. *United States FDA Medical Devices Control and Regulation Handbook. Vol. 1 Strategic and Practical Information.* 2010. International Publisher Publications, Washington, USA. 4th edition.
17. FDA website http://www.fda.gov/MedicalDevices/DeviceRegulationand Guidance/Overview/default.htm (accessed November 2010).
18. ISO website http://www.iso.org (accessed December 2010).
19. ASTM website http://www.astm.org (accessed November 2010).
20. FDA website http://www.accessdata.fda.gov/scripts/cdrh/cfdocs/cfStandards/search.cfm (accessed December 12, 2010).
21. FDA website http://www.fda.gov/GLP (accessed December 2010).
22. Peterson, W.A 2003. FDA/GLP Regulations. In *Good Laboratory Practice Regulations* 3rd edition, S. Weinberg, ed., New York: Marcel Decker.
23. FDA website http://www.fda.gov/GCP (accessed December 2010).

24. International Conferences on Harmonization website http://www.ich.org (accessed December 2010).
25. Feldman, M.D., et al. 2008. Who is responsible for evaluating the safety and effectiveness of medical devices? The role of Independent Technology Assessment. *J Gen Inter Med* 23(suppl 1): 57–63.
26. Dhruva, S.S. Bero, L.A., Redberg, R.F. 2009. Strength of the evidence examined by the FDA in premarket approval of cardiovascular devices. *JAMA* 302(24): 2679–2685.
27. Brown, S.L., Bright, R.A. and Travis, D.R. 2004. Medical device epidemiology and surveillance: Patient safety is the bottom line. *Expert Rev Med Dev* 1(1): 1–2.
28. Medical Device Regulations. 2003. Global overview and guiding principles. World Health Organization Report. Available at http://www.who.int/whr/2003/en/whr03_en.pdf.
29. Shelby, R.D., Shea, A.M., Schulman, K.A. 2007. Economic implications of potential changes to regulatory and reimbursement policies for medical devices. *J. Gen. Inter. Med.* 23(suppl 1): 50–56.
30. FDA website http://www.accessdata.fda.gov/scripts/cdrh/cfdocs/cfpcd/315.cfm?GMPPart=878#start (accessed December 2010).
31. FDA website http://www.fda.gov/MedicalDevices/DeviceRegulationand Guidance/Overview/default.htm.
32. FDA website www.fda.gov/cdrh/devadvice.
33. Saviola, J. 2005. The FDA role in medical devices clinical studies of human subjects. *J. Neural Eng* 2: S1–S4.
34. Clinical Trials Gov website http://www.clinicaltrials.gov (accessed December 2010).
35. Abded-Aleem, S.A. 2009. *Design, Execution and Management of Clinical Trials.* Haboken, NJ: Wiley.
36. Doval, H., et al. 2010. *Handbook of Clinical Trials and Statistics (Manual de Ensayos Clinicos y Estadistica).* Buenos Aires. Ed GEDIC.
37. Sedrakyan, A., et al. 2010. A framework for evidence evaluation and methodological issues in medical device development. *Med Care* 48 S121–S128.
38. Lawrence M.F., Furberg, C.D., and De Mets, D.L. (2010). *Fundamentals of Clinical Trials.* 4th ed. pp. 1–5. New York: Springer.
39. Thuesen, L., and Holm, N.R. 2010. Late coronary stent thrombosis. *Minerva Med* 101(1): 25–33.
40. Iancu, A., Grosz, C. and Lazar, A. 2010. Acute carotid stent thrombosis: Review of the literature and long-term follow-up. *Cardiovas Revas Med* 11(2): 110–113.
41. FDA website http://www.fda.gov/MedicalDevices/Safety/ReportaProblem/default.htm (accessed December 2010).
42. Reed, S.D., Shea, A.M. and Schulman, K.A. 2007. Economic implications of potential changes to regulatory and reimbursement policies for medical devices. *J Ger Int Med* 23(suppl 1): 50–56.
43. Findlay, R.J. 1999. Originator drug development. *Food and Drug Law Journal* 54: 227.
44. Turk, B. 2006. Targeting proteases successes, failures and future prospects. *Nature Review Drug Discovery* 5: 785–799.
45. Staples, M., et al. 2006. Application of micro- and nano-electromechanical devices to drug delivery. *Pharmaceutical Research* 23(5): 847–863.
46. Annas, G.J., and Elias S. 1999. Thalidomide and the Titanic: Reconstructing the technology tragedies of the twentieth century. *Am. J. Public Health* 89: 98–101.

47. Kaitin, K.I., Di Masi, J. 2011. Pharmaceutical Innovation in the 21st century: New drugs approvals in the first decade, 2000–2009. *Clin. Pharmacol. Ther* 89(2): 183–188.
48. Schoonmaker, M. 2005. The U.S. approval process for medical devices: Legislative issues and comparison with the drug model. CRS Report for Congress.
49. Depuy Panel Pack Complete Acetabular Hip System Orthopedic and Rehabilitation Devices Panel, August 18, 2009. Available at http://www.fda.gov/downloads/AdvisoryCommittees/CommitteesMeetingMaterials/MedicalDevices/MedicalDevicesAdvisoryCommittee/OrthopaedicandRehabilitationDevicesPanel/UCM178164.pdf (accessed April 2011).
50. Voskerician, G., et al. 2003. Biocompatibility and biofouling of MEMS drug delivery devices. *Biomaterials* 24: 1959–1967.
51. Bajaj, P., et al. 2007. Ultrananocrystalline diamond film as an optimal cell interface for biomedical applications. *Biomed Microdevices* 9(6): 787–794.
52. Matheny, M.E., Machado, L.O. and Resnic, F.S. 2006. Monitoring device safety in interventional cardiology. *Am Med Inform Assoc.* 13: 180–187.
53. Beekman, R.H., et al. 2009. Pathways to approval of pediatric cardiac devices in the united states: Challenges and solutions. *Pediatrics* 124: 155–162.
54. Florath, I., et al. 2005. Midterm outcome and quality of life after aortic valve replacement in elderly people: Mechanical versus stainless steel biological valves. *Heart* 91: 1023–1029.
55. Holmes, B.L., et al. 2001. The teratogenicity of anticonvulsant drugs. *N. Engl. J Med* 344(15): 1132–1138.
56. Jeejeebhoy, F.M. 2009. Prosthetic heart valves and management during pregnancy. *Canadian Family Physician* 55: 155–157.
57. Urizar, G.G., et al. 2004. Psychosocial intervention for a geriatric patient to address fears related to implantable cardioverter defibrillator discharges. *Psychosomatics* 45(2): 140–144.
58. Vidi, V., et al. 2009. Learning curve in the use of star close vascular closure device: an analysis of national cardiovascular registry data. Poster presented at the i2Summit, Georgia.
59. Felten, R.P., et al. 2005 Food and Drug Administration medical device review process: Clearance of a cot retriever for use in ischemic stroke. *Stroke* 34: 404–406.
60. FDA website http://www.accessdata.fda.gov/cdrh_docs/pdf7/P070005a.pdf (accessed December, 2010).
61. FDA website http://www.fda.gov/downloads/MedicalDevices/NewsEvents/WorkshopsConferences/UCM240789.pdf (accessed December, 2010).
62. Hunter, T.B. 2003. Medical devices and foreign bodies: An introduction. *RadioGraphics* 23: 193–194.
63. Hunter, T.B., Taljanovic, M.S., Tsau, P.H., Berger, W.G., Standen, J.R. 2004. Medical devices of the chest. *RadioGraphics* 24: 1725–1746.
64. Hunter, T.B., and Taljanovic, M.S. 2003. Glossary of medical devices and procedures: abbreviations, acronyms, and definitions. *RadioGraphics* 23: 195–213.
65. Hunter, T.B., and Taljanovic, M.S. Foreign bodies. 2003. *RadioGraphics* 23: 731–757.
66. Scott, W.W., Jr, Wheeler, P.S. 1986. Markers for implanted devices: Need for standardization. *AJR* 146: 387–390.
67. Wheeler, P.S. 1986. Device identification: Deficient marking systems. *AJR* 146: 418–419.

68. Wheeler, P.S., and Scott W.W., Jr. 1994. Medical devices and their radiologic visibility. In *Radiologic Guide to Medical Devices and Foreign Bodies*, T.B. Hunter and D.G. Bragg, eds. Waltham, MA: Elsevier.
69. National Nanotechnology Initiative website http://www.nano.gov/html/facts/whatIsNano.html (accessed December 2010).
70. Rocco, M. Nanotechnology Research Directions for Societal Need in 2020. Report. Available at http://www.wtec.org/nano2/Nanotechnology Research_Directions_to_2020/Nano_Research_Directions_to_2020pdf.
71. Vauthier, C. and Couvreur, P. 2007. Nanomedicines: A new approach for the treatment of serious diseases. *Journal of Biomedical Nanotechnology* 3: 1–12.
72. Helland, A., et al. 2007. Reviewing the environmental and human health knowledge base of carbon nanotubes. *Environ Health Perspect* 115: 1125–1131.
73. Rosen, Y., and Elman, N. 2009. Carbon nanotubes in drug delivery: Focus on infectious disease. *Expert Opin Drug Del* 6(5): 517–530.
74. Bawa, R. 2008. Nanoparticle-based therapeutics in humans: A survey. *Nanotechnology Law and Business* 5(2): 135–155.
75. Oberdoster, G., Oberdoster, E. and Oberdoster, J. 2005. Nanotoxicology: An emerging discipline evolving from studies of ultrafine particles. *Environ Health Perspect* 113: 823–839.
76. Bennett Woods, D. 2008. *Nanotechnology: Ethics and Society*. Boca Raton, FL: CRC Press.
77. Nanotechnology: A report of the Food and Drug Administration Nanotechnology Task Force 2007. Available at http://www.fda.gov/Science Research/Special Topics/Nanotechnology/Nanotechnology Task Force/default.html.
78. Code of Federal Regulations website http://www.gpoaccess.gov/cfr/.
79. Drobovolskaia, M.A., and McNeil, S.S. 2007. Immunological properties of nanoscale materials. *Nat. Nanotechnol.* 2: 469–478.
80. Woodrow Wilson Center. Project on Emerging Nanotechnologies website http://www.nanotechproject.org/ (Accessed December 2010).
81. Center for Disease Control and Prevention website http://www.cdc.gov.
82. Beatty, M.N., and Blumenthal, P.D. 2009. The levonorgestrel-releasing intrauterine system: Safety, efficacy and patient acceptability. *Therapeutics and Clinical Risk Management* 5: 561–574.
83. Irving. S. 1993. Another look at the Dalkon Shield: Meta-analysis underscores its problems. *Contraception* 48(1): 1–12.
84. Tatum, H.J., Schmidt M.A., and Philips, D.M. Morphological studies of Dalkon Shield tails removed from patients. *Contraception* 11(4): 465–477.
85. Tanne, J.H. 2006. FDA approves silicone breast implants 14 years after their withdrawal. *BMJ* 333: 1139.
86. Krumholz, H., et al. 2007. What have we learnt from Vioxx? *BMJ* 334: 120–123.
87. Finkelstein, J.B. 2008. Members of new FDA Board tied to industry. *JNCI* 100(5): 296–297.
88. Deyo, R.A. 2004. Gaps, tensions and conflicts in the FDA approval process: Implications for clinical practice. *J Am Board Fam Pract* 17(2): 142–149.
89. Avorn, J. 2007. Keeping science on top of drug evaluation. *NEJM* 357(7): 633–635.
90. Critical Path Initiative website http://www.fda.gov/ScienceResearch/specialtopics/criticalpathinitiative.

11

Innovative Product Development and Technology Adoption for Medical Applications

Stephen M. Jarrett

CONTENTS

Introduction

The rapid prototyping of innovative technology solutions into products for a wide range of customers is an essential business requirement for success in the technology development marketplace. Many technologies are available from different industries and from different applications that can be quickly applied to the medical area. Accomplishing this function requires research to find the emerging technologies that are complementary and in coordination with the many different disciplines. Few organizations are well positioned to accomplish this technology transfer and insertion function. Being a technology transition intermediary and a supplier of advanced systems to many customers from the Department of Defense and other federal agencies and even to the local hospital gives us a view of the requirements and the potential insertion points for new developments. A strong relationship with these customers is essential to position the intermediary to discuss the possibilities of technology insertion. Participation in the area of technology transition could be improved by a coordinated effort to find the technology kernels at the research and development sites and to actively team with other developers. Also, a formal concept development group to actively propose prototyping programs that could provide leap-ahead technology insertion projects for our customer base would accelerate the transition of advanced technology. Since the life blood of all small developers is the funding stream, multiple paths of transition should be the rule, not the exception.

The development of technology for future medical applications is a growing segment of our economy, and a number of companies, universities, and federal laboratories are striving to get products into the field and accepted for use by the medical community. The path to commercialization and commercial sales in this area is a tortuous route with many pitfalls that have the potential to derail a project before it reaches its objective. The ability to navigate this route effectively requires skill, persistence, and a good map. This chapter defines the basic steps in this route to assist in the successful transition of technologies from the idea phase to the operational phase of the testing program. The marketing of these technology products must also address the adoption of innovation criteria for the customers.

> Many technologies fail not because of the technical skills of their proponents, nor because of the market to which they are targeted. They fail simply because no one got sufficiently interested in them at the right time. (V. K. Jolly, 1997)

This quote from Jolly's book, *Commercializing New Technologies*, applies directly to the case of medical advancements. As we go through this process, we'll look at a number of examples and try to distill the factors that make the inventor more successful in the end. There is no silver bullet, and there is a lot of technology in the laboratories and even in the patent portfolio of many universities and companies that could be successful if it were ever to meet and marry the right complementary critical partner. There is a great benefit for developers of medical technology to examine technology from other fields and disciplines. There is also great power in the naming of the technology. Just because something was invented to move imagery around in the military from unmanned aerial vehicles doesn't mean that it has any difficulty in moving magnetic resonance imagery from the hospital to the radiologist's computer. We'll discuss some examples of this cross-disciplinary approach to development.

The naming of a technology has a great effect on the uses that are considered for its adoption. For this reason we need to look at other technology areas for those critical component technologies that can assist us in transitioning a laboratory development to a capability to fill the customer's request. We frequently hear about the "not invented here" syndrome. In my experience in the development of military systems, it has consistently come into focus that the naming of a system is critical to the future use of that system's capabilities in the field. It's only human nature for the inventor and program manager want to have the legacy system named after their service or group. The military gets around that roadblock usually by naming it "Joint ...," which is sometimes not totally correct. However, the thought process is sound. If you develop an imagery distribution system that is going to be used to distribute the magnetic resonance imagery (MRI) to the radiologist and the doctor, it is natural for you to name it an MRI distribution system. The issue comes up when in the future we want to use that same imagery system to distribute documents that have been scanned into the system and recorded in the same imagery format as the MRI data. Believe it or not, the software can't tell one JPEG file from another. It is sometimes difficult to come up with a name that describes the capabilities of the system without using the initial application in the title. It's helpful to think of systems or components in terms of what they do as a capability. This naming criterion is also important when the inventor is looking for other critical technologies to complete the system design. In the military acquisition system, this process is called the *analysis of alternatives*.

In competitive technology intelligence, these are called *technical options*. When we are working in the technology development of new products, such as a laptop, one alternative is to make a better laptop to generate competitive advantage. It can be lighter, last longer on a battery charge, and have a brighter screen or some other factor. It could also take a totally different approach and be an iPad, which has no ability to print, connects through a 3G connection to the Internet, and can automatically be linked to a bookseller to download a book. By the way, that same book cannot be read on a laptop.

One of my first developments was a microbolometer infrared detector. This detector was based on a new uncooled focal plane that was developed by an industry partner under funding from the Defense Advanced Research Projects Agency (DARPA). I was asked to transition this technology to the combat forces. The detector prototype that was presented to me was enormously expensive and was encased in a Plexiglas box. For this technology to transition, it was critical to get it into the hands of the customer and to generate a valid concept of operation that gave it an advantage over the current night-vision goggles that were in extensive use. When we first entered the caves of Afghanistan, that concept became clear. The microbolometer infrared detectors were called "pocketscopes." They weighed 12 ounces and operated for five hours on two double-A batteries. The principle of operation of the night-vision goggles was light intensification. In caves there is simply no light to intensify. The principle of operation of the infrared detector is that they don't see light, they see heat. A person is easily visible even in total darkness. However, a new advantage emerged when we determined that a handprint on the wall was visible for many minutes or even hours. We could tell not only if someone was there, but also if that person had been there for hours before we arrived. The light-intensification goggles had no capability in this area. Even with this advantage we had to adjust the development. We had to drive the cost down. I sent the commercial developer to Walmart to buy a Sportsman flashlight. I directed them to put the guts of their unit in the case of the flashlight. Our primary market was not combat forces but law enforcement personnel, and the cost was critical to generate volume.

Looking at this infrared technology from a medical sense, the question emerges, "What medical use could it have?" I took it to a dentist and experimented with looking in a patient's mouth to determine whether there was any gum irritation or infection. Since the unit can detect .0057 degrees of differential temperature, this was easily accomplished. Subsequent to this, I have found an infrared detector connected to a computer that can recognize the vein structure in the palm of your hand and is used to secure a computer from access by anyone other than the authorized user. The operator hovers his hand over the mouse, which has an IR camera facing up. The recorded vein structure of the palm of the user is as unique as the iris or the fingerprint.

We need to look at other technology areas for potential capabilities. This one factor is the reason that I rarely recommend a single technology for an application. It is important to look for technology options instead of looking for a single-technology solution. There are frequently a number of different technologies able to provide the capabilities that are requested by the customer. Not all of these technologies are at the same maturity level. Some of them are laboratory experiments that have exceptional potential but need several more years of development. Some are commercial products that may provide an 80% solution immediately. There may also be more than one way to accomplish the task requested by the customer. Looking at various options frequently yields new viewpoints on how to accomplish the outcome that is desired.

We also need to look at the motivation for the change in technology from the existing path being used to the new path being requested. If a new sampling technique for bacteria enables the user to do immediate triage at the site of the sample rather than sending the sample into the laboratory and getting results back in three days, then the motivation might be a great savings in time for the emergency technician. These factors are defined in several theories that delineate the reasons controlling the adoption of innovations. We frequently think it is only the technology that the decision is based on, but many times the decision criteria are based on much more subtle factors, such as peer or supervisor pressures. There is also a great difference in the reasoning between individuals. Therefore, the terms *early adopter, change agent,* and *innovator* were coined by Everett Rogers (Rogers 1995). When we transition technology from one industry or usage to another market, we frequently find a totally different group of acceptance factors. Even though the imaging technology may be used in both veterinary science and in human medical diagnosis, the factors affecting its adoption are not the same.

Connecting with the Medical Institutions, Doctors, and Nurses and Learning Their Needs

When you have an idea for a medical device that might benefit the doctor or nurse in the diagnosing or treatment of specific ailments, how do you get the specific requirements defined? Many medical institutions have their own research departments. This also holds for universities with medical teaching professors who are interested in spinning off their research projects into startup companies or licensing their inventions for royalty profits. This involves primarily the transition of technology from a medical research laboratory to a medical research product. However, much technology is developed in one area of application and transitioned to a totally different area of medical application.

Just as in other fields, the development of technology doesn't always happen in the field of use that the inventor thought he or she was working on at the time. This cross-fertilization is evident in many historical and current developments. The key question is how to determine whether the technology meets the requirements that the users will adopt? There are a number of adoption criteria in the theoretical works of Rogers, Ajzen, Fishbein, and Davis that relate the adoption of innovation to the medical areas. By examining some of these theoretical factors and by looking at some other techniques used by leading-edge thinkers such as Edison and da Vinci, we can form a process that gives us some clues about the development of critical medical devices that will be used in clinical practices on a wide scale.

Spending Time with the Customers in Their Environment

Who has a better feel for the challenges and issues faced by the customer than the blue-collar technician who listens every day to the complaints about "if they had only put this handle on the left rather than the right of the machine, it would have taken half the time to do this procedure"? Listening to the customer is not always a major part of the researcher's day in the laboratory. I'd be very surprised to find out that a researcher came up with the idea of using an automotive manufacturing robot to do brain surgery. So how did a system such as the CyberKnife get developed? I use this example because none of the parts of the CyberKnife came from medical research directly. I first became acquainted with the CyberKnife at Roper St. Francis Hospital in Charleston, South Carolina. It was demonstrated to our group on a tour as part of the ThinkTEC Innovation group at the Charleston Metro Chamber of Commerce. The head of the CyberKnife was originally used to inspect bridges for flaws in the concrete. The base of the CyberKnife was an automobile manufacturing robot. So how did this marriage ever occur?

One of the reasons that so much technology is left on the table in research facilities is their lack of the ability to develop "concepts of operation" for the new technologies. Another reason is the lack of teaming with other developers of critical components for a finished product owing to a misguided philosophy of "protecting intellectual property" that eventually becomes worthless as it is overcome by other research. In his book *The Innovator's Dilemma*, Christensen demonstrated this explicitly in the development of hard disc storage technology in the computer industry. In that example, in every instance of new technology being inserted in the miniaturization of new hard disc storage devices, the leaders could not visualize the benefits of the new technology. In most instances the new technology in its nascent form was not as capable as the existing product. The key missing concept was that computers were destined to become more mobile and smaller to fit the future market based primarily on laptops. In this mode the desktop unit didn't need a miniature disc with a high storage capacity. To enable the market transition to laptops, the miniature disc was a critical technology requirement.

In the medical technology industry sector, how do we team and find synergistic developments that will enable new products such as the CyberKnife?

"To invent, you need a good imagination and a pile of junk." —Thomas Edison

I'm not insinuating by this quote from Edison that all new technology is a pile of junk. What I am proposing is that many researchers get too close to their own work and need to step back and leverage the developments of others in other technology areas to move their own projects along faster to a reasonable conclusion. I don't know the CyberKnife developers, but I image the discussion may have gone something like this. "Who makes robots that have extreme accuracy?" "Don't they use them for very accurate welds and manufacturing in the automobile industry?" "Why don't we use our medical expertise to apply the technology they developed to the challenge of very precise manipulation of a device in the medical field?"

My friend Bob Miller and I have collaborated on a number of technology developments. We laugh about finishing each other's sentences sometimes. I'll have part of an idea defined and while we're discussing it he will comment, "Why don't we take this other technology that we found and put those two together?" About ten years ago industry went through a time when many companies put groups of expertise together in close proximity in their development areas. The leading-edge display group was in one area, the transmitter and receiver group was in another area, and the software developers were all together. This was to enable the easy transfer of information and ideas between these groups and to foster "over-the-lunch-table" discussions. These centers of technical excellence have been modified in many instances by "cross-functional teams." The Integrated Product Development Team (IPT) has supplanted the previous organizational model. In my opinion there needs to be both of these types of collaboration. When the develop calls for robotics, we call in a robotics expert. When the software needs modification, we bring in the software developer. But functionally the hardest part to put together for many teams is the operational concept of applying the product to the real world application. In the medical development we need that doctor or nurse who has the experience to know what factors are important and how to design the product to bring the most value to the user. Edison called it "master-mind collaboration" (Gelb 2007).

Integration into the Clinical Process (Concepts of Operations)

The integration of new technology into the medical field parallels that of the introduction and adoption of new developments in other technology-based applications. In military terminology this is termed the development of the concept of operations. This simply means "How will the user insert this technology into the current process, daily routine or method of servicing the customer?"

From the theoretical base we find ten factors that affect the adoption of new technology innovation. These factors are drawn from a combined model taking into account the diffusion of innovation theory of Everett Rogers, the theory of reasoned action by Ajzen and Fishbein, and the technology acceptance model of Davis.

1. Relative advantage
2. Compatibility
3. Image
4. Ease of use
5. Results demonstrability
6. Trialability
7. Perceived voluntariness

8. Subjective norm for customers

9. Subjective norm for peers

10. Subjective norm for community (Jarrett 2003)

Here is a chart that links these factors together in a realistic format that describes the decision-making criteria for explaining the behavior of the adopter of technology. Having dealt with researchers for many years, I've been accustomed to the idea of "if we make it they will come." In that frame of mind the researcher develops a technology or maybe even a product and "pitches it over the fence" and hopes that someone buys it. I've found it to be much more efficient and more of a winning strategy to take the adoption factors into consideration before the development than in the marketing phase of the product. Working with the user early in the process is a way to find out the real issues in the adoption of new technology. It always comes back to the generation of value in the eyes of the customer. If the nurse spends countless hours charting the vital signs of patients and we can use technology to automatically record the blood pressure, pulse rate, and temperature from our equipment, it's a positive factor in the adoption by the nurse of the new equipment. If this also generates more accuracy in the recording of the data and provides the nurse more time for patient interaction rather than data logging, it is beneficial to the patient and to the doctor.

Without getting too theoretical, the simple answer to the question of which factors affect the adoption of medical technology is to address how

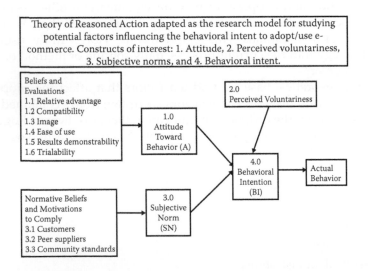

FIGURE 11.1
Combined adoption model (From Jarrett, S.M. 2003. *Factors Affecting the Adoption of E-Business in the Aerospace Industry.* Dissertation for Nova Southeastern University. Ann Arbor: ProQuest)

the customer will use the technology to his advantage in the medical application area. Let's look at an example and determine some of these factors. Currently, if you go in a leading hospital, you may find the nurses rolling a laptop computer around on a cart as they make their rounds. This computer is probably tied to the hospital network by wireless connectivity. What were some of the factors involved in the adoption of this system? First, the previous system was a clipboard with all of the patient's information on treatments, doses of medications, and other information annotated on it at the foot the patient's bed. This required someone to transcribe this information to the record-keeping system and required the nurse to go to the patient's room and read the current status to the doctor if he called in to inquire about the patient before making rounds. The nurse had to record the information, so recording it into the computer versus into a written chart is not much different, and it is a real time system so it only needs to be done once. The nurse doesn't need to chart the patient's status at the end of the shift because it's done concurrently with the visit and the administering of the medication. It can also be linked to the accountability of the drugs. So the nurse saves time, eliminates redundancy of recording the information, and can get a running inventory of the drugs on the floor, and the doctor can log onto the system from wherever he is to get an updated status.

So just from this short examination the nurse sees the system as compatible with her routine since she is recording the same data. The ease of use is that the data is recorded the same as on the chart but in an electronic format. The ease of use was probably also demonstrated by a simple fill in the blanks template for inputting the data on the computer. The results demonstrability is that they probably paralleled the hand charting and the electronic charting for some period of time to demonstrate the advantages. Not all of the factors will be active in every instance, and the importance of each may change as the product progresses through the development phase.

Mapping Out the Technology Insertion Strategy

The mapping out of the technology insertion is a technique that has been used by a number of prominent innovators, including Leonardo da Vinci and Thomas Edison. I think one of the factors in the use of mapping techniques is that many inventors and entrepreneurs are visually thinking people. This also may explain such "back of the napkin" ideas for many new products. I have to admit that I'm a strong advocate of this approach and rarely attend any meeting without some kind of a pad to write or draw pictures on during the meeting. It's not that I dislike meetings so much but that I've generally found after many years that two-hour meetings often yield 20 minutes worth of beneficial information and 100 minutes of people who like to listen to themselves talk.

In applying this technique to the medical field you need to have a few basics. First is to define the central theme for the new product. This sounds like a simple step, but it can prove to be very problematic. The "naming" of a technology frequently has a lot to do with the ability of the developers to transition the technology across technology boundaries into new applications. It's important not to limit the potential solutions too early in the process. In Gelb's book on innovation, this is called "kaleidoscope" thinking. I usually make two different maps. One is the technology version, and the second is the operational or developmental version. The first maps out all of the potential technologies and all the options that might provide solutions. The second maps out all the developmental paths that the technologies might take and ends with the contacts that should be followed in each technical area.

Let's look at an example of the technology version. We place the automated medical chart at the center and look for other technology connections. What benefits can we provide and what other enabling factors come to light that might provide value to the user? If we involve the user in this brainstorming session, we can come up with a number of other "what if?" benefits. What if everyone who needed the data had access to it after the nurse recorded it only once? What other data would make it more valuable? How does this technology fit into the current process? Does this save time, money, or lives by its adoption?

What can we learn from this map? As we populated the map, a few other factors came to the thought process. We originally were planning to automate the charting process to aid the nurse in time management. As we built this map it also came to light that by automating the chart we also enabled the doctor to view the chart in real time remotely. This would be another adopter who benefited. We also found that in the automation we could also provide links to the patient history, laboratory tests results, and even X-rays

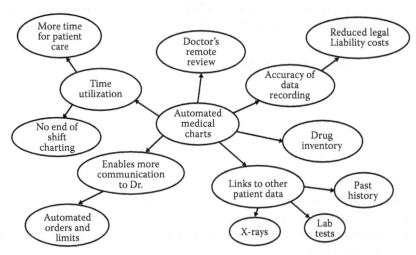

FIGURE 11.2
Medical Technology Mind Mapping example.

or MRIs. All of these factors would also be beneficial. If we made the charts more accurate it would also affect liability issues. By automating the doctor's review we could also provide automated communications of new doctor's orders and track the drug inventory as it was given instead of requiring extensive inventories after each shift.

Getting to the Prototype Stage of Development

In a number of technology transitions I've seen the most difficult part of the path to be getting from the idea stage to the prototype stage. We have also seen this from the medical laboratories at the universities and at the small businesses. Many groups get the initial research done, file for a patent, and file it in a locked drawer. Patents don't necessarily ensure that the developer will make any money whatsoever on the technology idea. Getting to the prototype for the application that will generate a product must be visualized. That's one reason that mind-mapping techniques work well for developers of advanced technology. Sometimes in this developmental cycle we get the chicken and egg syndrome. The developers don't have the funds to get the prototype built and the customer won't fund the development until they evaluate the prototype unit in operation. So how do we get to the next phase?

It's unbelievable how much technology we have in the research laboratory that never sees the light of day. I think the major hindrance to much of this develop is our legal system. We concentrate so much on the protection of intellectual property that we lose sight of the value-developing potential of collaboration. I've seen a number of innovation projects that were proposed in the PowerPoint stage. In the military we have another term for those projects: they are called "vapor ware." The success rate for developing a prototype and getting the customer on your team greatly improves when a rough prototype is connected with a visionary presentation of "you see this 50% solution, now think about if I could get funding to do the next 30%?" You might also note that 50% and 30% doesn't add up to 100%. It only adds up to 80%. Researchers frequently gold plate their proposals, especially to government sources, and ask for unrealistic amounts. The customer isn't ready to fund you to the third generation until you have demonstrated value at every step along the developmental path.

In the Pocketscope development we actually sold the prototypes. Several interesting facts emerged from that process. The first was that it was easier to get a prospective customer to test our units if they were not free. It probably gets back to the old adage of having some skin in the game. So the key is to get the customer on your developmental team as early in the process as possible. My recommendation would be as soon as any level of prototype is reasonably available. This generates buy-in by the customer of the final product early in the process.

Bringing the Customer on Board during the Design Rather than as a Buyer

When the customer is brought into the development process in the design phase rather than the marketing phase, you probably will end up changing the prototype. If you are developing a handheld reader for a microfluidics chip to detect malaria in the third world, you must take into account the operating conditions of that application. This may not be a laboratory environment that has a sterile sampling source and electrical power. In the development of a new sampling system for contamination in oysters for the National Oceanic and Atmospheric Administration (NOAA), the key issue was not the extreme level of accuracy but instead a go-no-go solution and in-field readout capability. The return of samples to the laboratory for analysis after several days was a key cost issue and shut down the shellfish industry harvesting of the seafood until it was completed. The customer knows the market the best. So interface with them early through direct interaction and through the organizations and groups that discuss their challenges.

Availability of Funds Drives Development: Diversifying with Multiple Customers

In most of the developments of advanced technology the funding stream has a great influence. So it makes sense to concentrate on diversifying the funding early in the project. Many projects are started as a result of some seed funding from a research and development source. The National Institutes of Health is one of the largest federal sources of R&D funding for new projects. If the funding comes as a Small Business Innovative Research (SBIR) grant, it has a relatively short life span and no guarantee for follow-on funding. Information from the AAAS shows the magnitude of federal funding that is devoted to the National Institutes of Health. It is much larger than the Department of Defense share of the R&D.

As we progress through developments in the medical area, just like in other sectors the availability of funding either enables or slows the process. As a professional warrior I prefer the "win early win often" approach to funding. Many researchers that I've worked with in other sectors prefer the "loyalty to a single source" approach. Multiple customers give you not only more options for funding but also more than one perspective on applying your research to the marketplace. Regulations for equipment in the veterinary practice are not the same as for the medical research area. Infrared imagery can be used in border security as well as finding infection in soft tissue. We need to look for the other use that we haven't thought of in the first round of research.

One approach that has proven successful in developing advanced technology is to propose a further development of an existing prototype for a different application. The key here is in the leveraging of other technology. In the federal government there is a form of research funding called Small Business Innovative Research (SBIR). This is a phased program where funding increases as you progress through the phases from I through III. There is a provision in the program that even though you got Phase I funding from the Department of Energy, you may go for Phase II funding from the Department of Defense or any other agency that you find that is interested. The second source in effect leverages the funds spent on Phase I but does not have to pay for them. This should be the attitude of the researcher in the search for funding. New sources get the benefit of previous development and only have to pay for the development of the new prototype. I've bounced developments from one agency to another several times and then back to the original source to their surprise and benefit.

Getting a Foothold in the Medical Market

The medical market is very competitive and very difficult to enter by a small business or lab-based developer. We should always look for alternative ways to get the technology proven. We seem to sometimes have blinders on when we are too deep into the technology mix. Being able to look through the development to the logical consequences in the future of the technology insertion is the key to finding the new product that will be a game-changer in the future. Setting up a program briefing plan to identify common technology needs and to raise the visibility of multiple projects is a way to identify alternative customers and other uses for the technology that you are developing. This is where the mind mapping techniques really shine. When we are able to visualize the development and see what it really does, those other applications naturally flow into the discussion.

Let's look at a case study to walk through the process. We'll go back to the pocketscope. The infrared capability of the scope gives us a capability to view heat differentials of very minute amounts. Where would this be applicable to the medical field? Since I'm an engineer and a business doctor and not a medical doctor, we'll go with that limited approach and look at only a couple of areas. I know from experience that infections cause the areas to be inflamed and therefore "hotter." Muscle pulls and strains also cause irritation that might yield inflammation. A similar unit is the portable handheld ultrasound unit that was used on our dog by the vet to detect a muscle strain after an agility trials injury. We probably need the handheld unit to display on a laptop or other screen instead of on an eyepiece. Since this pocketscope is actually a digital camera with a video out, that is not a problem—it's a cable. Next we look at what the concept of operation would be. It would probably parallel the handheld ultrasound. One difference would be that the ultrasound cannot detect infection. Also, the ultrasound probe must be in contact with the skin, whereas the infrared unit can be several feet away. The ultrasound looks under the skin while the infrared looks only at the surface temperature of the skin. Tumors might cause increased irritation and might be visible with the infrared. There seems to be some good similarities in the usage of the two units, and that could indicate that the market might accept it easily.

How would we go about getting this prototype into the field for evaluation by customers? Since the ultrasound is in frequent use in both medical and veterinary practices, we might want to look at both of those areas. The key is that developing the unit only in the human medical area runs into a number of certifications and regulations that the veterinary prototype avoids. The technology is basically the same. The process is the same but shorter. The legal issues are much different. I'd pick the vet area to prove the technology and demonstrate the concepts as an inexpensive route to getting to the market sooner. It's important not to name it an "animal evaluation tool" unless you have no plans to address the medical field later. Have you ever tried to get a vet medicine into the people market?

Building databases on technology sources, contacts and project requirements can be a good start toward identifying the markets to address with a technology. This falls in line with the second type of mind mapping that I mentioned before. I first used this application during my doctoral program, and it was in relation to the process to get a bill through Congress. In that application it is called Sayre Wheel Analysis. In this analysis the bill or issue is placed at the center, and the constituents who might be interested or influenced in the bill are all delineated around the wheel. An example of a Sayre Wheel is below. In a political sense if you were discussing a Social Security issue that would go in the center. Then we'd try to identify all of the potential players that we needed to influence or touch to get the bill passed. A good illustration of Sayer Wheel analysis is provided in a paper by McCarthy and Aronson (2001).

Looking at this from a technology viewpoint is very similar except we place the capability or technology in the center and work from there. Let's look at a medical example.

So in this example we should look for customers in the World Health Organization (WHO) and the Center for Disease Control (CDC). Developmental funds might be available from those organizations. Also there are a number of biotechnology firms working in this area globally, so they might become either partners or competitors. Rabies is a huge issue in the Third World countries that don't routinely vaccinate their pets. It's a people issue there, not just a pet issue, because of the many cases of humans who have been bitten; according to the WHO, there are 50,000 to 60,000 deaths per year from rabies. In the United States this would be both a person and a pet issue depending on how the vaccine is developed.

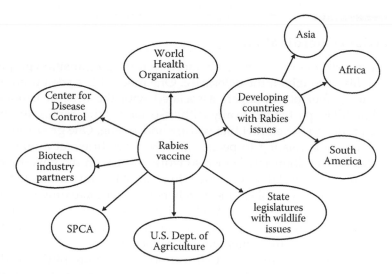

FIGURE 11.3
Medical Sayre Analysis model.

Business Plans

Write business plans for the development of advanced technologies to examine the path of transition and the evolutionary path for inserting these technologies into multiple customer product markets. It's important to get the process down for development, but it's even more important to determine the business model and the funding streams. Most developments that I've seen fail were based on funding issues. That is one reason for multiplying the potential sources of funds early in the cycle. There are a number of sources that can be tapped to get through the initial prototype developmental stages, but my recommendation would be to use as many as possible in the current fiscal environment. When we bring in partners that have critical technologies that are already developed, we diversify the risks. Don't multiply the risk by attempting to develop the entire system from scratch. Using someone else's display and technology platform to move your sensor technology to the higher level is a smart option.

The compatibility and ease of use factors can also be accentuated if the developer stays away from proprietary systems. The reason so many iPhone applications are available is that they are compatible with every iPhone. If you make a medical sensor and it has to have a proprietary display, it will probably be harder to sell than one that can be displayed on a standard personal laptop with some easy-to-load software application.

Consortium Coordination

Set up a consortium to coordinate advanced technology transition proposals with visibility throughout the enterprise. One method I've seen used effectively by a number of developers associated with the Federal Laboratory Consortium (FLC) partners is to ally with associations that represent a number of the groups in the market that you are addressing. One group addressing a disease issue relating to peanut growers gave briefs at a number of conferences for that industry. They also worked through the governmental agencies at the state level that controlled this part of the agriculture industry. They also brought in a number of researchers from the state universities. All of these actions were targeted toward building acceptance for the end product during the developmental phase. This group was also able to move this to the global level and enlist international participants.

When you are developing advanced technology, you are pitting yourself and your organization against the program manager network for the legacy systems. A program manager that has been working on a system for fifteen years becomes very attached to the legacy technology. The advanced technology developer is seen as attacking his personal legacy of the existing system. All program managers are evaluated on three factors: cost, schedule, and performance of the system. None of these factors is positive for inserting new "untested" technology. The fact that your technology may generate a ten-fold improvement in the capability is not in the decision-making process.

Many of the consortiums for industry segments have open forum discussions about serious issues that are common to the entire sector. When a new solution to this challenge is being discussed, the members don't want to be left out in the competitive market, so they are encouraged to adopt the technology to maintain their market share. The U.S. Department of Agriculture (USDA) was faced with an issue in the sugar industry over the use of an enzyme called dextranase in the processing of sugar beets. They solved this challenge with some innovated chemists and researchers, but the interesting part of the solution was the process they used. They allied with two industry partners who ran sugarcane factories. They got research grants from the American Sugar Cane League. Presentations were made to the American Society of Sugar Cane Technologists and the American Chemical Society. They wrote journal articles, industrial bulletin publications, and book chapters and held workshops on the new process in factories that would use it. Then after perfecting the process, they went global and briefed the International Commission for Uniform Methods in Sugar Analysis (ICUMSA). Being an engineer, I'm amazed that these organizations even exist. However, the process is what needs to be mirrored for medical technology development (SEFLC 2010). There are similar organizations for every segment of the medical industry, whether it is cancer research or magnetic resonance imagery interpretation. Hiding a technology under layers of legal paperwork is not my idea of an aggressive development. You can win a race by eliminating the competition, or by running faster than the competition. Getting your appropriate organizations and allies all lined up and running in the same direction is a good way to get to the finish line with customers and allies in tow.

Medical Materials and Advanced Applications

A great example of the insertion of advanced technology in the medical field is the current ability of doctors to replace knees and other joints. From an article at the BoneSmart website, we get this quote: "Total knee replacement offers the greatest quality of like improvement of all operations. It has one of the highest success rates and one of the best outcomes" (BoneSmart 2010). In the adoption of this procedure and the artificial knee joint, there were some significant factors involved. Obviously the design of the knee joint itself probably would be patterned on the natural knee. In research we'd probably term that as biomimicry. The knee was designed as a "replacement" for a natural knee that was damaged by accident or other factors such as age or arthritis. It was manufactured from materials that would make it durable and acceptable to the body's immune system. Other factors emerge as we get into the design phase before we get to actually trying it on a patient. How will it be attached to the bones of the leg? Can you use the existing tendons and muscles? Will the surgery be an hour and a half or ten hours to install it? Will the patient be severely limited in his or her abilities after the surgery?

From the article "Polymeric Medical Materials" (Merl 2010) we get this description of these materials.

> Polymeric materials are used in medical devices because of their unique properties, ease of manufacture, flexibility in design and low cost. Unique properties include:
>
> - Flexibility (tubing, seals, vascular ...)
> - Sorption/diffusion (drug delivery, contact lens ...)
> - Formability (bone cement, dental fillings ...)
> - Wear resistance (prosthetic joints ...)
> - Specific stiffness of composites (crutches, wheelchairs ...)

Replacement knees also involve metal parts. The BoneSmart article references metal components made from either cobalt chrome alloy or titanium alloy that can be fixed in place by either cement or bone "ingrowth." Similarly, a PRODISC-L Total Disc Replacement recently approved by the FDA describes the device this way:

> The PRODISC-L Total Disc Replacement is an artificial intervertebral disc made from metal and plastic that is used to treat pain associated with degenerative disc disease (DDD). (FDA 2010)

This replacement disc is composed of three parts:

- Two metal (cobalt-chrome alloy) endplates that are anchored to the top and bottom surfaces of the spinal bones (vertebrae)
- A plastic (ultra-high-molecular-weight polyethylene, or UHMWPE) inlay that fits between the two endplates (FDA 2010)

From this information, how do we analyze the insertion of materials in these new medical devices in relation to the adoption of innovation factors described in the previous sections?

These are the common factors from the theoretical work. If we can make several of these very positive or if we can eliminate any gross negative factors, we can more easily move these technologies to the medical field. It was very important to use materials that were acceptable to the body. When we look at some other medical issues we see that extensive testing is a key acceptance criterion. Whether it is a procedure or a device a major factor in acceptance is "do no harm in the long run." The Lasik eye procedure has become very accepted in the correction of vision. This procedure involves the reshaping of the cornea with a laser device. The success of this procedure and its acceptance in the marketplace required the patients to report success, and not just advertising by the doctors doing the procedure. Obviously there are extreme technology issues here in the laser power and accuracy, and in the precise control of the mechanical and computer devices. Similarly in the adoption of artificial knees, there was a huge factor in the demonstrating of results. Prosthetic limbs have been around for years, and many were and are still in use long after the limbs were amputated. The artificial knee restores natural functionality, so it yields a relative advantage. There is virtually no maintenance, and

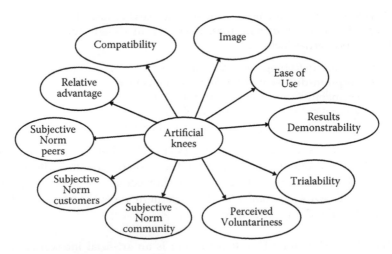

FIGURE 11.4
Mind map of artificial needs adoption criteria.

it is very easy to use after the surgery. It is compatible to our daily routine and to the body generally because after the insertion the materials are compatible to the immune system and require no continuing drugs to prevent rejection. The procedure is totally voluntary, but has very good success rates. I have an artificial knee replacement and it has worked well. I'm also acquainted with numerous other friends who have had both single and double knee replacements. Very few issues have surfaced with any of those acquaintances.

Evaluating all of these factors would yield these results for the artificial knee. Green indicates a positive factor. One negative would be the cost of the surgery.

If we took these same factors and applied them to the lumbar replacement disc, we should see a similar result. The success of the knee replacement surgery would also be a positive influence on that adoption owing to similar materials, procedures, and recovery. It should be noted that trialablity would probably be a negative. You just can't try it on like a pair of shoes and walk around in it for a while before you buy it. But there may be a factor here in letting other people you know try it for you. You know people who have had the surgery, and they were pleased with the results. You trust their opinion, and it becomes a positive factor in your adoption of the technology.

Summary of the Factors Affecting the Adoption of Innovative Medical Technology

When new technology is developed in any industry, it must demonstrate its value over the existing legacy systems. So the main issues of adoption of

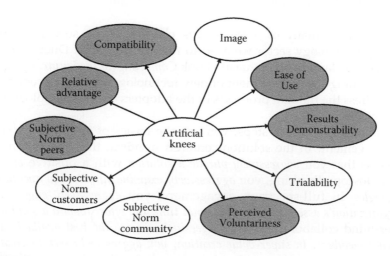

FIGURE 11.5
Mind map of artificial knees with criteria evaluated.

new technologies in the medical field must tap into the value stream to be candidates for adoption. When we address the processes that are in use we need to analyze the benefits of time reduction, cost reduction and improved capability to determine what these values are and how we can demonstrate them. The theoretical framework of innovation theory gives us a good start on identifying the potential factors that should be examined.

Adoption of a new medical technology also involves the subjective norms of the entire market sector, so logically we should try to influence this sector as a whole. The briefing and involvement of consortium organizations, industry partners and other research organizations will provide beneficial discussions and may prepare the market for your introduction. The mind-mapping technique of analyzing both the technology roadmap and the introduction market provides a visual way to generate discussion and to find the associated applications and complementary technologies that may enable your development. Using these techniques for linking the players, the technologies, and the users in the market provides a way to visualize the outcome and to guide the development down a successful transition path.

In his book *Innovate Like Edison*, Gelb identified the five competencies of innovation used by Thomas Edison to build new products and to invent new technologies. We must aggressively pursue these competencies if we are to be successful.

1. Solution-centered mindset
2. Kaleidoscopic thinking
3. Full-spectrum engagement
4. Master-mind collaboration
5. Super-value creation

I've identified many of these in this chapter. To be successful in the emerging technology sector we need to reward the "Wild Ducks" that are referenced in Hamal and Prahalad's book *Competing for the Future*. By aggressively pursuing the development of new technology into new products, we must address the thought processes of the adopters of this technology in the marketplace.

In summarizing these steps, you get the basics of aggressive technology engagement. In the solution-centered mindset, the primary thought process is that you *aggressively pursue a solution* with a positive attitude. In kaleidoscope thinking, you *aggressively examine all potential avenues and technologies*. In full-spectrum engagement, you *aggressively engage all of the organization's assets and your personal time and effort toward a solution*. In master-mind collaboration, *you aggressively engage the best available minds and team members*. In super-value creation, *you aggressively seek to create the most value for your customer and push your product to the optimum configuration* (Jarrett 2009).

The adoption of new technologies requires the generation of value in the eyes of the customer. We need to pay attention to the factors affecting this thought process and we need to actively pursue positive results.

Bibliography

BoneSmart. 2010. *The Basics of Total Knee Replacement Surgery*. www.bonesmart.org/knee_replacement.php.

Christensen, C. M. 2003. *The Innovator's Dilemma*. New York: Harvard Business School Press.

Federal Laboratory Consortium Southeast Region 2010 Awards for Excellence in Technology Transfer 2010.

Gelb, Michael J. 2007. *Innovate Like Edison: The Success System of America's Greatest Inventor*. Penguin Group (US) Inc.

Hamel, G., Prahalad, C. K. 1996. *Competing for the Future*. Boston: Harvard Business School Press.

Jarrett, S. M. 2009. *Aggressive Technology Engagement*. Published at the 2009 Charleston Defense Contractor's Association C5ISR conference in Charleston, SC.

Jarrett, S. M. 2003. *Factors Affecting the Adoption of E-Business in the Aerospace Industry*. Dissertation for Nova Southeastern University. Ann Arbor: ProQuest Information and Learning Company. UMI Microform 3096349.

Jolly, V. K. 1997. *Commercializing New Technologies: Getting from Mind to Market*. Boston: Harvard Business School Press.

McCarthy, Richard V., Aronson, Jay. 2001. *Analyzing the Balance between Consumer, Business and Government: The Emergent Internet Privacy Framework*. IACIS.

MERL Ltd. 2010. *Polymeric Medical Materials*. www.merl-ltd.co.uk/2003_industry/medical01.shtml.

Rogers, E. M. 1995. *Diffusion of Innovations*, 4th ed. New York: The Free Press.

U.S. Food and Drug Administration 2010. *PRODISC-L Total Disc Replacement – P050010*. www.fda.gov/MedicalDevices/ProductsandMedicalProcedures/Device Approval.

Bibliography

Appendix: Some Examples of FDA-Approved Products

Pablo Gurman

CHAPTER 2

Principles of Clinical and Engineering Integration in Hemocompatibility

COMPANY[1]	PRODUCT DESCRIPTION	TRADE NAME	APPROVAL PATHWAY	APPROVAL YEAR±
Abbott Vascular	Drug-eluting stent	Xience nano™ Everolimus Eluting Coronary Stent System	PMA	2011
St. Jude Medical Cardiovascular Division	Heart valve	St. Jude Medical Trifecta Valve™	PMA	2011
Boston Scientific Corporation	Drug-eluting stent	ION™ Paclitaxel-Eluting Coronary Stent System	PMA	2011
Medtronic, Inc.	Cryo-ablation catheter for arrhythmia therapy	Arctic Front CryoCatheter System™	PMA	2010
Thoratec Corporation	A ventricular-assisted device used in patients undergoing a heart transplant to avoid left ventricular failure as a bridge until transplantation has been performed	Thoratec HeartMate Ventricular Assist System II™ (LVAS)	PMA	2008

CHAPTER 3

Medical Applications of Micro-Electro-Mechanical-Systems (MEMS) Technology

COMPANY[1]	PRODUCT DESCRIPTION	TRADE NAME	APPROVAL PATHWAY	APPROVAL YEAR±
CardioMEMS	Implantable intra-aneurism pressure system	Cardiomems™ EndoSure Wireless Pressure Measurement System	510k	2008
Medtronic, Inc.	MEMS accelerometer for pacemaker	In Sync™	PMA	2011
Telecardia, Inc.	MEMS pressure sensor and MEMS pH sensor to detect early signs of acute myocardial infarction. The device evaluates ventricular wall tension.	Cardioguard™	PMA submission	2011
Cleveland Medical Devices Inc.	MEMS accelerometer and gyroscope for motion sensing in movement disorders such as tremor	Kinesia™	510k	2007
iSTAT Corporation (now Abbott Laboratories)	Cartridge for in vitro diagnostics test used in the i-STAT point of care device	i-STAT™ System	510k	1999

CHAPTER 4

Nanoparticles to Cross Biological Barriers

COMPANY[1]	PRODUCT DESCRIPTION	TRADE NAME	APPROVAL PATHWAY	APPROVAL YEAR±
Ortho Biotech	Doxorubicin hydrochloride liposomes	Doxil™	NDA 050718	1995
Abraxis Bioscience	Paclitaxel-bound albumin nanoparticles	Abraxane™	NDA 021660 – fast approval	2005
Graceway	Estradiol emulsion	Estradiol Topical Emulsion™	Approved drug products with therapeutic equivalence (Orange Book)	2003
Wyeth Pharmaceuticals	Nanocrystalline sirolimus	Rapamune™	NDA021110	2000
Abbott Laboratories	Nanocrystalline fenofibrate	Tricor™	NDA 021656	2004
Merck and Co., Inc.	Nanocrystalline aprepitant	Emend™	NDA 022023	2008
Zeneus Pharma	Liposomal doxorubicin	Myocet™		
Astellas	Liposomal amphotercin B	Ambisome™	NDA 050740	1997
Feridex	Iron oxide superparamagnetic nanoparticles	Feridex™	NDA020416	1996
Three Rivers Pharmaceuticals	Amphotericin B	Amphotec™	NDA 050729	1996
Biowave Corporation	Microneedle array for electrical delivery through skin	Deepwave Percutaneous Neuromodulation Pain Therapy System™	510k	2006

CHAPTER 5

Biomaterials, Dental Materials, and Device Retrieval and Analysis

COMPANY[1]	PRODUCT DESCRIPTION	TRADE NAME	APPROVAL PATHWAY	APPROVAL YEAR±
3M	Polymer-based dental cement	Rely X™ Luting Plus Automix Resin Modified Glass Ionomer Cement	510k	2011
Oroscience Inc.	Dental cement	Periogenix™	510k	2009
AIDI Biomedical, LLC	Endosseus dental implant and endosseuss dental implant abatement	AIDI Dental Implant System™	510k	2010

CHAPTER 6

Biomaterials and the Central Nervous System: Neurosurgical Applications of Materials Science

COMPANY[1]	PRODUCT DESCRIPTION	TRADE NAME	APPROVAL PATHWAY	APPROVAL YEAR±
Medtronic, Inc.	Implantable multiprogrammable quadripolar deep brain stimulation system	Medtronic Activa Dystonia Therapy™	HDE	2003
EISAI, Inc.	Intracranial implant drug delivery system releasing carmustine	Gliadel™	NDA	1996
Codman and Shurtleff, Inc.	Antimicrobial ventricular catheter for central nervous system fluid shunt	Codman Bactiseal EVD Catheter Set™	510K	2009

CHAPTER 7

Biomaterials in Obstetrics and Gynecology

COMPANY[1]	PRODUCT DESCRIPTION	TRADE NAME	APPROVAL PATHWAY	APPROVAL YEAR±
Ortho Pharmaceutical Corp	Copper-based intrauterine device (IUD)	Paragard™	PMA	1986
Bayer Healthcare	Levonorgestrel release intrauterine device	Mirena™	NDA	2000
Conceptus Inc	Transcervical contraceptive device	Essure™	PMA	2002
Hologic Inc	Permanent contraception system	Adiana™	PMA	2007
Ethicon Women Health and Urology	Absorbable adhesion barrier for prevention of post-surgical adhesions	Interceed™	510K	1993
FzioMed, Inc.	Absorbable adhesion barrier gel for prevention of post-surgical adhesions	Intercoat™	PMA	
Allergan	Silicone-filled breast implants	Inamed™	PMA	2006
Genzyme Corp.	Absorbable adhesion barrier for pelvic and abdominal surgery	Seprafilm™	PMA	1996
Controlled Therapeutics Inc. (currently distributed by Forest Pharmaceuticals)	Prostaglandin vaginal controlled release insert	Cervidil™ *	NDA	1995
Merocel Corp.	Osmotic cervical dilatators for labor work	Lamicel™	PMA	1996
Cook Inc.	Polyvinyl alcohol particles for embolization of symptomatic uterine fibroids	Polyvinyl Alcohol, Foam Embolization Particles™	510k	2008
Boston Scientific Corporation	Polymeric surgical mesh sling package with a delivery device for the treatment of stress urinary incontinence	Surgical Mesh (SIS)™	510k	2008
W.L. Gore and Associates Inc.	Polytetrafluoroethylene loaded with antimicrobial surgical mesh with a modified texturing pattern in the ingrowth surface intended for the reconstruction of hernias and soft tissues deficiencies and temporary bridging of fascial defects	Gore-Tex™ Dual Mesh™ PLUS Biomaterial	510k	2000

CHAPTER 8

Tissue Engineering: Focus on the Cardiovascular System

COMPANY[1]	PRODUCT DESCRIPTION	TRADE NAME	APPROVAL PATHWAY	APPROVAL YEAR±
SyntheMed, Inc.	PLGA poly(lactic-co-glycolic) bioadhesion barrier to avoid cardiac adhesion in pediatric patients undergoing repeated sternotomies	REPEL-CV™	PMA	2009
Nycomed Danmark ApS	Fibrin sealant patch that serves as adjunct for hemostasis in cardiac surgery	TachoSil™	BLA	2010

CHAPTER 9

Tissue Engineering: Focus on the Musculoskeletal System

COMPANY[1]	PRODUCT DESCRIPTION	TRADE NAME	APPROVAL PATHWAY	APPROVAL YEAR±
DePuy Orthopaedics, Inc.	Bone void filler based on hydroxiapatite + collagen type I	Healos Fx Bone Graft Substitute™	510k	2006
Globus Medical, Inc.	Bone void filler based on PLGA microspheres	Micro Fuse Bone Void Filler™	510k	2008
Medtronic, Inc.	BMP-2 +collagen type I	Infuse™	PMA	2007
Orthovita, Inc.	βTCP	Vitoss™	510k	2003
Stryker	Genetic engineered morphogenetic bone protein (rhBMP-7) + collagen type I for inducing posterolateral spinal fusion	OP-1 Putty™	Humanitarian Device Exemption (HDE)	2008
Synthes, Inc.	Resorbable calcium salt bone void filler	Norian SRS™	510k	2004
Wright Medical Technologies	Bone void filler based on calcium sulphate	Osteoset BVF Kit™	510k	2001

1 Manufacturer: It should be considered that changes in the manufacturer or distributor of the product could have taken place from the date of submission to the FDA. Readers are advised to consult the FDA website for an update of this information. Trade names other than those specified in these tables could be found for the same product, since many manufacturers may be producing the same product under different PMA, NDA, or 510k numbers (e.g., absorbable adhesion barriers).

± Approval Year: Approval year for PMA and NDA could differ significantly from those described in other parts of the book owing to the different dates that apply to the original document and its supplements.

β Cervidil in U.S., Canada, Australia, and New Zealand; Propess in most of the other countries.

These tables are designed with an educational purpose only and by no means are a comprehensive list of the products that appear in this book. They are not intended to replace the information contained in the FDA website, where readers are referred to for more detailed and complete information regarding the regulatory status of each product.

References

FDA website http://www.fda.gov
TELECARDIA website http://www.telecardiacorp.com
http://www.electroic.com/articles/stm/2001/07/bfda-panel-recommends-approval-medtronics-mems-pacemaker-b.html
http://glneurotech.com/PDFs/kinesia_specs.pdf
http://www.seprafilm.com/patients/about/seprafilm.aspx
http://www.cytokinepharmasciences.com/cervidil-propess.shtml

Index

A

Abraxane, 87, 88
Adipose-derived stem cells (ADSCs),
    200–201
Adriana, 153–154
Adult skeletal muscle stem cells, 171–172
ALOX5AP, 11
American Society for Testing and
    Materials (ASTM), 125,
    234–235
Amphiphilic peptides, 213–214
Analysis of alternatives, 273
Anisotropic etching, 37, 38–39, 70. *See
    also* Bulk micromachining
Anodic bonding, 45
Antiphospholipid syndrome (APS),
    15–17
Atherosclerosis risk factors, 10

B

Belmont Report, 226
BiCMOS, 54
Bilirubin, 10
Bioanalyzer, 76
Biocompatibility
    MEMS devices, 62–63
Biodegredation phenomena, 130
Biomaterials science
    growth in field, 224
    historical overview, 225–226
    integrated approach to, 1, 2
    neurosurgical applications; *see*
        Neurosurgical applications of
        biomaterials
    obstetric and gynecology
        applications; *see* OB/GYN
        applications of biomaterials
    overview of field, 1
Biomimicry, 197, 287
Biomolecules, 106
Bone marrow stem cells, 169–171
BoneSmart, 287

Brain-derived neurotrophic factor
    (BDNF), 136
Brain-machine-interface (BMI)
    technology, 136
Breast implants, 157–158
Bulk micromachining, 32–33. *See also*
    Micromachining
    anisotropic etching, 37, 38–39, 70
    chemical wet etching, 35–37
    etch stops, 39–40
    technology of, 34; *See also*
        Micro-electro-mechanical
        systems (MEMS)

C

Carboxymethylcellulose (CMC), 155
Cardiac stem cells, resident, 171
Cardiac stents
    bioabsorbable, 18, 19, 20–21
    biodegradable, 18, 19
    biomaterials used in, 18, 19
    drug-eluting stents (DES), 12, 13, 19, 20
    genetic polymorphism, relationship
        between; *see* Genetic
        polymophism, cardiac stents
CardioGene Study, 10
Cardiomyocyte (CM) population
    propagation, 166, 168, 170
Cardiovascular use of biomaterials
    adult skeletal muscle stem cells,
        171–172
    bone marrow stem cells, 169–171
    cardiac stem cells, resident, 171
    cardiomyocyte (CM) population
        propagation, 166, 168, 170
    cell aggregation, 175
    cell sheets, 175–176
    decellularized matrix, 176–177
    human embryonic stem cells
        (hESCs), 166, 168
    hydrogels; *see* Hydrogels
    induced pluripotent stem cell (iPS),
        166, 169

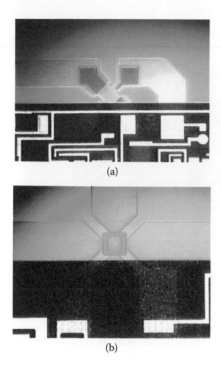

(a)

(b)

FIGURE 3.27
(a) Optical photograph of top surface of Freescale Semiconductor pressure sensor that employs the original "X-ducer." An "X" has been added to show the transducer layout. (b) The newer "picture frame" piezoresistor configuration. (Reprinted with permission, copyright Freescale Semiconductor, Inc.)

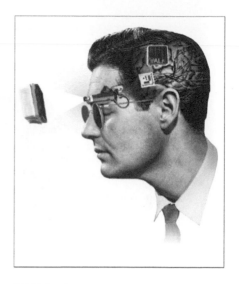

FIGURE 3.35
Illustration of one implementation of a cortical implant for visual prosthesis.

1024 sites and 64 data channels on 400μm centers

FIGURE 3.36
Photograph of MEMS microelectrode array for visual prosthesis.

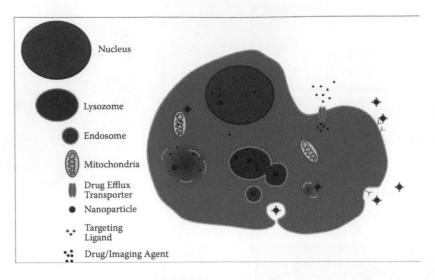

FIGURE 4.2
Specific cellular or intracellular targets may increase nanoparticle treatment efficacy. To avoid the degradation of the payload, nanoparticles may need to actively escape endosomal or lysosomal compartments. Delivery of the payload to specific organelles may be mediated by the nanocarrier.

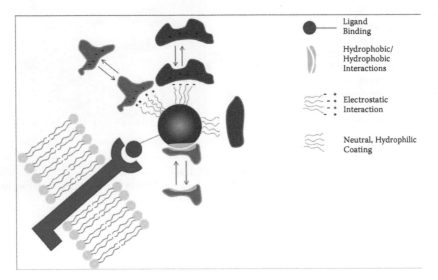

FIGURE 4.3
Once within the body, nanoparticle fate will be determined by interaction with proteins. The physiochemical properties and surface engineering of a nanoparticle will determine the particle/protein interaction. Electrostatic interactions between the nanoparticle surface charge and the positive and negative domains of proteins, along with non-polar interactions between hydrophobic regions, will determine protein adsorption. Hydrophilic coatings, such as PEG, can confer "stealth" properties. Various ligands attached to the nanoparticle surface can target cell-specific moeities and mediate cellular processes.

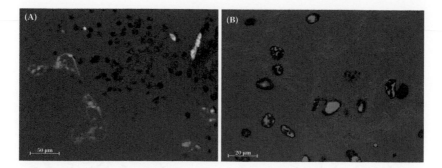

FIGURE 8.1
In-vitro construction of engineered vascularized cardiac muscle using a multi-cellular strategy of hES-CMs, endothelial cells, and embryonic fibroblasts seeded within a porous polymer scaffold of PLLA/PLGA. (A) The endothelial cells (vWF, green) within the scaffold self-organized to lumen vessel structures located in close proximity to the hES-CMs (troponin I red). (B) Higher magnification reveals that the hES-CMs matured to a certain degree, presenting developed cytoplasm and sarcomeric pattering (troponin I, red). Nuclei are stained with DAPI (blue).

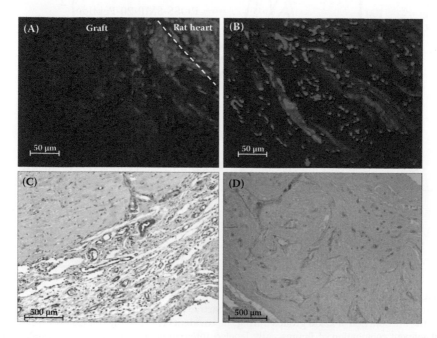

FIGURE 8.2
Transplantation of the engineered human vascularized cardiac tissue demonstrating localization of the hES-CMs (troponin I, red) in the graft area next to the myocardium (A), and structural maturation of the CMs in the graft area (B). The graft area was occupied with intense vascularization, as detected by staining with aSMA antibody (host and human derived vessels, brown, C) and with human specific endothelial CD31 antibody (human implanted vessels, brown, D). Nuclei were stained with DAPI (A, B in blue) or with hematoxylin (C, D in blue).

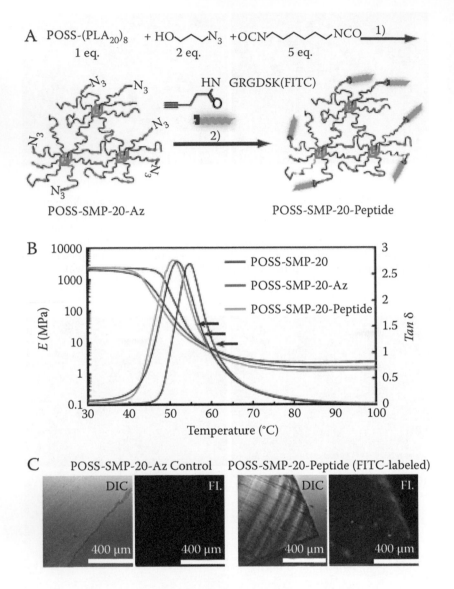

FIGURE 9.2
Chemical modification of POSS-SMP with a bioactive peptide: (A) Synthetic scheme illustrating the introduction of azido groups during the covalent cross-linking of POSS-(PLA20)8 and subsequent conjugation of fluorescently labeled integrin-binding peptide to POSS-SMP via "click" chemistry. (B) Storage modulus (E′)-temperature curves and loss angle (Tan δ)-temperature curves (denoted by black arrows) of POSS-SMP-20, POSS-SMP-20-Az, and POSS-SMP-20-Peptide. (C) Differential interference contrast (DIC) and fluorescent (Fl) micrographs confirming the covalent conjugation of the fluorescently labeled peptide via click chemistry. (From *PNAS* 107:7652–7657, 2010. With permission from the Publisher.)

Printed and bound by CPI Group (UK) Ltd, Croydon, CR0 4YY
23/10/2024
01778263-0013